AF581652

Functional Nanomaterials and Polymer Nanocomposites: Current Uses and Potential Applications

Functional Nanomaterials and Polymer Nanocomposites: Current Uses and Potential Applications

Editor

Raghvendra Singh Yadav

MDPI • Basel • Beijing • Wuhan • Barcelona • Belgrade • Manchester • Tokyo • Cluj • Tianjin

Editor
Raghvendra Singh Yadav
Centre of Polymer Systems
Tomas Bata University in Zlin
Zlín
Czech Republic

Editorial Office
MDPI
St. Alban-Anlage 66
4052 Basel, Switzerland

This is a reprint of articles from the Special Issue published online in the open access journal *International Journal of Molecular Sciences* (ISSN 1422-0067) (available at: www.mdpi.com/journal/ijms/special_issues/nanomaterials_polymernanocomposites).

For citation purposes, cite each article independently as indicated on the article page online and as indicated below:

LastName, A.A.; LastName, B.B.; LastName, C.C. Article Title. *Journal Name* **Year**, *Volume Number*, Page Range.

ISBN 978-3-0365-6587-3 (Hbk)
ISBN 978-3-0365-6586-6 (PDF)

© 2023 by the authors. Articles in this book are Open Access and distributed under the Creative Commons Attribution (CC BY) license, which allows users to download, copy and build upon published articles, as long as the author and publisher are properly credited, which ensures maximum dissemination and a wider impact of our publications.
The book as a whole is distributed by MDPI under the terms and conditions of the Creative Commons license CC BY-NC-ND.

Contents

About the Editor

Raghvendra Singh Yadav

Dr. Raghvendra Singh Yadav is a Senior Scientist and Leader of a sub-research group in the field of two-dimensional nanostructures (graphene, MXene, borophene, transition metal dichalcogenides, etc.) -based Innovative Functional Nanocomposites and its various Applications at Centre of Polymer Systems, Tomas Bata University in Zlin, Czech Republic, from 1 October 2017.

Dr. Yadav has been involved as an Editorial Board Member in several international high-impacted journals, namely (1) *Crystals* (I. F. = 2.670, MDPI, Switzerland), (2) *International Journal of Molecular Sciences* (I.F.=6.208, MDPI, Switzerland), (3) *Nanomaterials* (I.F. = 5.719, MDPI, Switzerland), (4) *Frontiers in Materials* (I.F =3.985, Switzerland), etc. Dr. Yadav has also worked as a Guest Editor of Special Issues in several international journals on material science, nanoscience, and nanotechnology. Dr. Yadav received several projects as "Principal Investigator" such as from (i) Standard Project-Grant Agency of Czech Republic, Prague, Czech Republic, (ii) a Fast Track Young Scientist Project-Department of Science and Technology, New Delhi, India, and (iii) Beam time allotment, Inter-University Accelerator Centre (IUAC), New Delhi, India.

Dr. Yadav has published more than 65 publications in reputed international journals that have received 2493 citations with an h-index 28. Dr. Yadav has also published five books and two book chapters in the field of materials science and nanotechnology.

International Journal of
Molecular Sciences

MDPI

Editorial

Functional Nanomaterials and Polymer Nanocomposites: Current Uses and Potential Applications

Raghvendra Singh Yadav

Centre of Polymer Systems, Tomas Bata University in Zlin, Trida Tomase Bati 5678, 76001 Zlin, Czech Republic; yadav@utb.cz; Tel.: +420-576-031-725

In the present Special Issue "Functional Nanomaterials and Polymer Nanocomposites: Current Uses and Potential Applications", two review articles and nine original research articles are published. The published review article by M.S.A. Darwish et al. [1] presents the research advances on polymeric nanocomposites for environmental and industrial applications. Further, C.V. Rocha et al. [2] provide a review article on current advances in the development and biomedical applications of PLGA-based materials.

In a published research article in this Special Issue, M. Aviv et al. [3] describe the behavior of the double-fluorinated Fmoc-Phe derivatives, Fmoc-3,4F-Phe and Fmoc-3,5F-Phe, and the influence that the position of single fluorine has on the self-assembly process and physical characteristics that the material produces. Moreover, S. Stojanov et al. [4] reported the incorporation of vaginal lactobacilli into electrospun nanofibers to achieve a prospective solid vaginal delivery system, and further, the fluorescent proteins were incorporated to differentiate them and allow their tracking in the future probiotic-delivery investigations. C. Miyamaru et al. [5] developed $CaCO_3$-coated vesicles by biomineralization and further their utilization as carriers of drug-delivery systems. In this Special Issue, M.B. Stie et al. [6] described mucoadhesive electrospun nanofiber-based hybrid system with the controlled and unidirectional release of desmopressin. Anju et al. [7] investigated highly efficient electromagnetic interference shielding of $Cu_xCo_{1-x}Fe_2O_4$ (x = 0.33, 0.67, 1) magnetic nanoparticle-based polyurethane nanocomposites with reduced graphene oxide. Further, C.-Y. Wu et al. [8] report conductive supramolecular polymer nanocomposites with tunable characteristics to manipulate cell growth and functions. J. Fèvre et al. [9] describe, in this Special Issue, chelating polymers for targeted decontamination of actinides and application of PEI-MP to Hydroxyapatite-Th(IV). Furthermore, Z. Wang et al. [10] describe mono-sized anion-exchange magnetic microspheres for protein adsorption. In addition, M. Park et al. [11] investigated the impact of hexagonal boron nitride insulating layers on the driving ability of ionic electroactive polymer actuators for lightweight artificial muscles.

Funding: We thank the financial support of the Ministry of Education, Youth, and Sports of the Czech Republic-DKRVO (RP/CPS/2022/007).

Acknowledgments: I, as a guest editor, appreciate all the authors for their review/research articles in this Special Issue. Further, I also thank the reviewers for their help in the evaluation of review/research articles.

Conflicts of Interest: The author declares no conflict of interest.

Citation: Yadav, R.S. Functional Nanomaterials and Polymer Nanocomposites: Current Uses and Potential Applications. *Int. J. Mol. Sci.* **2022**, *23*, 12713. https://doi.org/10.3390/ijms232112713

Received: 18 October 2022
Accepted: 19 October 2022
Published: 22 October 2022

Publisher's Note: MDPI stays neutral with regard to jurisdictional claims in published maps and institutional affiliations.

Copyright: © 2022 by the author. Licensee MDPI, Basel, Switzerland. This article is an open access article distributed under the terms and conditions of the Creative Commons Attribution (CC BY) license (https://creativecommons.org/licenses/by/4.0/).

References

1. Darwish, M.S.A.; Mostafa, M.H.; Al-Harbi, L.M. Polymeric Nanocomposites for Environmental and Industrial Applications. *Int. J. Mol. Sci.* **2022**, *23*, 1023. [CrossRef] [PubMed]
2. Rocha, C.V.; Gonçalves, V.; da Silva, M.C.; Bañobre-López, M.; Gallo, J. PLGA-Based Composites for Various Biomedical Applications. *Int. J. Mol. Sci.* **2022**, *23*, 2034. [CrossRef] [PubMed]
3. Aviv, M.; Cohen-Gerassi, D.; Orr, A.A.; Misra, R.; Arnon, Z.A.; Shimon, L.J.W.; Shacham-Diamand, Y.; Tamamis, P.; Adler-Abramovich, L. Modification of a Single Atom Affects the Physical Properties of Double Fluorinated Fmoc-Phe Derivatives. *Int. J. Mol. Sci.* **2021**, *22*, 9634. [CrossRef] [PubMed]
4. Stojanov, S.; Plavec, T.V.; Kristl, J.; Zupančič, Š.; Berlec, A. Engineering of Vaginal Lactobacilli to Express Fluorescent Proteins Enables the Analysis of Their Mixture in Nanofibers. *Int. J. Mol. Sci.* **2021**, *22*, 13631. [CrossRef] [PubMed]
5. Miyamaru, C.; Koide, M.; Kato, N.; Matsubara, S.; Higuchi, M. Fabrication of $CaCO_3$-Coated Vesicles by Biomineralization and Their Application as Carriers of Drug Delivery Systems. *Int. J. Mol. Sci.* **2022**, *23*, 789. [CrossRef] [PubMed]
6. Stie, M.B.; Gätke, J.R.; Chronakis, I.S.; Jacobsen, J.; Nielsen, H.M. Mucoadhesive Electrospun Nanofiber-Based Hybrid System with Controlled and Unidirectional Release of Desmopressin. *Int. J. Mol. Sci.* **2022**, *23*, 1458. [CrossRef] [PubMed]
7. Anju; Yadav, R.S.; Pötschke, P.; Pionteck, J.; Krause, B.; Kuřitka, I.; Vilčáková, J.; Škoda, D.; Urbánek, P.; Machovský, M.; et al. $Cu_xCo_{1-x}Fe_2O_4$ (x = 0.33, 0.67, 1) Spinel Ferrite Nanoparticles Based Thermoplastic Polyurethane Nanocomposites with Reduced Graphene Oxide for Highly Efficient Electromagnetic Interference Shielding. *Int. J. Mol. Sci.* **2022**, *23*, 2610. [CrossRef] [PubMed]
8. Wu, C.-Y.; Melaku, A.Z.; Ilhami, F.B.; Chiu, C.-W.; Cheng, C.-C. Conductive Supramolecular Polymer Nanocomposites with Tunable Properties to Manipulate Cell Growth and Functions. *Int. J. Mol. Sci.* **2022**, *23*, 4332. [CrossRef] [PubMed]
9. Fèvre, J.; Leveille, E.; Jeanson, A.; Santucci-Darmanin, S.; Pierrefite-Carle, V.; Carle, G.F.; Den Auwer, C.; Di Giorgio, C. Chelating Polymers for Targeted Decontamination of Actinides: Application of PEI-MP to Hydroxyapatite-Th(IV). *Int. J. Mol. Sci.* **2022**, *23*, 4732. [CrossRef] [PubMed]
10. Wang, Z.; Wang, W.; Meng, Z.; Xue, M. Mono-Sized Anion-Exchange Magnetic Microspheres for Protein Adsorption. *Int. J. Mol. Sci.* **2022**, *23*, 4963. [CrossRef] [PubMed]
11. Park, M.; Chun, Y.; Kim, S.; Sohn, K.Y.; Jeon, M. Effects of Hexagonal Boron Nitride Insulating Layers on the Driving Performance of Ionic Electroactive Polymer Actuators for Light-Weight Artificial Muscles. *Int. J. Mol. Sci.* **2022**, *23*, 4981. [CrossRef] [PubMed]

International Journal of *Molecular Sciences*

MDPI

Article

Mono-Sized Anion-Exchange Magnetic Microspheres for Protein Adsorption

Zhe Wang [1,2], Wei Wang [1], Zihui Meng [1] and Min Xue [1,*]

1 School of Chemistry and Chemical Engineering, Beijing Institute of Technology, Beijing 102488, China; wangzhe@bit.edu.cn (Z.W.); wangwei2020_ripp@126.com (W.W.); mengzh@bit.edu.cn (Z.M.)
2 Academy of National Food and Strategic Reserves Administration, Beijing 100037, China
* Correspondence: minxue@bit.edu.cn

Abstract: In this study, mono-sized anion-exchange microspheres with polyglycidylmethacrylate were engineered and processed to introduce magnetic granules by penetration–deposition approaches. The obtained magnetic microspheres showed a uniform particle diameter of 1.235 μm in average and a good spherical shape with a saturation magnetic intensity of 12.48 emu/g by VSM and 12% magnetite content by TGA. The magnetic microspheres showed no cytotoxicity when the concentration was below 10 μg/mg. The magnetic microspheres possess respective adsorption capacity for three proteins including Bovine albumin, Hemoglobin from bovine blood, and Cytochrome C. These magnetic microspheres are also potential biomaterials as targeting medicine carriers or protein separation carriers at low concentration.

Keywords: magnetic microspheres; surface embedding; magnetic separation; protein adsorption

Citation: Wang, Z.; Wang, W.; Meng, Z.; Xue, M. Mono-Sized Anion-Exchange Magnetic Microspheres for Protein Adsorption. *Int. J. Mol. Sci.* **2022**, *23*, 4963. https://doi.org/10.3390/ijms23094963

Academic Editors: Raghvendra Singh Yadav and Ángel Serrano-Aroca

Received: 14 March 2022
Accepted: 27 April 2022
Published: 29 April 2022

Publisher's Note: MDPI stays neutral with regard to jurisdictional claims in published maps and institutional affiliations.

Copyright: © 2022 by the authors. Licensee MDPI, Basel, Switzerland. This article is an open access article distributed under the terms and conditions of the Creative Commons Attribution (CC BY) license (https://creativecommons.org/licenses/by/4.0/).

1. Introduction

The purified proteins play a crucial role in the research of protein on life activities, such as catalytic metabolic reactions and growth control. However, current separation and purification methods are tedious and time-consuming [1–3], such as affinity chromatography, dialysis, salting, and ultrafiltration. Magnetic separation technology has potential in protein purification due to its advantages of easy operation and rapid separation [4–6]. Magnetic microspheres are composite material particles [6,7] consisting of both inorganic magnetic materials providing magnetism and organic active functional groups carrying affinity ligand to target on the surface.

Magnetic microspheres have been successfully used for the separation of proteins [8–10] based on the interaction between protein and functional groups or special ligands on the microspheres, including electrostatic adsorption and specific adsorption. Moreover, magnetic microspheres modified with affinity ligands may have high selectivity to the target proteins, but the available ligands are limited and relative expensive [11,12]. Some commercial magnetic beads modified with monoclonal antibodies were successfully used for target substances identification, especially in diagnosis. However, the high cost, the tedious modification process, and the difference in separation effect severely limit their widespread application.

Magnetic polymer microspheres could be synthesized by several methods. The embedding method [13] is typically applied in the preparation of magnetic microspheres with a magnetic shell, which is simple and easy to carry out, but results in the magnetic particles with irregular shapes and polydisperse states. The emulsion polymerization method [14] provides monodispersed magnetic microspheres, but the small grain size beads below 1.0 μm that exhibit higher separation efficiency under magnetic field are hard obtain. An in situ method [15] is a way to obtain the magnetic nanocomposite materials by binding nanoscale magnetic materials on the pre-synthetic polymer surface. During the magnetization process, the particle size and distribution of the monodisperse polymer microspheres could be maintained. Each microsphere, containing the same concentration of magnetic particles, ensures that it has uniform magnetic response in the magnetic field.

In this study, two kinds of anion-exchange microspheres were prepared by an in situ method and applied for protein adsorption study. A novel method for modification of amino-microsphere to carboxyl-microsphere by EDC, NHS, and sodium carboxymethyl cellulose was proposed. The particle size, functional groups, and magnetic properties of the resultant magnetic particles were characterized. The maximum binding capacity was relatively high compared with similar research [16–18].

2. Results and Discussion

2.1. Synthesis of Anion-Exchange Magnetic Microspheres

The functional magnetic microspheres were synthesized by an in situ synthesis method with moderate size and functional group beneficiation on the surface for further modification. Glycidyl methacrylate (GMA) was selected as the basic monomer to structure monodisperse polymeric microspheres. The PGMA microsphere surface was rich in amino group after reacting with EDA (Figure 1a). The amino group is a strong polar group, which can form an ionic bond and a complex coordinate bond with a metal ion, thereby reducing the probability of collision between the particles and preventing excessive aggregation of the particles. Magnetic microspheres were prepared when Fe_3O_4 nanoparticles were precipitated in the surface and the internal of PGMA microspheres through the interfacial stripping precipitation method (see Figure 1b). Furthermore, a novel method in which carboxymethyl cellulose was bonded to the surface of magnetic microspheres by the EDC method was applied to prepare carboxyl magnetic microspheres (Figure 1c).

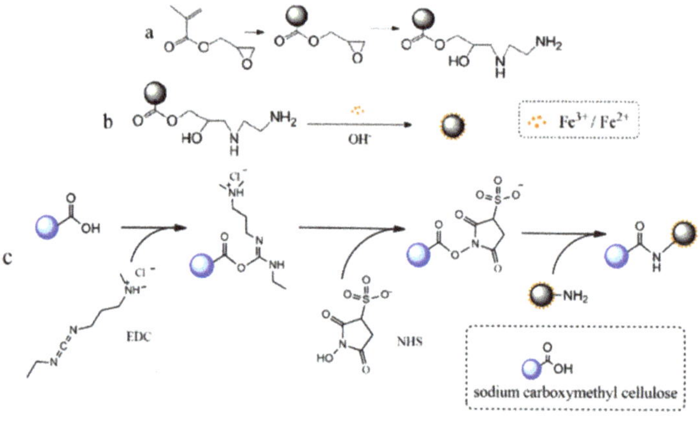

Figure 1. Schematic diagram of the preparation of magnetic anion-exchanged microspheres. (**a**) Preparation and amino modification of PGMA Polymer Microspheres. (**b**) Synthesis of amino magnetic microspheres. (**c**) Carboxyl coating on the surface of amino magnetic microspheres.

2.2. Characterization of the Magnetic Microspheres

The morphologies of microspheres were studied by SEM. It could be observed that the PGMA microspheres were mono-sized microspheres with very smooth surfaces, while the magnetic microspheres were relatively rough on the surface (Figure 2). The average diameter of the magnetic microspheres was 1.235 ± 0.017 μm according to the 100 microspheres selected from SEM images randomly, and the diameter of the Fe_3O_4 particles coated by sodium carboxymethyl cellulose on the microspheres was 30~50 nm. The size of these magnetic microspheres is relatively smaller compared with a similar polymerization method [19]. Generally, smaller diameter means larger specific surface area and greater adsorption capacity for the target.

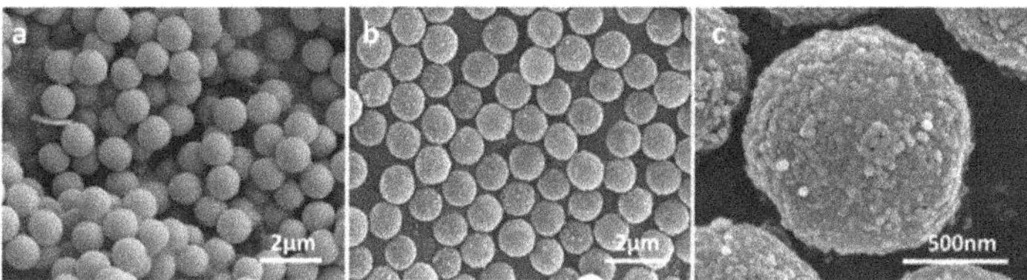

Figure 2. SEM images of microspheres. (**a**) PGMA microspheres. (**b**) Carboxyl magnetic microspheres. (**c**) The close-up of carboxyl magnetic microspheres.

The presence of the functional groups of PGMA and amino-PGMA microspheres was verified by FT-IR in Figure 3. PGMA microspheres were obtained by the polymerization of GMA monomer with DVB as cross-linker; therefore, epoxy groups and carbonyl group should distribute throughout the microspheres surface. The strong adsorption peak at 1727 cm^{-1} corresponds to C=O stretching vibration. Compared with Figure 3a, the characteristic bands at 848 cm^{-1}, 908 cm^{-1}, and 1250 cm^{-1} belong to the epoxy group which disappeared in Figure 3b, indicating that the epoxy groups transformed into an amino group after the amino modification.

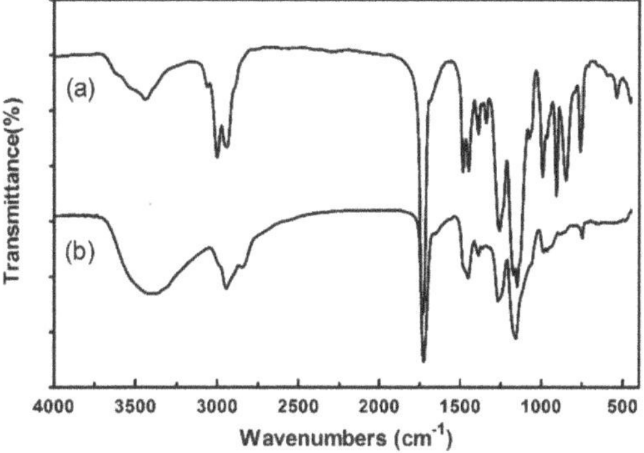

Figure 3. IR spectrum of microspheres. (**a**) PGMA microspheres. (**b**) Amino microspheres.

Controlling the magnetite content of the microspheres is important for realizing the rapid response to external magnetic fields for efficient adsorption. TGA measurement showed that the main weight loss of all microspheres was in the range of 200~450 °C, indicating that the microspheres should have stable thermal performance in the adsorption condition (Figure 4A). By comparing the residual mass after full burning, it can be calculated that the magnetite content of amino magnetic spheres and carboxyl magnetic spheres is about 12%. The magnetic properties of the carboxyl magnetic microspheres were measured by vibrating sample magnetometry (VSM). The magnetization curve shows that the saturation magnetization of the microspheres reached 12.48 emu/g, while the residual magnetism and coercive force were almost zero (Figure 4B). It means the magnetic microspheres are of high enough magnetism such that they can be separated from a suspension quickly. Moreover, they are superparamagnetic, which could disperse in the solution uniformly when there is no magnetic field. Compared with previous reports, magnetic intensity lowering of sodium carboxymethyl cellulose coated magnetic decrease may be caused by the embedding of the coating.

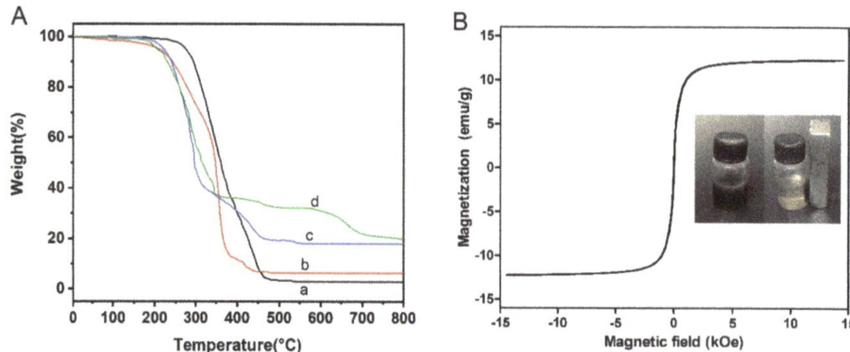

Figure 4. (**A**) TGA of (a) PGMA microspheres, (b) Amino microspheres, (c) Amino magnetic microspheres, and (d) Carboxyl magnetic microspheres. (**B**) Hysteresis curve of carboxyl magnetic microspheres. The inset shows photos of the carboxyl magnetic microsphere in aqueous solution without (left) and with (right) a magnet.

The density of amino groups and carboxyl groups on corresponding microspheres measured by the titration method was 2.75 mmol/g and 1.32 mmol/g respectively. It could be inferred from the result that amino groups incompletely reacted with carboxymethyl cellulose, so both amino groups and carboxyl groups exist on the surface of carboxyl magnetic microspheres.

Investigation of the biological safety of magnetic microspheres is critical for biomolecular separation medium. Herein, human pulmonary epithelial cells were used to study the in vitro cytotoxicity of carboxyl magnetic microspheres solution and its supernatant measured using a water-soluble tetrazolium cell proliferation assay. It could be inferred from Figure 5 that the supernatant of suspension showed no cytotoxicity to the cell, and the carboxyl magnetic microspheres showed no cytotoxicity when their concentrations were lower than 10 μg/mg. Therefore, it could be said that the magnetic microspheres are potential targeting medicine carriers or cell separation carriers at low concentrations.

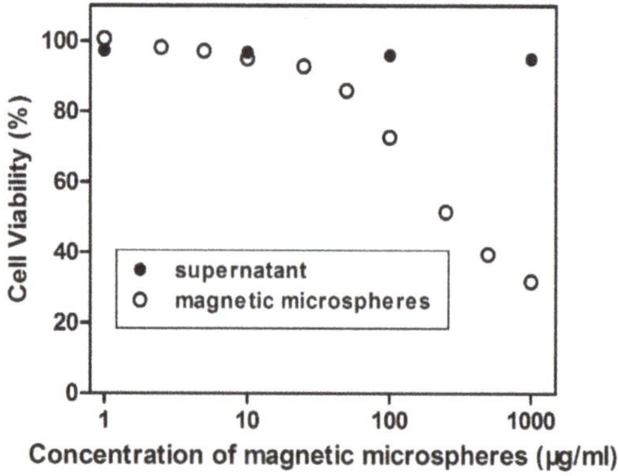

Figure 5. Relation curve of cell viability with the concentration of magnetic microspheres.

2.3. Binding Capability of Magnetic Microspheres

The adsorption capability of magnetic microspheres to proteins is mainly affected by the properties of the protein, the functional groups content on the surface of the microspheres, and the adsorption conditions. Generally, proteins are more likely to precipitate around their isoelectric point. It is comprehensible that pH value has a great effect on the adsorption of proteins by the microspheres. The maximum adsorption capacity of three proteins to amino magnetic microspheres is in close proximity to the isoelectric point (pI) in Figure 6a. Amino magnetic microspheres were positively charged while BSA was negatively charged at pH 5, which means BSA (pI 4.6) would be more easily absorbed on the magnetic surface of the microspheres due to its electrostatic interaction. The absorption mechanism of Hb (pI 7.0) was similar to BSA with the optimal amount of adsorption at pH 7. However, the adsorption for Cyt C (pI 10.65) was not very significant and the adsorption amount differences between each pH value were slight.

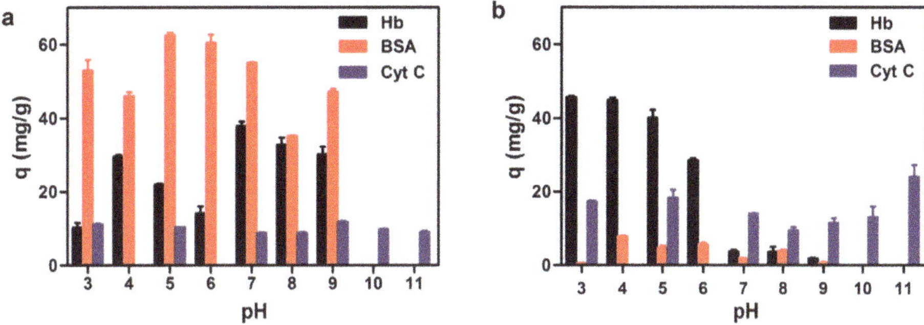

Figure 6. Adsorption capacity of three proteins to magnetic microspheres with different pH values. (**a**) Amino magnetic microspheres. (**b**) Carboxyl magnetic microspheres.

The carboxyl magnetic microsphere was negatively charged on the protein surface. Due to electrostatic adsorption, binding capacities of Hb to the microspheres were greater when the pH value was below 7. Since there were both carboxyl groups and amino groups existing on the carboxyl magnetic microsphere surface, the adsorption of carboxyl magnetic microspheres was more complicated than the adsorption of amino magnetic microspheres. Binding experimental results (Figure 6b) suggested the overall adsorption of BSA was not much with the maximum adsorption when the pH value was close to its isoelectric point; greater adsorption of Hb is observed in an acidic environment; maximum absorption of Cyt C is also near the isoelectric point. According to our experiment, the optimal adsorption conditions Hb, BSA, and Cyt C are pH 5, pH 7, and pH 9 for amino microspheres, pH 4, pH 3, and pH 11 for the carboxyl microspheres, respectively. The adsorption of Hb was also recorded in the concentration ranging from 0 to 10 mg/mL at 25 °C (Figure 7). With the increase of the initial concentration of Hb, the equilibrium adsorption amount generally increased and eventually became saturated.

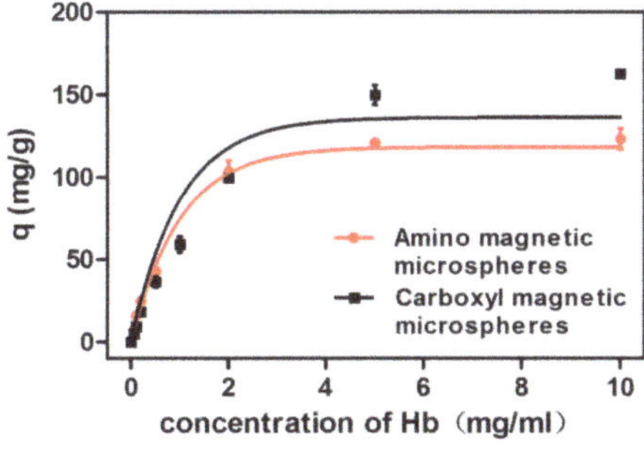

Figure 7. Adsorption isotherm of magnetic microspheres on Hb at 25 °C.

The adsorption capacities of several magnetic microsphere adsorption materials were compared in Table 1. We can see that the adsorption capacity of the prepared microspheres is comparable to that of the magnetic microspheres prepared by the in situ method. Meanwhile, they have the advantages of simple preparation and low cost. The immunoaffinity magnetic microspheres are highly specific to the target protein, however, sacrificing part of the adsorption capacity. The magnetic microspheres prepared by the imprinting technology have ordered imprinting pores and excellent specificity for target protein, realizing an efficient adsorption rate, which is higher than immunomagnetic microspheres. In addition, the surface mesoporous structure and the enrichment effect of metal particles are also considered to be the factors that improve the ability to adsorb proteins.

Table 1. Comparison of the adsorption capacity of several magnetic microspheres for proteins.

Magnetic Microspheres	Protein	Maximum Adsorption Capacity (mg/g)
Thepreparedmagnetmicrospheres	Hb	217
Chitosan-based magnetic beads byin situmethod [20]	BSA	240.5
Cu^{2+}-cooperated magnetic imprinted nanomaterial [21]	Hb	116.3
surface-imprinted polyvinyl alcohol microspheres [22]	papain	44
magnetic immunoaffinity beads by dispersion polymerization [23]	Anti-Tf	2.0
Fe_3O_4@PMAA@Ni microspheres with flower-like Ni nanofoams [7]	Hb	2660

The adsorption isotherm describes the distribution of the adsorbed molecules between the liquid phase and the solid phase when the adsorption process reaches equilibrium. To study the adsorption mechanism of amino magnetic microspheres and carboxyl magnetic microspheres to Hb, whose adsorption capacity is larger than that of other proteins, the isotherm data were analyzed based on the Langmuir and Freundlich models respectively. The expressions, adsorption constantsm and correlation coefficients of the Langmuir and Freundlich models at 25 °C were calculated and are presented in Table 2.

Table 2. The adsorption isotherm parameters of Hb by amino magnetic microspheres and carboxyl magnetic microspheres.

Adsorbents	Langmuir Adsorption Isotherm $\frac{c_e}{q_e}=\frac{c_e}{q_m}+\frac{1}{Kq_m}$			Freundlich Adsorption Isotherm $\ln q_e=\ln K_F+\frac{\ln c_e}{n}$		
	q_m (mg/g)	K	R^2	K_F (mg/g)	$1/n$	R^2
amino magnetic microspheres	131.27	1.4837	0.9966	52.55	0.5342	0.9427
carboxyl magnetic microspheres	215.74	0.4282	0.9997	48.41	0.6603	0.9750

Where c_e is the equilibrium concentration (mg/L), q_e is the adsorption amount at equilibrium (mg/g), K is the Langmuir adsorption equilibrium constant, q_m is the Langmuir constant, which represents the saturated monolayer adsorption capacity (mg/g), K_F is a Freundlich constant related to the adsorption capacity (mg/g), and n is a Freundlich adsorption equilibrium constant relevant to the adsorption intensity.

Comparing the adsorption constants and correlation coefficients (R^2) of the Langmuir and Freundlich isotherms, it is suggested that the adsorption of the magnetic microspheres to Hb is in accordance with the Langmuir equation and the adsorption process is chemical monolayer adsorption. In Langmuir model, the maximum capacity q_mof the carboxyl microspheres in the Langmuir constant monolayer was 215.74 mg/g at 25 °C, indicating that the carboxyl magnetic microspheres have better adsorption capacity than the amino magnetic microspheres.

The kinetic curves (Figure 8) showed that the adsorption of the microspheres saturated quickly in 15 min. It is because electrostatic force between magnetic microspheres and proteins occurs mainly on the surface of the magnetic microspheres without the inward diffusion phenomenon. In the initial stage, there are a large number of active sites on the surface of the magnetic microspheres, and proteins were easily adsorbed. As the adsorption increases, the surface active sites of the magnetic microspheres decrease, and the adsorption becomes slower. Therefore, the microspheres would be more efficient for protein purification.

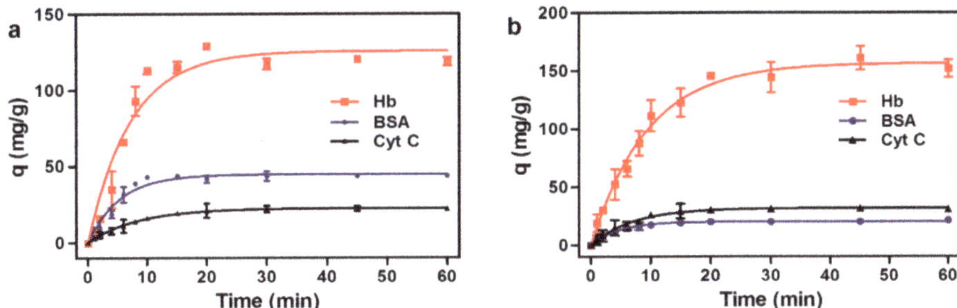

Figure 8. Adsorption kinetics of three proteins to magnetic microspheres. (**a**) Amino magnetic microspheres. (**b**) Carboxyl magnetic microspheres.

The adsorption data of the microspheres to Hb were respectively fitted to the pseudo first-order and pseudo-second-order kinetic models (Table 3). The correlation coefficients of the pseudo second-order kinetic model for amino magnetic microspheres and carboxyl magnetic microspheres were 0.9981 and 0.9911, and the maximum adsorption amounts obtained were 119.05 mg/g and 172.41 mg/g, respectively, which are consistent with the experimental results. It is suggested that the adsorption process is a pseudo second-order kinetic adsorption, which is consistent with the mentioned adsorption mechanism above.

Table 3. Adsorption kinetics parameters of HB by amino magnetic microspheres and carboxyl magnetic microspheres.

	Amino Magnetic Microspheres			Carboxyl Magnetic Microspheres		
	k	q_e/(mg/g)	R^2	k	q_e/(mg/g)	R^2
Lagergren first-order rate kinetics $\ln(q_e - q_t) = \ln q_e - kt$	0.2660	166.05	0.9209	0.1143	153.33	0.9767
Lagergren second-order rate kinetics $\frac{t}{q_t} = \frac{1}{kq_e^2} + \frac{t}{q_e}$	2.352	119.05	0.9981	0.0010	172.41	0.9911

Where q_e and q_t signify the amount adsorbed at equilibrium and at any time t, k is a Lagergren constant.

3. Materials and Methods

3.1. Materials

Glycidyl methacrylate (GMA) was obtained from TCI Company (Shanghai, China). Divinylbenzene was obtained from J&K chemical (Beijing, China). 1-Ethyl-3-(3-dimethyllaminopropyl) carbodiimide hydrochloride (EDC·HCl) and N-Hydroxysuccinimide (NHS) were purchased from Shanghai Medpep corporation (Shanghai, China). Bovine albumin (BSA), Hemoglobin from bovine blood (Hb), Cytochrome C (Cyt C) were obtained from Aladdin chemical corporation (Shanghai, China). Polyvinyl pyrrolidone (PVP K-30), 2,2′-azobis-(isobutyronitrile) (AIBN), ethylenediamine (EDA), anhydrous morpholine ethanesulfonic acid, and other chemicals were received from Beijing Chemical Factory (Beijing, China). Cell Counting Kit-8 (CCK-8) was received from Dojindo Laboratories, Kumamoto, Japan. Human pulmonary epithelial cells were purchased from InvivoGen (San Diego, CA, USA) and grown in Dulbecco's modified Eagle's medium (Sigma-Aldrich, St. Louis, MO, USA) containing 10% (*v/v*) fetal bovine serum, 50 units/mL penicillin, 50 mg/L streptomycin, 100 μg/mL normocin, and 10 μg/mL blasticidin. All chemicals were used without further treatment. Deionized water used in polymerization and characterization was distilled and purified by Aqua Pro (Chongqing, China).

3.2. Synthesis of PGMA Microspheres

PGMA microspheres are fabricated by dispersion polymerization method [24]. The polymerization was carried out under nitrogen in the three-necked flask equipped with a condenser. PVP K-30 (2.4 g) and GMA (8.0 g) dissolved in ethanol (67.0 g) was stirred at 300 rpm under nitrogen at room temperature for 15 min. After the initiator AIBN (0.16 g/5 g ethanol) was added, the polymerization was carried out at 70 °C for 2 h. Thereafter, Divinylbenzene (DVB, 0.24 g) was added into the flask smoothly, keeping the reaction going on for 5 h. Then the microspheres were centrifuged and washed with ethanol and water several times and dried under vacuum.

3.3. Synthesis of Amino Magnetic Microspheres

The amino magnetic microspheres were synthesized according to the reported method [15]. The dry PGMA microspheres (2.0 g) were added into a mixture of ethylene diamine (EDA, 50 mL) and water (50 mL) while stirring at 80 °C for 6 h. The microspheres were centrifuged and washed with water, and then dried under vacuum. The EDA functionalized microspheres (1.0 g) were added into water (100 mL), which was cooled to 0 °C under nitrogen for 30 min. Afterwards, $FeCl_3 \cdot 6H_2O$ (0.41 g) and $FeSO_4 \cdot 7H_2O$ (0.24 g) dissolved in water (10 mL) were added to the mixture respectively, and stirred for 3 h below 5 °C. After adding the ammonia solution (10 mL) smoothly, the ice bath was removed and the temperature was raised to 80 °C for 1.5 h. The resulting microspheres were centrifuged and washed with 0.5 M HCl three times and followed by pure water. The magnetic microspheres were dried by lyophilization and reserved.

3.4. Synthesis of Carboxyl Magnetic Microspheres

The amino magnetic microspheres were modified with sodium carboxymethyl cellulose to the synthesis of the amino-microspheres. EDC (1.94 g) and NHS (0.58 g) were dissolved in MES solution (100 mL, 0.1 M) together with sodium carboxymethyl cellulose solution (100 mL, 2.5 g/L). Then the dry amino magnetic microspheres (1.0 g) were added and stirred at room temperature for 2 h. Finally, the microspheres were washed with pure water and dried by lyophilization.

3.5. Characterizations of the Magnetic Microspheres

The morphology of magnetic microspheres was observed by scanning electron microscopy (SEM, S-4800, HITACHI, Tokyo, Japan). The sample powders were sputter-coated with gold before examination. The magnetic properties of magnetic microspheres were measured by a vibrating sample magnetometer (VSM, 9600-1, LDJ Electronics, Troy, MI, USA) at room temperature. TGA was performed with a thermal gravimetric analyzer (DTG-60H, Shimadzu, Kyoto, Japan) in the temperature range from room temperature to 800 °C with a scanning rate of 10 °C/min under nitrogen stream. The presence of certain functional groups was detected by Fourier Transform infrared spectrometer (FT-IR, ALPHA, Bruker, Billerica, MD, USA). The densities of amino groups and carboxyl groups on the microspheres were measured by a titration method.

3.6. Cytotoxicity Test of the Carboxyl Magnetic Microspheres

The cytotoxicity of the carboxyl magnetic microspheres was investigated using a CCK-8 method in vitro. The 96-well plates were seeded with a suspension of 5000 human pulmonary epithelial cells for 24 h to allow the cells to adhere. Then serial dilutions of carboxyl magnetic microspheres solution, the supernatant and medium alone (control) were added into the wells. After incubation at 37 °C for 24 h in an atmosphere of 5% CO_2, 10 μL CCK-8 solution was added to each well and the cells were incubated for another three hours. Absorbance at 450 nm was determined using a microplate reader using a microplate reader (MTP-880 Lab, Corona Electric, Ibaraki, Japan). Cytotoxicity was expressed as a percentage of viable cells compared with untreated control ones.

3.7. Binding Experiment

The binding properties of magnetic anion-exchange microspheres to the proteins were studied by HPLC with a diode array detector and the C8 column at 40 °C. The standard curves and adsorption capability for the proteins were measured with the corresponding conditions. For BSA, the mobile phase was acetonitrile/water (2/8–8/2, *v/v*), using a linear gradient elution at the wavelength of 280 nm, and the injection volume was 10 μL. For Hb, the mobile phase was acetonitrile/water (5/5, *v/v*), using isocratic elution at the wavelength of 400 nm, and the injection volume was 20 μL. For Cyt C, the mobile phase was acetonitrile/water (3/7–5/5, *v/v*), using linear gradient elution at the wavelength of 400 nm, and the injection volume was 20 μL.

The adsorption of proteins by magnetic microspheres was carried out in phosphate buffers (100 mM) of different pH values ranging from 3–11, adjusted with phosphoric acid solution or sodium hydroxide solution. The following experiment was performed in triplicate. The dry magnetic polymer microspheres (5 mg) were dispersed in 1 mL buffer solution followed by the adsorption experiment while the initial concentration of protein was determined from 0 to 10 mg/mL and the equilibrium time was 60 min. In the adsorption kinetics experiment, protein solution with initial concentration (BSA 5 mg/mL, Hb 5 mg/mL, Cyt C 1 mg/mL) was added, and the mixture was incubated at 25 °C for different times (0~60 min) respectively. Then the tubes were placed in the magnetic separation rack for 2 min, and the supernatant was extracted carefully for HPLC detection.

The protein binding quantity q (mg/g) of magnetic microspheres could be calculated from Formula (1).

$$q = \frac{(C_0 - C) \times V}{W} \quad (1)$$

where C_0 and C are the protein concentrations (mg/mL) before and after adsorption; V is the volume of protein solution (mL); W is the weight of magnetic microspheres (g).

4. Conclusions

A new process to obtain the carboxymethyl cellulose surface-coated magnetic polymer microspheres by EDC method was performed in this study. The superparamagnetism and no significant cytotoxicity of the magnetic microspheres attribute to their potential application in vivo. The adsorption capacity of three proteins (BSA, Hb, and Cyt C) on amino magnetic microspheres and carboxyl magnetic microspheres was evaluated, wherein maximum adsorption capacity of Hb on carboxyl magnetic microspheres reached 215.74 mg/g within sufficient binding time at appropriate pH value. However, further studies based on the increase of the stability of magnetic microspheres, specific adsorption of a certain protein, and desorption of protein are required. This paper provides an idea for the preparation of magnetic microspheres for protein separation, which is expected to be a fast and efficient new way of protein separation in the future.

Author Contributions: Original draft writing and validation, Z.W.; methodology, W.W.; conceptualization, Z.M.; supervision, review, and editing, M.X. All authors have read and agreed to the published version of the manuscript.

Funding: This research was funded by the Natural Science Foundation of China, grant number 21874009.

Institutional Review Board Statement: Not applicable.

Informed Consent Statement: Not applicable.

Data Availability Statement: Not applicable.

Acknowledgments: We are grateful for the support of the SEM test from Analysis &Testing Center in Beijing Institute of Technology and cytotoxicity test from National Institute for Materials Science (NIMS), Japan.

Conflicts of Interest: The authors declare no conflict of interest.

References

1. Schmelter, C.; Funke, S.; Treml, J.; Beschnitt, A.; Perumal, N.; Manicam, C.; Pfeiffer, N.; Grus, F. Comparison of Two Solid-Phase Extraction (SPE) Methods for the Identification and Quantification of Porcine Retinal Protein Markers by LC-MS/MS. *Int. J. Mol. Sci.* **2018**, *19*, 3847. [CrossRef] [PubMed]
2. Karakus, C.; Uslu, M.; Yazici, D.; Salih, B.A. Evaluation of immobilized metal affinity chromatography kits for the purification of histidine-tagged recombinant CagA protein. *J. Chromatogr. B* **2016**, *1021*, 182–187. [CrossRef] [PubMed]
3. Keilhauer, E.C.; Hein, M.Y.; Mann, M. Accurate Protein Complex Retrieval by Affinity Enrichment Mass Spectrometry (AE-MS) Rather than Affinity Purification Mass Spectrometry (AP-MS). *Mol. Cell. Proteom.* **2015**, *14*, 120–135. [CrossRef] [PubMed]
4. Ma, Y.; Chen, T.; Iqbal, M.Z.; Yang, F.; Hampp, N.; Wu, A.; Luo, L. Applications of magnetic materials separation in biological nanomedicine. *Electrophoresis* **2019**, *40*, 2011–2028. [CrossRef]
5. Schwaminger, S.P.; Blank-Shim, S.A.; Scheifele, I.; Pipich, V.; Fraga-García, P.; Berensmeier, S. Design of Interactions Between Nanomaterials and Proteins: A Highly Affine Peptide Tag to Bare Iron Oxide Nanoparticles for Magnetic Protein Separation. *Biotechnol. J.* **2019**, *14*, 1800055. [CrossRef]
6. Tavakoli, Z.; Rasekh, B.; Yazdian, F.; Maghsoudi, A.; Soleimani, M.; Mohammadnejad, J. One-step separation of the recombinant protein by using the amine-functionalized magnetic mesoporous silica nanoparticles; an efficient and facile approach. *Int. J. Biol. Macromol.* **2019**, *135*, 600–608. [CrossRef]
7. Wang, Y.; Wei, Y.; Gao, P.; Sun, S.; Du, Q.; Wang, Z.; Jiang, Y. Preparation of Fe_3O_4@PMAA@Ni Microspheres towards the Efficient and Selective Enrichment of Histidine-Rich Proteins. *ACS Appl. Mater. Interfaces* **2021**, *13*, 11166–11176. [CrossRef]
8. Yang, Q.; Wu, Y.; Lan, F.; Ma, S.; Xie, L.; He, B.; Gu, Z. Hollow superparamagnetic PLGA/Fe3O4 composite microspheres for lysozyme adsorption. *Nanotechnology* **2014**, *25*, 085702. [CrossRef]
9. Yang, S.; Zhang, X.; Zhao, W.; Sun, L.; Luo, A. Preparation and evaluation of Fe_3O_4 nanoparticles incorporated molecularly imprinted polymers for protein separation. *J. Mater. Sci.* **2015**, *51*, 937–949. [CrossRef]
10. Bae, S.W.; Kim, J.I.; Choi, I.; Sung, J.; Hong, J.-I.; Yeo, W.-S. Zinc Ion-immobilized Magnetic Microspheres for Enrichment and Identification of Multi-phosphorylated Peptides by Mass Spectrometry. *Anal. Sci.* **2017**, *33*, 1381–1386. [CrossRef]
11. Lai, Z.; Zhang, M.; Zhou, J.; Chen, T.; Li, D.; Shen, X.; Liu, J.; Zhou, J.; Li, Z. Fe_3O_4@PANI: A magnetic polyaniline nanomaterial for highly efficient and handy enrichment of intact N-glycopeptides. *Analyst* **2021**, *146*, 4261–4267. [CrossRef]
12. Shen, L.H.; Zhou, J.; Wang, Y.X.; Kang, N.; Ke, X.B.; Bi, S.L.; Ren, L. Efficient Encapsulation of Fe3O4Nanoparticles into Genetically Engineered Hepatitis B Core Virus-Like Particles Through a Specific Interaction for Potential Bioapplications. *Small* **2015**, *11*, 1190–1196. [CrossRef]
13. Wang, Y.; He, J.; Chen, J.; Ren, L.; Jiang, B.; Zhao, J. Synthesis of Monodisperse, Hierarchically Mesoporous, Silica Microspheres Embedded with Magnetic Nanoparticles. *ACS Appl. Mater. Interfaces* **2012**, *4*, 2735–2742. [CrossRef]
14. Liu, Y.; Song, L.; Du, L.; Gao, P.; Liang, N.; Wu, S.; Minami, T.; Zang, L.; Yu, C.; Xu, X. Preparation of Polyaniline/Emulsion Microsphere Composite for Efficient Adsorption of Organic Dyes. *Polymers* **2020**, *12*, 167. [CrossRef]
15. Yan, B.; Zeng, C.; Yu, L.; Wang, C.; Zhang, L. Preparation of hollow zeolite NaA/chitosan composite microspheres via in situ hydrolysis-gelation-hydrothermal synthesis of TEOS. *Microporous Mesoporous Mater.* **2018**, *257*, 262–271. [CrossRef]
16. Liao, M.H.; Chen, D.H. Fast and efficient adsorption/desorption of protein by a novel magnetic nano-adsorbent. *Biotechnol. Lett.* **2002**, *24*, 1913–1917. [CrossRef]
17. Heebøll-Nielsen, A.; Dalkiær, M.; Hubbuch, J.J.; Thomas, O.R.T. Superparamagnetic adsorbents for high-gradient magnetic fishing of lectins out of legume extracts. *Biotechnol. Bioeng.* **2004**, *87*, 311–323. [CrossRef]
18. Zhang, H.P.; Bai, S.; Xu, L.; Sun, Y. Fabrication of mono-sized magnetic anion exchange beads for plasmid DNA purification. *J. Chromatogr. B* **2009**, *877*, 127–133. [CrossRef]
19. Jiang, X.Y.; Bai, S.; Sun, Y. Fabrication and characterization of rigid magnetic monodisperse microspheres for protein adsorption. *J. Chromatogr. B* **2007**, *852*, 62–68. [CrossRef]
20. Mahdavinia, G.R.; Soleymani, M.; Etemadi, H.; Sabzi, M.; Atlasi, Z. Model protein BSA adsorption onto novel magnetic chitosan/PVA/laponite RD hydrogel nanocomposite beads. *Int. J. Biol. Macromol.* **2018**, *107*, 719–729. [CrossRef]
21. Shi, L.; Tang, Y.; Hao, Y.; He, G.; Gao, R.; Tang, X. Selective adsorption of protein by a high-efficiency Cu^{2+}-cooperated magnetic imprinted nanomaterial. *J. Sep. Sci.* **2016**, *39*, 2876–2883. [CrossRef] [PubMed]

22. Gao, B.; Chen, T.; Cui, K. Constituting of a new surface-initiating system on polymeric microspheres and preparation of basic protein surface-imprinted material in aqueous solution. *Polym. Adv. Technol.* **2018**, *29*, 575–586. [CrossRef]
23. Saçlıgil, D.; Şenel, S.; Yavuz, H.; Denizli, A. Purification of transferrin by magnetic immunoaffinity beads. *J. Sep. Sci.* **2015**, *38*, 2729–2736. [CrossRef] [PubMed]
24. Ma, Z.; Guan, Y.; Liu, H. Synthesis and characterization of micron-sized monodisperse superparamagnetic polymer particles with amino groups. *J. Polym. Sci. Part A Polym. Chem.* **2005**, *43*, 3433–3439. [CrossRef]

International Journal of *Molecular Sciences*

MDPI

Article

Effects of Hexagonal Boron Nitride Insulating Layers on the Driving Performance of Ionic Electroactive Polymer Actuators for Light-Weight Artificial Muscles

Minjeong Park [1], Youngjae Chun [2], Seonpil Kim [3], Keun Yong Sohn [1,*] and Minhyon Jeon [1,*]

1 Department of Nanoscience and Engineering, Center for Nano Manufacturing, Inje University, Gimhae 50834, Korea; mjpark9121@gmail.com

2 Department of Industrial Engineering, Bioengineering, University of Pittsburgh, Pittsburgh, PA 15261, USA; yjchun@pitt.edu

3 Department of Military Information Science, Gyeongju University, Gyeongju 38065, Korea; seonpil@gu.ac.kr

* Correspondence: ksohn@inje.ac.kr (K.Y.S.); mjeon@inje.ac.kr (M.J.); Tel.: +82-55-320-3714 (K.Y.S.); +82-55-320-3672 (M.J.); Fax: +82-320-3963 (M.J.)

Abstract: To improve the energy efficiency and driving performance of ionic electroactive polymer actuators, we propose inserting insulating layers of 170 nm hexagonal boron nitride (h-BN) particles between the ionic polymer membrane and electrodes. In experiments, actuators exhibited better capacitance (4.020×10^{-1} F), displacement (6.01 mm), and curvature (35.59 m^{-1}) with such layers than without them. The excellent insulating properties and uniform morphology of the layers reduced the interfacial resistance, and the ion conductivity (0.071 S m^{-1}) within the ionic polymer improved significantly. Durability was enhanced because the h-BN layer is chemically and thermally stable and efficiently blocks heat diffusion and ion hydrate evaporation during operation. The results demonstrate a close relationship between the capacitance and driving performance of actuators. A gripper prepared from the proposed ionic electroactive polymer actuator can stably hold an object even under strong external vibration and fast or slow movement.

Keywords: insulators; interface structure; electrochemistry; mechanical properties; ionic electroactivepolymers; acuators; artificial muscle; capacitance

Citation: Park, M.; Chun, Y.; Kim, S.; Sohn, K.Y.; Jeon, M. Effects of Hexagonal Boron Nitride Insulating Layers on the Driving Performance of Ionic Electroactive Polymer Actuators for Light-Weight Artificial Muscles. *Int. J. Mol. Sci.* **2022**, *23*, 4981. https://doi.org/10.3390/ijms23094981

Academic Editor: Raghvendra Singh Yadav

Received: 4 April 2022
Accepted: 28 April 2022
Published: 29 April 2022

Publisher's Note: MDPI stays neutral with regard to jurisdictional claims in published maps and institutional affiliations.

Copyright: © 2022 by the authors. Licensee MDPI, Basel, Switzerland. This article is an open access article distributed under the terms and conditions of the Creative Commons Attribution (CC BY) license (https://creativecommons.org/licenses/by/4.0/).

1. Introduction

Ionic electroactive polymer (IEAP) actuators have garnered significant attention for use in applications such as intelligent robots, biomedical devices, and micro-electro-mechanical systems [1–3]. They are lightweight, operate at low voltages, and are suitable for a wide range of environments [4–8]. IEAP consists of two electrodes and a polymer membrane. Various types of electrodes [9–11] and polymer materials [12–17] have been considered for improving IEAP actuator performance, but further research is required to increase energy efficiency, reliability, and durability. Recently, the interface between the electrode and polymer membrane has attracted research attention with regard to improving the actuation performance, stability, and electrochemical properties of the IEAP actuator. For the study of the actuator interface, the basic structure of IEAP should be noted.

The structure of an IEAP actuator is similar to that of a capacitor, which stores energy in the form of electrical charge [18,19]. This has led to some researchers in the IEAP field to report the capacitance results of actuators [13]. For example, Akle et al. [20] reported a linear correlation between the strain response and capacitance of ionomeric materials. However, the capacitance and strain values of actuators require further improvement, and there have been few studies of the relationship between the driving performance and capacitance of IEAP actuators. Known methods of improving capacitance include the use of electrolyte polymers, insulators, and carbon materials [21–23]; we suggest that the electrochemical and driving properties of IEAP actuators could be improved by placing insulating layers between the polymer membrane and electrodes, forming a structure similar to that of a traditional capacitor. As the insulator, we suggest hexagonal boron nitride (h-BN). This chemically stable material, which has a two-dimensional structure, excellent thermal conductivity (~400 W m^{-1} K^{-1}), and a wide band gap of ~6 eV [24–29], has been reported to block large ions under an applied voltage [30]; these characteristics indicate that it would be a suitable insulating material for use in IEAP actuators.

IEAP actuators work because ion hydrates and water molecules move within the ionic polymer membrane upon application of an external voltage. Unfortunately, as the actuator is operated, the ion hydrates evaporate because of the heat generated by the electrodes; this reduces durability. However, water molecules are larger (>0.28 nm) than the lattice constant of h-BN ($a_{\text{h-BN}}$ = 0.25 nm); therefore, their evaporation from the ionic polymer can be efficiently blocked by h-BN insulating layers.

In this study, we developed and tested an IEAP actuator with insulating layers of h-BN particles between the ionic polymer membrane and the top and bottom electrodes. The insulating properties of h-BN particles of different sizes were systematically tested to determine the optimal performance. We observed the changes in the driving and electrochemical properties according to the h-BN particle size and compared the results with those obtained using a conventional IEAP actuator without insulating layers. The IEAP actuator with h-BN insulating layers (particle size = 170 nm) exhibited improved capacitance and ion conductivity compared to the actuator without h-BN insulating layers; thus, the driving characteristics were improved. In addition, the layered actuator had high energy and power densities and exhibited stable gripping abilities even when subjected to external vibration and movement, suggesting that IEAP actuators with insulating layers could be utilized in soft robotics and the flexible actuator field as well as in energy storage.

2. Results

We designed an IEAP actuator (Figure 1) with an ionic polymer (Nafion-117) membrane, P/GO–Ag electrodes, and insulating layers between the polymer membrane and electrodes composed of h-BN powders with particle sizes of 1 μm, 800 nm, 300 nm, or 170 nm (denoted h-BN−1 μm, h-BN−800 nm, h-BN−300 nm, and h-BN−170 nm, respectively), in a Nafion matrix. The P/GO–Ag paper electrode and h-BN insulating layers were included to increase the mobility of ions in the Nafion membrane and decrease the interface resistance between the electrodes and Nafion membrane, respectively.

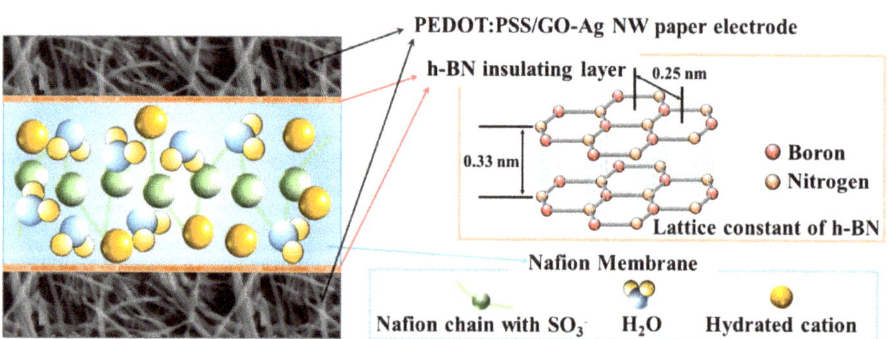

Figure 1. Schematic of ionic electroactive polymer (IEAP) actuator structure with hexagonal boron nitride (h-BN) insulating layers.

Figure S1 shows the morphological properties of the P/GO–Ag paper electrode. The Ag NWs within the electrode were well connected, as shown in Figure S1a,c. The average thickness of the P/GO–Ag electrode was ~13 μm (Figure S1b), and the surface was smooth, with an average surface roughness of ~88 nm (Figure S1c). The sheet resistance was uniform across the entire electrode surface, with a value of ~200 mΩ sq^{-1}. The h-BN powder was well dispersed without aggregation within the insulating layer (Figure 2a–d), particularly for smaller h-BN particle sizes; the roughness of the insulating layer decreased with decreasing h-BN particle size (Figure S2). The P/GO–Ag electrode and Nafion-117 membrane became more smoothly attached as the h-BN particle size decreased (Figure 2e–h). These results suggest that the h-BN−170-nm insulating layer, which was thin and uniform on both sides of the Nafion membrane, may have reduced ion intercalation to the PEDOT:PSS layers of the electrode. This would increase the capacitance and water uptake (WUP) of the actuator by blocking the leakage of ions from the Nafion membrane through the P/GO–Ag electrodes.

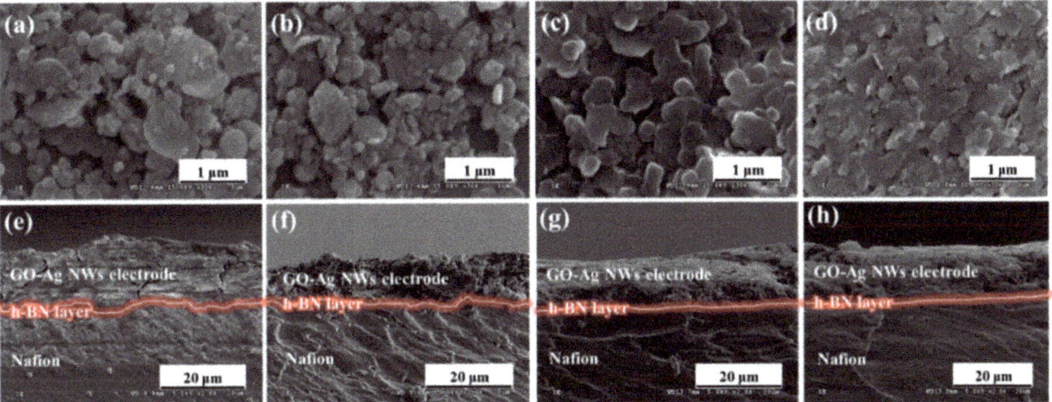

Figure 2. Top and cross-sectional scanning electron microscopy (SEM) images of IEAP actuators with h-BN insulating layers with different particle sizes: (**a**,**b**) 1 µm, (**c**,**d**) 800 nm, (**e**,**f**) 300 nm, and (**g**,**h**) 170 nm.

We also investigated the effect of these layers on the ion capacitance and driving performance. An IEAP actuator with insulating layers between the electrodes and ionic polymer membrane can be regarded as a capacitor. We measured the electrochemical properties of the IEAP actuators with and without h-BN insulating layers and observed the effect of the h-BN insulating layers on the current–potential curves and capacitance. Cyclic voltammetry (CV) was conducted in the applied voltage range of −1.0 to +1.0 V at a scan rate of 50 mV s^{-1}, revealing quasi-ideal parallel-plate capacitive behavior.

The CV curves differed with the size of the h-BN particles. The actuators without h-BN layers produced rectangular CV curves (Figure 3a and Figure S3a). No redox peak was observed and the specific capacitance was 0.5 mF g^{-1} at a scan rate of 50 mV s^{-1}. By contrast, the actuators with h-BN layers produced redox peaks in both the negative and positive regions (−0.4 to +0.4 V) because of the redox reaction at the h-BN surface. The area of the CV curve was much larger for the IEAP actuators with h-BN insulating layers than for those without them; it increased as the h-BN particle size decreased.

The total area of the CV curve corresponds to the capacitance. The average capacitances are presented in Figure S3, and the average specific capacitances considering the weight of the IEAP actuators are presented in Figure 3b. The IEAP actuators with h-BN insulating layers had much higher specific capacitances than those without, and the specific capacitance increased with decreasing h-BN particle size. In particular, the IEAP actuator with h-BN−170 nm insulating layers exhibited greater capacitance (4.020×10^{-1} F) and specific capacitance (5.83 F g^{-1}) than the IEAP actuator without insulating layers. These results are sufficient to show that the IEAP actuator has excellent capacity performance and high energy efficiency within the Nafion membrane [31–33].

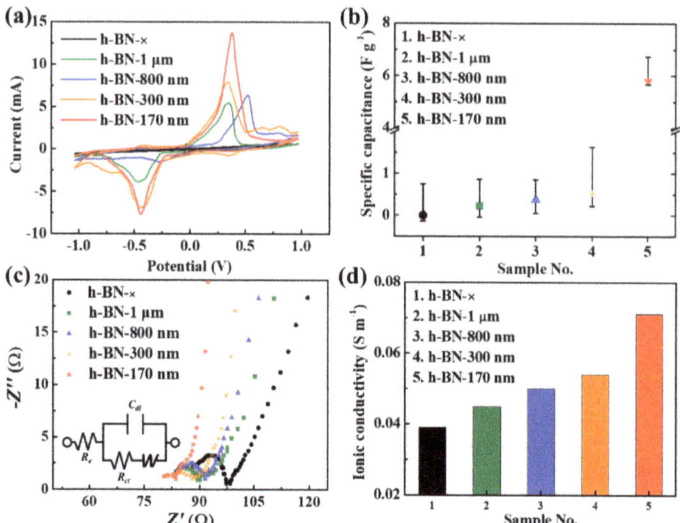

Figure 3. Electrochemical properties of IEAP actuators without insulating layers (h-BN−×) and with h-BN insulating layers with different particle sizes (1 μm, 800 nm, 300 nm, and 170 nm): (**a**) Cyclic voltammetry (CV) curves; (**b**) Specific capacitance; (**c**) Nyquist plots (inset: diagram of equivalent electrical circuit; W represents the Warburg diffusion element, R_e is the electrolyte resistance, R_{ct} is the polarization resistance, and C_{dl} is the double-layer capacitance); (**d**) Ionic conductivity.

The electrochemical impedance of the fabricated IEAP actuators was measured in the frequency range from 1 Hz to 100 kHz. Figure 3c shows Nyquist plots of the IEAP actuators. Each Nyquist plot comprises a semicircle and line. The presence of the semicircle demonstrates that the system has an equivalent electrical circuit (inset in Figure 3c) with electrolyte resistance R_e and double-layer capacitance C_{dl} in parallel at high frequencies [34,35]. The ion conductivity (σ_i) was calculated from the electrolyte resistance of the IEAP actuator (Equation (S2)); it increases as the electrolyte resistance of the actuator decreases. The electrolyte resistance represents the interface resistance between the electrode and ion polymer film, and a decrease in interface resistance can enhance the ion conductivity of the ion polymer film. The ion conductivities of the IEAP actuators with h-BN insulating layers increased with decreasing h-BN particle size (Figure 3b). The ion conductivity of the IEAP actuator with h-BN−170-nm insulating layers was approximately twice that of the IEAP actuator with none.

We measured the driving performances of the IEAP actuators according to the particle size of the h-BN powder. The driving performance was also evaluated as a function of the capacitance. In addition, the weights of the IEAP actuators before and after the ion-substitution process were measured to assess the WUP within the Nafion membrane (Equation (S3)), thus obtaining information on the amount of water molecules and ion hydrates contained in the Nafion membrane. The weights and WUPs of the IEAP actuators are presented in Table 1. The IEAP actuator with h-BN−170-nm insulating layers had the highest WUP of the fabricated actuators (32.69%), because the uniform formation of the insulating layers across the electrode/Nafion membrane interface reduced the ion intercalation of the PEDOT:PSS layer and ensured the ions were retained inside the Nafion membrane. Table 1 also presents the weights of the IEAP actuators before and after actuation. The weight loss of the IEAP actuator with h-BN−170-nm insulating layers was ~2.9%, which is one-ninth that of the IEAP actuator without insulating layers (27.54%). The results suggest that the 170 nm h-BN particles form a uniform interface layer, improving the durability of the IEAP actuator because they are smaller than water molecules and therefore inhibit evaporation from the Nafion membrane during operation.

Table 1. Water uptake (WUP) and weight loss (WL) of IEAP actuators without and with h-BN insulating layers.

Sample	W_d [1] [g]	W_s [2] [g]	W_f [3] [g]	WUP [4] [%]	WL [5] [%]
h-BN−× [6]	0.056	0.069	0.050	23.21	27.54
h-BN−1 μm [7]	0.059	0.074	0.063	25.42	14.86
h-BN−800 nm [7]	0.056	0.071	0.065	26.79	8.45
h-BN−300 nm [7]	0.054	0.069	0.065	27.78	5.80
h-BN−170 nm [7]	0.052	0.069	0.067	32.69	2.90

[1] W_d: Initial (dry) weight of IEAP actuator. [2] W_s: Weight of IEAP actuator after ion substitution of water molecules and ion hydrates in ionic polymer. [3] W_f: Weight of IEAP actuator after actuation. [4] WUP: Water uptake during ion substitution {WUP = $[(W_s - W_d)/W_d] \times 100$}. [5] WL: Weight loss during actuation {WL = $[(W_s - W_f)/W_s] \times 100$}. [6] IEAP actuator without insulating layers. [7] IEAP actuators with insulating layers comprising h-BN powder with a particle size of 1 μm, 800 nm, 300 nm, or 170 nm in a Nafion matrix.

The Nafion membrane of the IEAP actuator with h-BN−170-nm insulating layers contained 50% more ions than that of the actuator without insulating layers (see WUP in Table 1). Therefore, the insulating layers may improve the driving properties. The driving performances of the IEAP actuators were measured under a sinusoidal input with a peak voltage of 2.5 V and excitation frequency of 0.2 Hz (Figure 4a,b). The IEAP actuator with h-BN−170 nm insulating layers exhibited a larger displacement (3.21 mm) and peak-to-peak value (6.01 mm) than the other IEAP actuators. In agreement with the weight-loss results, the IEAP actuator with h-BN−170 nm insulating layers had better durability than the other IEAP actuators: it showed only slight degradation (within 4.5%) of the peak-to-peak value over 5 h, compared to the drastic 96.3% degradation over 5 h of the IEAP actuator without insulating layers (Figure 4b). Its curvature performance (35.59 m^{-1}) was also better than those of the other IEAP actuators and ~7.4 times greater than that of the actuator without insulating layers. Figure 4d shows the actual movement of the IEAP actuators after 60 s under 2.5 V_{DC} input. These results show that the h-BN insulating layers reduced the interface resistance, thus increasing the ion storage capacitance and ion conductivity in the Nafion-polymer membrane. The driving characteristics improved accordingly. Figure S4 shows the driving performance of the IEAP actuator with h-BN−170-nm insulating layers under various voltages and frequencies.

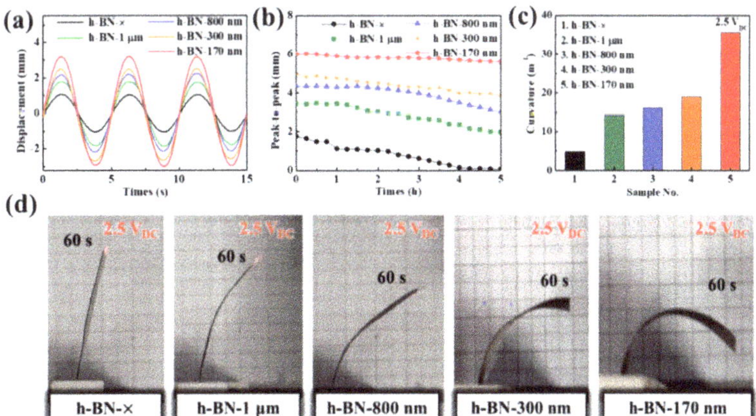

Figure 4. Performances of IEAP actuators without insulating layers (h-BN−×) and with h-BN insulating layers with different particle sizes (1 μm, 800 nm, 300 nm, and 170 nm): (**a**) Displacement at 0.2 Hz and 2.5 V_{AC}; (**b**) Peak-to-peak value at 0.2 Hz and 2.5 V_{AC}; (**c**) Curvature at 2.5 V_{DC}; (**d**) Photographs of movement at 2.5 V_{DC}.

3. Discussion

The surface resistance, capacitance, and driving characteristics of the actuators obtained in this study are compared with reported results [36–41] in Table 2. The maximum strain value of the IEAP actuator with h-BN−170-nm insulating layers was 8.31%, larger than those of the earlier actuators (the calculation details [42] are provided in the Supplementary Materials). In addition, the electrode surface resistance and capacitance of the developed actuator were superior to those of earlier actuators. Most previous studies did not investigate both the capacitance and driving characteristics of the actuator, and none mention the correlation between them reported here. Figure 5 shows that the displacement and curvature of the IEAP actuators increase with increasing capacitance. Moreover, the capacitance and driving performance of IEAP actuators increase with decreasing h-BN particle size. In particular, the capacitance of the IEAP actuator with h-BN−170-nm insulating layers is more than 10^5 times that of the IEAP actuator without insulating layers. These results highlight that the h-BN insulating layers increase the capacitance of the IEAP actuator, with a positive effect on the driving performance.

Table 2. Electrode, capacitance, and driving characteristics of various actuators from prior reports and the IEAP actuator with h-BN−170-nm insulating layers.

Actuator Types	Electrode Sheet Resistance [Ω sq.$^{-1}$]	Specific Capacitance [F g^{-1}]	Max. Displacement [mm] (at 2.5 VAC)	Curvature [m^{-1}]	Strain [%]	Ref.
P/GO-Ag IEAP [1] Actuator with h-BN−170-nm Insulating Layers	2.00×10^{-5}	5.83	6.01	35.59	8.31	This work
Au	<100	10^{-4}	-	-	1.04	[5]
Pt	10	7.5×10^{-4} [F cm^{-2}]	-	-	5	[10]
Ag nanopowder	0.12–0.15	-	5.00	-	-	[14]
Graphene	-	0.95 [meq/g]	0.145 (at 0.5 VAC)	-	-	[15]
Ionomer	-	-	5.00	-	-	[16]
IPMC	-	-	-	0.2×10^{-3}	-	[17]
Ionomeric-IL	-	(1–5) $\times 10^{-3}$ [F cm^{-2}]	-	-	2.44	[43]
PSS-*b*-PMB	-	0.12	4.00 (at 3.0 VAC)	-	4	[44]

[1] PEDOS:PSS/graphene oxide–Ag nanowire electrode-based ionic electroactive polymer.

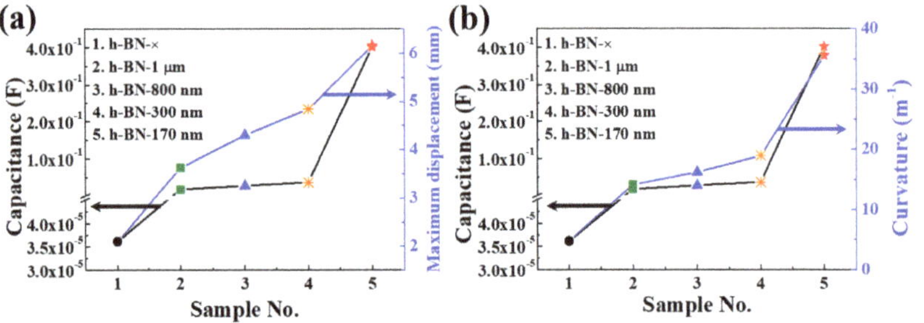

Figure 5. Effect of h-BN layer thickness on capacitance and driving performance of IEAP actuators without insulating layers (h-BN−×) and with h-BN insulating layers with different particle sizes (1 μm, 800 nm, 300 nm, and 170 nm): (**a**) Capacitance vs. maximum displacement; (**b**) Capacitance vs. curvature.

As shown in Figure 6a, different potential application sectors of IEAP actuators have distinct power-density and energy-density requirements; actuators without insulating layers do not satisfy any of them. We investigated the power densities (P) and energy densities (E) of the IEAP actuators with h-BN insulating layers (Figure 6a). The IEAP actuators with h-BN insulating layers had supercapacitor characteristics [41]. As the h-BN particle size decreased from 1 μm to 170 nm, P increased from 23.97 to 606.88 W kg^{-1}, whereas E increased from 0.2 to 5.06 Wh kg^{-1}. In particular, the P and E values of the IEAP actuator with h-BN−170 nm insulating layers were superior (P: 606.88 W kg^{-1}, E: 5.06 Wh kg^{-1}) to those of the IEAP actuator without insulating layers (P: 0.054 W kg^{-1}, E: 4.54 × 10−4 Wh kg^{-1}).

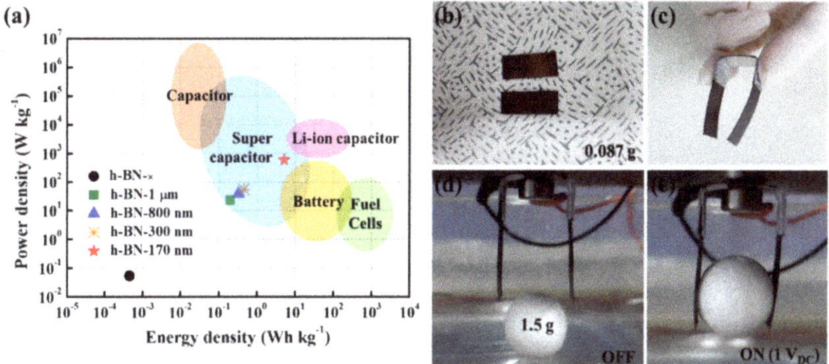

Figure 6. (**a**) Average power and energy densities of IEAP actuators without insulating layers (h-BN−×) and with h-BN insulating layers with different particle sizes (1 μm, 800 nm, 300 nm, and 170 nm) compared to requirements of various industrial fields. Images of (**b**) sheets of IEAP actuator with h-BN−170 nm used in (**c**) two-finger gripper. Images taken (**d**) before and (**e**) after grasping an object at 1.0 V_{DC}.

A simple device was developed to illustrate the applicability of the IEAP actuator with h-BN−170 nm insulating layers. A two-finger gripper composed of two IEAP cantilevers (IEAP-actuator sheets) was designed to imitate the grasping motion of fingers; its operation is shown in Figure 6d, e, as well as Video S1. Figure 6b shows the two metal plates used to create the gripper. The two-finger gripper had a length and width of 23 and 10 mm, respectively, and a weight of 0.087 g (Figure 6c). Given the importance of water molecules and cations inside the polymer membrane for actuator operation, working voltage conditions that would not cause water decomposition (i.e., <1.23 V) were required; therefore, the actuator was operated at 1 V. The performance of the two-finger gripper on a 1.5 g plastic ball is shown in Figure 6d. In fact, the two-finger gripper was able to maintain a stable hold on objects that weighed more than 8.6 times as much as the actuator device, despite strong external vibration and fast or slow movement (Video S1, Supporting Information).

The high capacitance, high driving performance, and minimal leakage of vaporized water molecules and cations during electrical stimuli exhibited by IEAP actuators with h-BN−170 nm insulating layers suggest other possible applications, including biomedical devices [42,45] and biomimetic soft robots [46,47]. Moreover, the power and energy density results show that these actuators could be used as supercapacitors [43].

4. Materials and Methods

4.1. Materials and Characterization

We used deionized (DI) water (resistivity: >18 MΩ cm at 25 °C) in all experiments. Poly(3,4-ethylenedioxythiophene):poly(styrenesulfonate) (PEDOT:PSS) (1.3 wt%) and Triton X-100 were purchased from Sigma-Aldrich (St. Louis, MO, USA). Graphene oxide (GO) was synthesized by the Hummers and Hoffman methods, and Ag nanowires (NWs) were purchased from DUKSAN Hi-Metal (Ulsan, Korea). Cellulose-membrane filter papers (pore size: 0.20 μm, diameter: 47 mm) were purchased from HYUNDAI MICRO., Ltd. (Anseong, Korea). The PEDOT:PSS, GO, Ag NWs, and cellulose filter papers were used to fabricate PEDOT:PSS/GO–Ag (P/GO–Ag) NW paper electrodes for the IEAP actuator. Sulfonated-tetrafluoroethylene-based fluoropolymer-copolymer membranes (thickness: ~183 μm; Nafion 117) and Nafion solution were purchased from Dupont (Wilmington, DE, USA). These were used as the polymer membranes of the IEAP actuator and the attachment resin between the polymer membranes and P/GO–Ag electrodes, respectively. Insulating layers for the IEAP actuators were formed from h-BN powders with different particle sizes (1 μm, 800 nm, 300 nm, and 170 nm), purchased from Ditto Technology (DT-BN-20PG, Korea). Lithium chloride (LiCl) purchased from Sigma-Aldrich (Seoul, Korea) and 1-ethyl-3-methylimidazolium trifluoromethanesulfanate (EMIM-OTf) purchased from Merck (Seoul, Korea) were used in the ion-substitution process to produce the actuator. We fabricated an h-BN/Nafion mixture using an ultrasonic probe sonicator (UW200, Bandelin, Germany) to ensure that the h-BN powder (0.5 wt%) was well dispersed in the Nafion solution. The morphological properties of the P/GO–Ag electrode, h-BN insulating layer, and IEAP actuator were investigated using field-emission scanning electron microscopy (S-4300, Hitachi, Japan) and atomic force microscopy (non-contact mode, XE-100, Parksystems, Suwon, Korea). The sheet resistance of the P/GO–Ag electrodes was measured using a four-point probe system (DASOL, ENG, FPP-HS 8). The electrochemical properties of the IEAP actuator were measured using cyclic voltammetry (CompactStat, HS Technology, Korea), and the current density, capacitance, and impedance were analyzed. The driving performance was measured using a laser displacement sensor (OMRON, ZS-LD80; beam length: 0.9 mm, beam thickness: 60 μm).

4.2. Fabrication of IEAP Actuator with Insulating Layers

First, we incorporated Li+ and EMIM-OTf cations into the Nafion-117 membrane by an ion exchange process using LiCl (LiCl: 16 g, DI water: 244 mL) and EMIM-OTf (EMIM-OTf: 55 g, methanol: 28 g) solutions. Second, we mixed GO and Ag NWs in a 1:2.5 volume ratio, and then fabricated the GO–Ag NW electrodes by filtration using the cellulose filter paper. The filtered GO–Ag NW electrode was put into acetone to peel off the filter paper and dried in a vacuum oven at 100 °C. Subsequently, we coated the Nafion-117 membranes with the h-BN/Nafion mixture to create a uniform insulating layer. The GO–Ag NW electrodes were attached to the Nafion-117 membrane by hot-pressing at 0.1 MPa and 100 °C for 5 min. Finally, a mixture of Triton X-100 (0.766 mL) and PEDOT:PSS (10 mL) [27] was spin-coated onto both surfaces of the IEAP actuator. The finished P/GO–Ag electrode-based IEAP actuators with h-BN insulating layers were rectangular with dimensions 0.5 × 3.5 cm. An actuator sample without h-BN insulating layers was fabricated in a similar manner but without the h-BN/Nafion mixture coating.

5. Conclusions

In this study, IEAP actuators with h-BN insulating layers between the ionic polymer membrane and the top and bottom electrodes were fabricated and tested. We demonstrated for the first time that the driving performance of an actuator improves as its capacitance increases, highlighting the strong relationship between the capacitance and actuator performance. The IEAP actuator with h-BN−170-nm insulating layers showed the highest capacitance (4.020×10^{-1} F) and largest bending value (6.01 mm) of the tested IEAP actuators. The durability of this IEAP actuator (displacement degradation within 4.5%) was also greatly improved (by a factor of ~21.4) over that of the IEAP actuator without insulating layers (displacement degradation of 96.3%). We attribute the improved durability to the lattice constant of the h-BN particles being smaller than the size of the hydrated ions; this suppresses evaporation from the Nafion membrane during operation. Our results suggest a new direction for IEAP actuator research and may to contribute to the commercialization of actuators in the fields of intelligent robotics and biomimetic medical devices.

Supplementary Materials: The following supporting information can be downloaded at: https://www.mdpi.com/article/10.3390/ijms23094981/s1.

Author Contributions: Conceptualization, M.J.; validation, M.P., Y.C. and S.K.; investigation, M.P., K.Y.S. and S.K.; data curation, M.P. and Y.C.; writing—original draft preparation, M.P.; writing—review and editing, Y.C., K.Y.S. and M.J.; visualization, K.Y.S. and M.P.; supervision, M.J.; project administration, M.P., S.K. and M.J. All authors have read and agreed to the published version of the manuscript.

Funding: This research was supported by Basic Science Research Program through the National Research Foundation of Korea (NRF) funded by the Ministry of Education (NRF-2016R1D1A1B01011724).

Institutional Review Board Statement: Not applicable.

Informed Consent Statement: Not applicable.

Conflicts of Interest: The authors declare no conflict of interest. The funders had no role in the design of the study; in the collection, analyses, or interpretation of data; in the writing of the manuscript; or in the decision to publish the results.

References

1. Park:, M.; Kim, J.; Song, H.; Kim, S.; Jeon, M. Fast and Stable Ionic Electroactive Polymer Actuators with PEDOT:PSS/(Graphene–Ag-Nanowires) Nanocomposite Electrodes. *Sensors* **2018**, *18*, 3126. [CrossRef]
2. Kong, W.; Wang, C.; Jia, C.; Kuang, Y.; Pastel, G.; Chen, C.; Chen, G.; He, S.; Huang, H.; Zhang, J.; et al. Muscle-Inspired Highly Anisotropic, Strong, Ion-Conductive Hydrogels. *Adv. Mater.* **2018**, *30*, 1801934. [CrossRef] [PubMed]
3. Terasawa, N.; Asaka, K. Superior Performance Hybrid (Electrostatic Double-Layer and Faradaic Capacitor) Polymer Actuators Incorporating Noble Metal Oxides and Carbon Black. *Sens. Actuators B* **2015**, *210*, 748–755. [CrossRef]
4. Chen, I.-W.P.; Yang, M.-C.; Yang, C.-H.; Zhong, D.-X.; Hsu, M.-C.; Chen, Y. Newton Output Blocking Force Under Low-Voltage Stimulation for Carbon Nanotube–Electroactive Polymer Composite Artificial Muscles. *ACS Appl. Mater. Interfaces* **2017**, *9*, 5550–5555. [CrossRef]
5. Biswal, D.K.; Nayak, B. Analysis of Time Dependent Bending Response of Ag-IPMC Actuator. *Procedia Eng.* **2016**, *144*, 600–606. [CrossRef]
6. Mirvakili, S.M.; Hunter, I.W. Multidirectional Artificial Muscles from Nylon. *Adv. Mater.* **2017**, *29*, 1604734. [CrossRef] [PubMed]
7. Chen, P.; Xu, Y.; He, S.; Sun, X.; Pan, S.; Deng, J.; Chen, D.; Peng, H. Hierarchically Arranged Helical Fibre Actuators Driven by Solvents and Vapours. *Nat. Nanotechnol.* **2015**, *10*, 1077–1783. [CrossRef]
8. Lima, M.D.; Li, N.; de Andrade, M.J.; Fang, S.; Oh, J.; Spinks, G.M.; Kozlov, M.E.; Haines, C.S.; Suh, D.; Foroughi, J.; et al. Electrically, Chemically, and Photonically Powered Torsional and Tensile Actuation of Hybrid Carbon Nanotube Yarn Muscles. *Science* **2012**, *338*, 928–932. [CrossRef]
9. Hong, W.; Almomani, A.; Montazami, R. Influence of Ionic Liquid Concentration on the Electromechanical Performance of Ionic Electroactive Polymer Actuators. *Org. Electron.* **2014**, *15*, 2982–2987. [CrossRef]
10. Tiwari, R.; Garcia, E. The State of Understanding of Ionic Polymer Metal Composite Architecture: A Review. *Smart Mater. Struct.* **2011**, *20*, 083001. [CrossRef]
11. Ma, S.; Zhang, Y.; Liang, Y.; Ren, L.; Tian, W.; Ren, L. High-Performance Ionic-Polymer–Metal Composite: Toward Large-Deformation Fast-Response Artificial Muscles. *Adv. Funct. Mater.* **2020**, *30*, 1908508. [CrossRef]
12. Kim, J.; Jeon, J.-H.; Kim, H.-J.; Lim, H.; Oh, I.-K. Durable and Water-Floatable Ionic Polymer Actuator with Hydrophobic and Asymmetrically Laser-Scribed Reduced Graphene Oxide Paper Electrodes. *ACS Nano* **2014**, *8*, 2986–2997. [CrossRef] [PubMed]
13. Kim, J.; Park, M.; Kim, S.; Jeon, M. Effect of Ionic Polymer Membrane with Multiwalled Carbon Nanotubes on the Mechanical Performance of Ionic Electroactive Polymer Actuators. *Polymers* **2020**, *12*, 396. [CrossRef] [PubMed]
14. Park, M.; Kim, J.; Song, H.; Kim, S.; Jeon, M. Effects of Thickness and Multi Walled Carbon Nanotube Incorporation on the Performance of Nafion Membrane–Based Ionic Polymer–Metal Composite Actuators. *Nanosci. Nanotechnol. Lett.* **2018**, *10*, 1121–1125. [CrossRef]
15. Yoo, S.; Park, M.; Kim, S.; Chung, P.S.; Jeon, M. Graphene Oxide-Silver Nanowires Paper Electrodes with Poly(3,4-Ethylenedioxythiophene)-Poly(Styrenesulfonate) to Enhance the Driving Properties of Ionic Electroactive Polymer Actuators. *Nanosci. Nanotechnol. Lett.* **2018**, *10*, 1107–1112. [CrossRef]
16. Park, M.; Yoo, S.; Bae, Y.; Kim, S.; Jeon, M. Enhanced Stability and Driving Performance of GO–Ag-NW-Based Ionic Electroactive Polymer Actuators with Triton X-100-PEDOT:PSS Nanofibrils. *Polymers* **2019**, *11*, 906. [CrossRef]
17. Nam, J.; Hwang, T.; Kim, K.J.; Lee, D.-C. A New High-Performance Ionic Polymer–Metal Composite Based on Nafion/Polyimide Blends. *Smart Mater. Struct.* **2017**, *26*, 035015. [CrossRef]
18. Vellacheri, R.; Zhao, H.; Mühlstädt, M.; Ming, J.; Al-Haddad, A.; Wu, M.; Jandt, K.D.; Lei, Y. All-Solid-State Cable-Type Supercapacitors with Ultrahigh Rate Capability. *Adv. Mater. Technol.* **2016**, *1*, 1600012. [CrossRef]
19. Yu, X.; Su, X.; Yan, K.; Hu, H.; Peng, M.; Cai, X.; Zou, D. Stretchable, Conductive, and Stable PEDOT-Modified Textiles Through a Novel In Situ Polymerization Process for Stretchable Supercapacitors. *Adv. Mater. Technol.* **2016**, *1*, 1600009. [CrossRef]

20. Akle, B.J.; Leo, D.J.; Hickner, M.A.; McGrath, J.E. Correlation of Capacitance and Actuation in Ionomeric Polymer Transducers. *J. Mater. Sci.* **2005**, *40*, 3715–3724. [CrossRef]
21. Lochmann, S.; Bräuniger, Y.; Gottsmann, V.; Galle, L.; Grothe, J.; Kaskel, S. Switchable Supercapacitors with Transistor-Like Gating Characteristics (G-Cap). *Adv. Funct. Mater.* **2020**, *30*, 1910439. [CrossRef]
22. Nketia-Yawson, B.; Noh, Y.-Y. Recent Progress on High-Capacitance Polymer Gate Dielectrics for Flexible Low-Voltage Transistors. *Adv. Funct. Mater.* **2018**, *28*, 1802201. [CrossRef]
23. Monajjemi, M. Metal-Doped Graphene Layers Composed with Boron Nitride–Graphene as an Insulator: A Nano-Capacitor. *J. Mol. Model.* **2014**, *20*, 2507. [CrossRef] [PubMed]
24. Wang, Y.; Chen, X.; Ye, W.; Wu, Z.; Han, Y.; Han, T.; He, Y.; Cai, Y.; Wang, N. Side-Gate Modulation Effects on High-Quality BN-Graphene-BN Nanoribbon Capacitors. *Appl. Phys. Lett.* **2014**, *105*, 243507. [CrossRef]
25. Shi, G.; Hanlumyuang, Y.; Liu, Z.; Gong, Y.; Gao, W.; Li, B.; Kono, J.; Lou, J.; Vajtai, R.; Sharma, P.; et al. Boron Nitride–Graphene Nanocapacitor and the Origins of Anomalous Size-Dependent Increase of Capacitance. *Nano Lett.* **2014**, *14*, 1739–1744. [CrossRef]
26. Ożçelik, V.O.; Ciraci, S. Nanoscale Dielectric Capacitors Composed of Graphene and Boron Nitride Layers: A First-Principles Study of High Capacitance at Nanoscale. *J. Phys. Chem. C* **2013**, *117*, 15327–15334. [CrossRef]
27. Izyumskaya, N.; Demchenko, D.O.; Das, S.; Özgür, Ü.; Avrutin, V.; Morkoç, H. Recent Development of Boron Nitride towards Electronic Applications. *Adv. Electron. Mater.* **2017**, *3*, 1600485. [CrossRef]
28. Coleman, J.N.; Lotya, M.; O'Neill, A.; Bergin, S.D.; King, P.J.; Khan, U.; Young, K.; Gaucher, A.; De, S.; Smith, R.J.; et al. Two-Dimensional Nanosheets Produced by Liquid Exfoliation of Layered Materials. *Science* **2011**, *331*, 568. [CrossRef]
29. Khan, U.; May, P.; O'Neill, A.; Bell, A.P.; Boussac, E.; Martin, A.; Semple, J.; Coleman, J.N. Polymer Reinforcement Using Liquid-Exfoliated Boron Nitride Nanosheets. *Nanoscale* **2013**, *5*, 571–581. [CrossRef]
30. Yoon, S.I.; Ma, K.Y.; Kim, T.-Y.; Shin, H.S. Proton Conductivity of a Hexagonal Boron Nitride Membrane and Its Energy Applications. *J. Mater. Chem. A* **2020**, *8*, 2898–2912. [CrossRef]
31. Aziz, M.A.; Shanmugam, S. Sulfonated Graphene Oxide-Decorated Block Copolymer as a Proton-Exchange Membrane: Improving the Ion Selectivity for All-Vanadium Redox Flow Batteries. *J. Mater. Chem. A* **2018**, *6*, 17740–17750. [CrossRef]
32. Huang, J.Q.; Zhang, Q.; Peng, H.J.; Liu, X.Y.; Qian, W.Z.; Wei, F. Ionic Shield for Polysulfides Towards Highly-Stable Lithium–Sulfur Batteries. *Energy Env. Sci.* **2014**, *7*, 347–353. [CrossRef]
33. Prakash, O.; Jana, K.K.; Manohar, M.; Shahi, V.K.; Khan, S.A.; Avasthi, D.; Maiti, P. Fabrication of a Low-Cost Functionalized Poly(Vinylidene Fluoride) Nanohybrid Membrane for Superior Fuel Cells. *Sustain. Energy Fuels* **2019**, *3*, 1269–1282. [CrossRef]
34. Ribeiro, D.V.; Souza, C.A.C.; Abrantes, J.C.C. Use of Electrochemical Impedance Spectroscopy (EIS) to Monitoring the Corrosion of Reinforced Concrete. *Rev. IBRACON Estrut. Mater.* **2015**, *8*, 529–546. [CrossRef]
35. Lin, Y.; Williams, T.V.; Xu, T.-B.; Cao, W.; Elsayed-Ali, H.E.; Connell, J.W. Aqueous Dispersions of Few-Layered and Monolayered Hexagonal Boron Nitride Nanosheets from Sonication-Assisted Hydrolysis: Critical Role of Water. *J. Phys. Chem. C* **2011**, *115*, 2679–2685. [CrossRef]
36. Yan, Y.; Santaniello, T.; Bettini, L.G.; Minnai, C.; Bellacicca, A.; Porotti, R.; Denti, I.; Faraone, G.; Merlini, M.; Lenardi, C.; et al. Electroactive Ionic Soft Actuators with Monolithically Integrated Gold Nanocomposite Electrodes. *Adv. Mater.* **2017**, *29*, 1606109. [CrossRef]
37. Jung, J.-H.; Jeon, J.-H.; Sridhar, V.; Oh, I.-K. Electro-Active Graphene–Nafion Actuators. *Carbon* **2011**, *49*, 1279–1289. [CrossRef]
38. Fallahi, A.; Bahramzadeh, Y.; Tabatabaie, S.E.; Shahinpoor, M. A Novel Multifunctional Soft Robotic Transducer Made with Poly(Ethylene-co-Methacrylic Acid) Ionomer Metal Nanocomposite. *Int. J. Intell. Robot. Appl.* **2017**, *1*, 143–156. [CrossRef]
39. Almomani, A.; Hong, W.; Hong, W.; Montazami, R. Influence of Temperature on the Electromechanical Properties of Ionic Liquid-Doped Ionic Polymer-Metal Composite Actuators. *Polymers* **2017**, *9*, 358. [CrossRef]
40. Akle, B.J.; Bennett, M.D.; Leo, D.J. High-Strain Ionomeric–Ionic Liquid Electroactive Actuators. *Sens. Actuators A* **2006**, *126*, 173–181. [CrossRef]
41. Kim, O.; Shin, T.J.; Park, M.J. Fast Low-Voltage Electroactive Actuators Using Nanostructured Polymer Electrolytes. *Nat. Commun.* **2013**, *4*, 2208. [CrossRef] [PubMed]
42. Sun, Z.; Zhao, G.; Guo, H.; Xu, Y.; Yang, J. Investigation into the Actuating Properties of Ionic Polymer Metal Composites Using Various Electrolytes. *Ionics* **2015**, *21*, 1577–1586. [CrossRef]
43. Yeom, S.-W.; Oh, I.-K. A Biomimetic Jellyfish Robot Based on Ionic Polymer Metal Composite Actuators. *Smart Mater. Struct.* **2009**, *18*, 085002. [CrossRef]
44. Rajapaksha, C.P.H.; Feng, C.; Piedrahita, C.; Cao, J.; Kaphle, V.; Lüssem, B.; Kyu, T.; Jákli, A. Poly (ethylene glycol) Diacrylate Based Electro-Active Ionic Elastomer. *Macromol. Rapid Commun.* **2020**, *41*, 1900636. [CrossRef] [PubMed]
45. Horiuchi, T.; Mihashi, T.; Fujikado, T.; Oshika, T.; Asaka, K. Voltage-Controlled IPMC Actuators for Accommodating Intra-Ocular Lens Systems. *Smart Mater. Struct.* **2017**, *26*, 045021. [CrossRef]
46. Wang, Y.; Liu, J.; Zhu, D.; Chen, H. Active Tube-Shaped Actuator with Embedded Square Rod-Shaped Ionic Polymer-Metal Composites for Robotic-Assisted Manipulation. *Appl. Bionics Biomech.* **2018**, *2018*, 4031705. [CrossRef]
47. Zhao, Y.; Xu, D.; Sheng, J.; Meng, Q.; Wu, D.; Wang, L.; Xiao, J.; Lv, W.; Chen, Q.; Sun, D. Biomimetic Beetle-Inspired Flapping Air Vehicle Actuated by Ionic Polymer-Metal Composite Actuator. *Appl. Bionics Biomech.* **2018**, *2018*, 3091579. [CrossRef]

International Journal of
Molecular Sciences

Article

Chelating Polymers for Targeted Decontamination of Actinides: Application of PEI-MP to Hydroxyapatite-Th(IV)

Jeanne Fèvre [1], Elena Leveille [1], Aurélie Jeanson [1], Sabine Santucci-Darmanin [2], Valérie Pierrefite-Carle [2], Georges F. Carle [2], Christophe Den Auwer [1] and Christophe Di Giorgio [1,*]

1 CNRS, ICN, Université côte d'Azur, 06108 Nice, France; jeanne.fevre@univ-cotedazur.fr (J.F.); elena.leveille@univ-cotedazur.fr (E.L.); aurelie.jeanson@univ-cotedazur.fr (A.J.); christophe.denauwer@univ-cotedazur.fr (C.D.A.)
2 CEA, TIRO-MATOs, Université côte d'Azur, 06100 Nice, France; sabine.santucci@univ-cotedazur.fr (S.S.-D.); valerie.peirrefite@unice.fr (V.P.-C.); georges.carle@unice.fr (G.F.C.)
* Correspondence: christophe.di-giorgio@univ-cotedazur.fr

Abstract: In case of an incident in the nuclear industry or an act of war or terrorism, the dissemination of plutonium could contaminate the environment and, hence, humans. Human contamination mainly occurs via inhalation and/or wounding (and, less likely, ingestion). In such cases, plutonium, if soluble, reaches circulation, whereas the poorly soluble fraction (such as small colloids) is trapped in alveolar macrophages or remains at the site of wounding. Once in the blood, the plutonium is delivered to the liver and/or to the bone, particularly into its mineral part, mostly composed of hydroxyapatite. Countermeasures against plutonium exist and consist of intravenous injections or inhalation of diethylenetetraminepentaacetate salts. Their effectiveness is, however, mainly confined to the circulating soluble forms of plutonium. Furthermore, the short bioavailability of diethylenetetraminepentaacetate results in its rapid elimination. To overcome these limitations and to provide a complementary approach to this common therapy, we developed polymeric analogs to indirectly target the problematic retention sites. We present herein a first study regarding the decontamination abilities of polyethyleneimine methylcarboxylate (structural diethylenetetraminepentaacetate polymer analog) and polyethyleneimine methylphosphonate (phosphonate polymeric analog) directed against Th(IV), used here as a Pu(IV) surrogate, which was incorporated into hydroxyapatite used as a bone model. Our results suggest that polyethylenimine methylphosphonate could be a good candidate for powerful bone decontamination action.

Keywords: actinides; decontamination; DTPA; chelating polymers; PEI-MP

Citation: Fèvre, J.; Leveille, E.; Jeanson, A.; Santucci-Darmanin, S.; Pierrefite-Carle, V.; Carle, G.F.; Den Auwer, C.; Di Giorgio, C. Chelating Polymers for Targeted Decontamination of Actinides: Application of PEI-MP to Hydroxyapatite-Th(IV). *Int. J. Mol. Sci.* **2022**, *23*, 4732. https://doi.org/10.3390/ijms23094732

Academic Editor: Raghvendra Singh Yadav

Received: 8 March 2022
Accepted: 22 April 2022
Published: 25 April 2022

Publisher's Note: MDPI stays neutral with regard to jurisdictional claims in published maps and institutional affiliations.

Copyright: © 2022 by the authors. Licensee MDPI, Basel, Switzerland. This article is an open access article distributed under the terms and conditions of the Creative Commons Attribution (CC BY) license (https://creativecommons.org/licenses/by/4.0/).

1. Introduction

Actinide contamination may result from an incident in the nuclear industry, a malicious act targeting a nuclear power plant, the explosion of a dirty bomb, or a degraded situation in a nuclear-powered naval vessel such as an icebreaker or submarine. Although rare in the history of nuclear facilities, radioelement contamination is particularly harmful because these elements, all alpha emitters, present both chemical and radiological toxicity. Among these, plutonium (Pu) is particularly involved in both civil and military nuclear activity. It is highly toxic whatever its isotopy and remains an emblematic radioelement linked to the nuclear industry for the general public.

Despite the panic it may inspire, one should still consider that Pu is only/mainly toxic once it has entered the organism. Indeed, alpha emitters exhibit a limited power of penetration (e.g., about 5 cm in air and only about 30 μm, representing a few cell diameters, in living tissues); however, the deposited energy is enormous (alpha particles are on the order of MeV energy). Unfortunately, human exposure to Pu can be efficiently widespread through the inhalation of small particles kicked up by wind and dust after the accident has occurred.

Inhalation is, thus, the most likely entry route into the organism. The retention and the fate of the inhaled particles depend on their size and physicochemical form [1]. Bigger particles are either filtered by the upper respiratory region and swallowed or transferred into the throat by the lung clearance process (elimination of the mucus layer and particles through the natural motion of the bronchial cilia). In both cases, these particles are directed into the gastrointestinal tract and mainly excreted. On the contrary, the smaller particles (10 nm to 1 μm) are capable of reaching the lung alveoli [2], where they are sequestered by alveolar macrophages [3] and eventually transferred to lymph nodes or into lung tissues, representing long-term storage (for years), which contributes to lung cancers. The soluble forms (nitrates, citrates, and certain oxides) of inhaled Pu are absorbed more easily, pass into the bloodstream [4], and are redistributed throughout the body. About 90% of this absorbed Pu is then equally deposited in the liver and bones, where it contributes to the distribution of the dose over very long periods causing, in particular, bone cancer. In case of ingestion of Pu, subsequent entry into the bloodstream from the digestive tract is very low (<1%) [5]. Most of the ingested Pu is then eliminated in the feces [6].

As for the absorption of Pu through the skin, this represents a risk only for workers in highly contaminated areas and/or in the case of wounding (cut or blast injury). In that case, Pu follows the same paths as described earlier in the case of its transfer into the blood, whereby about 90% of the Pu absorbed is retained, mostly in the liver and bones.

Therefore, these three major compartments, lung, liver, and bones, constitute a real sanctuary for sequestered Pu.

The current recommended and approved treatment for contamination with transuranic radionuclides (e.g., plutonium, americium, and curium) is chelate calcium- and zinc-diethylenetriaminepentaacetic acid (Ca and Zn-DTPA), administered intravenously (i.v.) or by nebulizer. However, DTPA exhibits a narrow biodistribution [7,8] and, thus, is eliminated very quickly through urine [9]. Moreover, DTPA is only active on the soluble forms of Pu [10]. Of course, new chelating agents, some of which are promising, such as hydroxypyridonates (e.g., 3,4,3-LI(1,2-HOPO)) are still being developed [11]. However, as with DTPA chelation therapy, their action is mainly directed toward the soluble forms of Pu such as nitrates.

To address the limitations of DTPA, special formulations such as aerosolized Ca-DTPA for pulmonary administration have been tested [12,13]. Despite its effectiveness on soluble forms of Pu deposited in the lungs, it remained ineffective on insoluble oxides (PuO_2) at the primary site of contamination. However, independently of the primary site of contamination or the DTPA treatment regimen (i.v. or by aerosolized Ca-DTPA), it reduced the systemic retention (skeleton and liver). Thus, there is still a need for a more effective treatment for the remaining Pu at the primary site of contamination.

Stealth liposomes encapsulating DTPA have also been tested [14] for their enhanced half-life in the blood and, hence, higher biodistribution. However, again, they showed improved efficiency when soluble forms such as ^{238}Pu-phytate were injected. As for the bones, to the best of our knowledge, only one team [15] reported a real in vivo decontamination, of uranium only, using a 3,2-hydroxypyridinone-based compound. Indeed, the lower observed actinide level in the liver and/or bones highlighted in most studies should be rather attributed to the chelation effect (subsequent to the i.v. administration) of the decontaminating agent rather than to actinide extravasation (e.g., real decontamination) from the liver and/or bones. In other words, once the Pu has been incorporated into a retention compartment, it is virtually inextricable.

In spite of its limitations (effectiveness mainly confined to circulating soluble forms of Pu), i.v. DTPA remains the most efficient treatment and constitutes a solid basis for comparison. It should be noticed that DTPA decontamination therapy is also quite well tolerated since, according to the National Council on Radiation Protection and Measurements, NCRP (Report No. 166), Bethesda, MD, 2011, the recommended dose (i.v. or nebulized inhalation) represents 1 g (in one shot) of chelate/day, while, in case of multiple treatments, total amounts as high as 500 g of DTPA could be administrated within several years. However, at the present time, no treatment fully meets all the required specifications. More specifically, no agent is currently capable, in addition to its action in the bloodstream, of (i) specifically targeting the three major biological Pu retention compartments, and (ii) extracting it from there.

Our challenge was then to propose a simple, effective, and affordable complementary method to DTPA therapy based on a polymeric platform. We postulated that a chelating polymer would indirectly target the main organs (liver and lungs) or the bones (provided an affinity for bone fixation sites would be implemented in that case). This strategy could, therefore, be entirely complementary to Ca-/Zn-DTPA therapy which, as previously seen, is the only one currently in use despite being mainly effective for the circulating (e.g., soluble) forms of Pu.

During the past years, our group has been developing a macromolecular approach based on a polyethyleneimine backbone for the decontamination of actinides [16–19]. This can provide, after adequate functionalization, polymeric analogs of DTPA, PEI-MC (polyethyleneimine methylcarboxylate, true polymeric analog) and PEI-MP (polyethyleneimine methylphosphonate, phosphonate analog) from a commercially available 25 kDa branched PEI (polyethyleneimine, see Figure 1). One fully functionalized polymer is most likely represented by a wide polydisperse population with a mass fraction ranging from ~5–150 kDa.

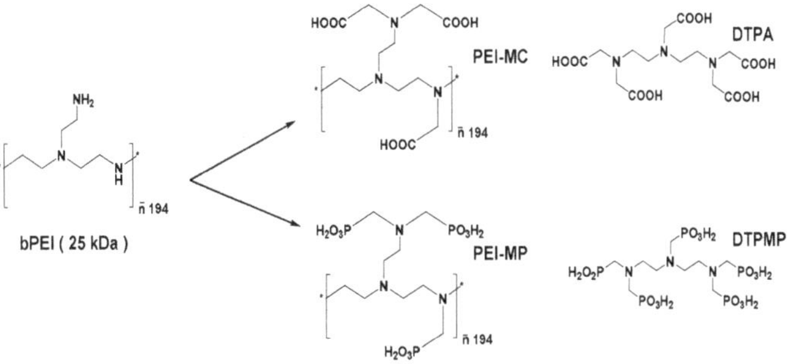

Figure 1. Chemical structures of the chelating polymers carrying carboxylic or phosphonic functions and their reference molecular analogs.

The syntheses of these chelating PEIs can be carried out in a single step, with very high yields, and are very cheap. Purification via ultrafiltration is straightforward. Too many chelating agents developed so far require a greater number of tedious synthesis and purification steps. Furthermore, as these compounds are polyelectrolytes salts, they readily dissolve in water, making them very easy to use.

We have clearly demonstrated the ability of PEI-based polymers to sequester both U(VI) (uranyl) and actinides(IV) (Pu and Th as a chemical surrogate of Pu). Indeed, under (pseudo)physiological conditions, the two polymers show an EC_{50} (50% effective concentration) comparable to DTPA taken as a reference. Complexation to Pu(IV) was also demonstrated by EXAFS spectroscopy with both polymers, PEI-MC and PEI-MP [18,19].

We present now, in this report, a first study on the decontamination of Th(IV) from a hydroxyapatite (HAp) matrix in which Th(IV) has been used as a chemical surrogate of Pu(IV) (see Section 2.4). Dose–response and kinetic profiles of decontamination associated with this model are provided. Furthermore, viability experiments realized in bones constitutive cells (osteoblasts and osteoclasts) are also reported for PEI-MC and PEI-MP compared to DTPA, taken as the gold standard.

2. Results and Discussion

2.1. PEI-MP as a Bone Seeker

The development of radiopharmaceuticals with specific bone-targeting abilities, such as ligand–radionuclide conjugates, has been carried out extensively considering metastatic bone cancer. In that case, the ligand should not only prevent radionuclide dissociation but also exhibit a strong affinity for the principal mineral phase of bone, hydroxyapatite (HAp). Indeed, HAp which is a calcium phosphate, $Ca_5(PO_4)_3(OH)$, constitutes 70% of the bone weight. HAp, then, undoubtedly constitutes a robust structural and chemical model for studying the actinide decontamination process. Phosphonates show a particular affinity for Ca^{2+}. As a consequence, lower and higher (di)phosphonate analogs of DTPA have been considered as ligands with bone-seeking properties for diverse radionuclides (^{153}Sm, ^{117m}Sn, ^{99m}Tc). For example, the beta emitter ^{153}Sm complexed with ethylenediaminetetramethylene phosphonate (Quadramet®) found a clinical use with osteoblastic skeletal metastases [20]. To improve radionuclide import, Zeevaart et al. showed that PEI-MP, loaded to some extent with radionuclides (^{99m}Tc, ^{153}Sm), could target the bone after i.v. injection in dogs [21]. Dormehl et al. demonstrated, in primates, that $t_{1/2}$ and percentage of uptake of the ^{99m}Tc–PEI-MP complexes, in diverse compartments, could be modulated by manipulating the mass fraction of the polymer. The same study also reported on the excretion of these complexes through the kidneys related to diverse mass fractions (kDa) of the polymer [22]. From these studies, they concluded that the fraction size 10–30 kDa seems to be the most suitable for radioisotopic therapy, with a lower uptake and shorter $t_{1/2}$ in liver and kidneys while having the highest bone uptake.

In the case of actinide contamination, particularly in the case of bone invasion by Pu(IV), we envisioned a similar approach with the use of chelating polymers as potential candidates. PEI-MP in particular should exhibit bone-seeking abilities. After adsorbing onto bone (i.e., HAp), it could behave as a decontaminant and extract the actinide(IV) provided the thermodynamic and kinetic profiles (adsorption/desorption of ligand and complexes) are compatible/tunable with the targeted biological utilization. As the liver or even the kidneys (in the case of contamination by U) could also be taken as relevant targets, our chelating polymers were not fractionated except for the smallest sizes (<5 kDa), which were eliminated via ultrafiltration (see Section 3).

2.2. Toxicity of PEI Chelates toward Bone Constitutive Cells

Bones undergo a constant cycle of construction and resorption, which must remain in equilibrium. Many types of cells and factors are involved in this process. Osteocytes and osteoblasts from mesenchymal stem cells are in charge of matrix formation and mineralization, whereas osteoclasts ensure the resorption process. The whole process constitutes the bone homeostasis. The potential decontamination of actinides through this chelation therapy should ideally be highly tolerated by the constitutive cells. Accordingly, viability was assayed onto SAOS-2 osteoblastic cells and MLO-A5 cells (osteocytes). The results are presented in Figure 2.

Interestingly, PEI-MP did not show any toxicity regardless of cell type or dose utilized, except at the highest dose, 50 mM, on SAOS-2 osteoblasts, where we observed a very slight effect (80%), despite the result at 48 h not showing a significant statistical difference with the untreated cells. On the contrary, DTPA was toxic from 10 mM toward both cell lines. This result, together with the evidence of PEI-MP excretion provided by Dormehl et al. [22], is very encouraging given that, when used in vivo, PEI-MP does not show any serious deleterious effects.

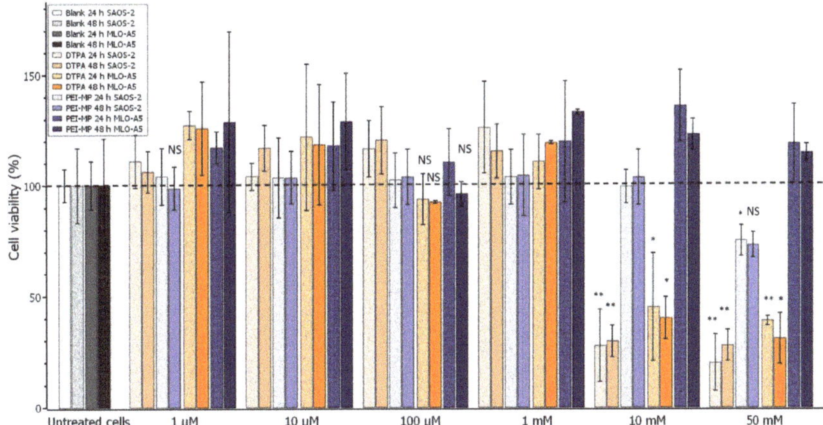

Figure 2. Graph representing the mean percentage of cell viability for MLO-A5 or SAOS-2 cells treated with increasing concentrations of chelates for 24 or 48 h. Error bars represent the SD. The mean and the SD were calculated from independent experiments ($n = 3$). * $p < 0.05$; ** $p < 0.005$, NS (not significant), according to *t*-test results (equal means) of DTPA or PEI-MP when inferior to their respective blank (untreated MLO-A5 or SAOS-2 cells at the corresponding time).

2.3. Affinity toward HAp

To assess the targeting abilities of the chelates for the bones, their affinity toward a hydroxyapatite powder was measured by thermogravimetric analyses. We, thus, monitored the mass percentage degradation of the organic matter, in the 250–500 °C range, due to PEI-MP, PEI-MC, and DTPA (as a control) after being immobilized on HAp.

Figure 3 shows the thermogravimetric analysis curves for each chelating molecule. For reading convenience, each thermogram was normalized so that 100% fit the initial mass loss plateau (around 200 °C). Keeping the initial ratio of chelating monomer to HAp constant, the HAp mass loss exhibited the expected trend with increasing loss from DTPA to its homolog PEI-MC and then to the phosphonate analog PEI-MP. Maximum adsorption onto HAp was observed with PEI-MP with a mass loss of about 4.7%, whereas the degradation due to PEI MC corresponded to about 1.8%. As for the DTPA, adsorption was negligible (0.3%). This experiment confirmed that the chelating polymers showed a higher affinity toward the target HAp. One can calculate that PEI-MC and PEI-MP exhibited sixfold and 15-fold higher affinity, respectively, for the HAp matrix as a bone model than DTPA. An affinity of 0.042 µmol of PEI-MP/µmol of HAp could be calculated.

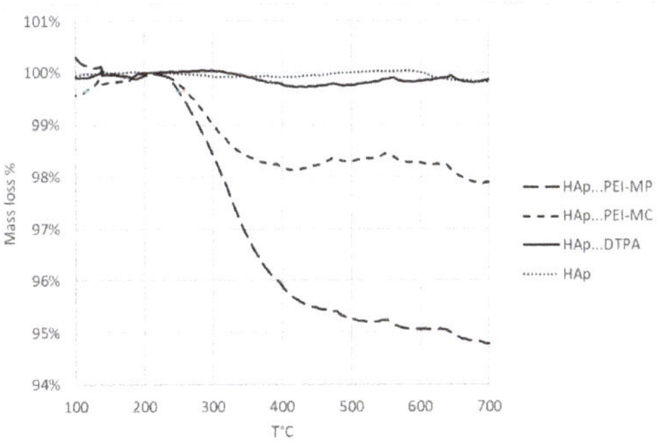

Figure 3. Thermograms of HAp, HAp...PEI-MP, HAp...PEI-MC, and HAp...DTPA in the 100–700 °C range showing the percentage of mass loss due to the organic part (chelates).

2.4. Contamination of HAp

In case of contamination, postmortem radiometry showed that 90% of plutonium activity in extrapulmonary sites was deposited in the liver and skeleton. Additionally, the fraction of plutonium deposited in the skeleton increased with time [23]. Decontamination of Pu(IV) from the bones remains of particular interest especially, for workers who have suffered from a wound while handling irradiated material.

To facilitate both handling and radioprotection, the experiments were carried out with Th(IV) as a chemical surrogate of Pu(IV), which is the major form of Pu in biological media. Th is indeed easier to manipulate (6 log lower specific activity for ^{232}Th compared to ^{239}Pu) while exhibiting the same +IV oxidation state with a comparable ionic radius [24,25]. This is very important since chelation and desorption of complexes from the HAp matrix strongly depend on these parameters. Nonetheless, this approach must be considered with some care, and test experiments with Pu will possibly be performed after full optimization with Th.

HAp was contaminated with Th(IV) through a reverse ionic exchange process. The Th(IV) source was provided by thorium carbonate, generated in situ from its nitrate form, in sodium carbonate excess, under the chemical form of the pentacarbonate complex $[Th(CO_3)_5]^{6-}$. A contact time of 48 h with HAp was set to ensure equilibrium was reached. After purification, the extent of Th^{4+} replacement at Ca^{2+} was measured by ICP-MS analyses. Thus, 35 $\pm$ 1.2 µg (n = 3 on six independent HAp-Th samples) of elementary Th (0.7% of initial Th) was found to be incorporated in 5 mg of HAp. Considering that 5 mg of HAp contained 49.75 µmol of Ca, this corresponds to a molar ratio Th/Ca of almost 0.003. Furthermore, the speciation of the Th(IV) cations into the HAp was investigated via EXAFS spectroscopy to make sure all the Th was fully incorporated (chemisorbed) into the matrix and not simply physisorbed as a carbonate complex. This spectrum constitutes a reference to be compared with those from the chelate–Th complexes, DTPA–Th, PEI-MC–Th, and PEI-MP–Th, after the decontamination process was carried out.

The EXAFS spectrum of HAp–Th is presented in Figure 4, and the corresponding best-fit parameters are displayed in Table 1.

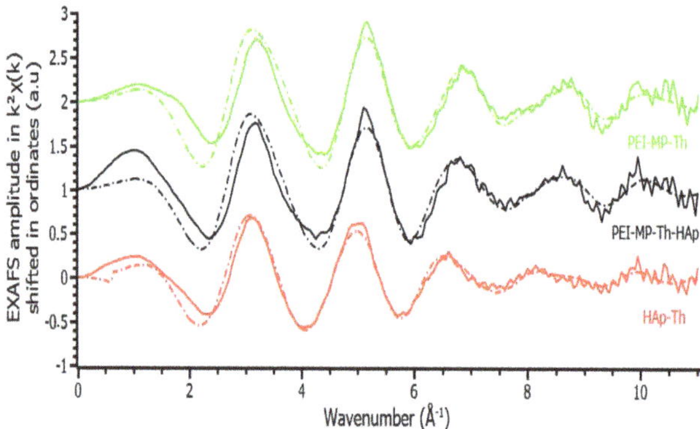

Figure 4. Th L_{III} edge experimental EXAFS spectra of PEI-MP–Th (green), PEI-MP–Th–HAp (black), and HAp–Th (red). Fits are shown in dotted line.

The fitting procedure included the contributions of oxygen and phosphorus atoms from the phosphate groups, showing that the Th atoms were incorporated in the phosphate matrix. The average distance of the oxygen contributions of the first coordination sphere (2.45 Å) was close to the average distance in the phosphate diphosphate $Th_4(PO_4)_4P_2O_7$ (2.44 Å) [26]. The short phosphorus distance (3.21 Å) is in agreement with bidentate phosphate groups, while the longer phosphorus contribution (3.82 Å) would correspond to monodentate phosphate groups as in the phosphate diphosphate phase. Adding a

Th–Th contribution did not significantly improve the fit, meaning that the Th atoms were dispersed in the apatite phase and did not form clusters.

Table 1. EXAFS best-fit parameters for HAp–Th, PEI-MP–Th, and PEI-MP–HAp–Th under (pseudo)physiological conditions [a].

Sample	First Coordination Sphere	Second Coordination Sphere	Fit Parameters
HAp–Th	2.0 (1) O at 2.32 (3) Å σ^2 = 0.0032 Å^2 4.0 (1) O at 2.45 (2) Å σ^2 = 0.0032 Å^2 2.0 (1) O at 2.59 (5) Å σ^2 = 0.0032 Å^2	*1* P at 3.21 (3) Å σ^2 = 0.0074 Å^2 *6* P at 3.82 (6) Å σ^2 = 0.0181 Å^2	S_{02} = *1.0* e_0 = 1.86 eV R_{factor} = 3.3% Q = 15
PEI-MP–Th	*9* O at 2.36 (1) Å σ^2 = 0.0109 Å^2	*2.8* P at 3.88 (4) Å σ^2 = 0.0927 Å^2 *1.8* Cl at 3.13 (3) Å σ^2 = 0.0117 Å^2	S_{02} = *1.0* e_0 = −0.13 eV R_{factor} = 5.4% Q = 88
PEI-MP–HAp–Th	*9* O at 2.37 (2) Å σ^2 = 0.0111 Å^2	*2.8* P at 3.82 (9) Å σ^2 = 0.0171 Å^2 *1.8* Cl at 3.14 (4) Å σ^2 = 0.0096 Å^2	S_{02} = *1.0* e_0 = −0.34 eV R_{factor} = 5.5% Q = 52

[a] σ^2 is the Debye–Waller factor of the considered scattering path. S_{02} is the global amplitude factor, e_0 is the energy threshold, R_{factor} is the agreement factor of the fit in percentage, and Q is the quality factor (reduced chi^2) of the fit. Uncertainties given in brackets are related to the last digit. Numbers in italics were fixed. For HAp–Th, the sum of the coordination numbers of the oxygen atoms was fixed to 8, and the coordination numbers for the second sphere were fixed to the corresponding crystallographic phase of $Th_4(PO_4)_4P_2O_7$. The coordination numbers for PEI-MP–Th were fixed to the values obtained in the previous study [17].

Surprisingly, a similarly good fit could be obtained by replacing one of the phosphorus contributions with C and O (distal) contributions of a carbonate anion (R_{factor} = 2.6%). Thus, the EXAFS spectrum did not allow us to conclude on the presence or not of carbonate anions in the Th coordination sphere within the apatite matrix. Indeed human bone may contain up to 8 wt.% carbonate ions that occupy phosphate and hydroxide positions in the HAp lattice [27]. Note, however, that the spectrum could not be adjusted with carbonate contributions only (like in the pentacarbonato complex [28]), which confirmed that thorium in HAp was not in the form of pure carbonate complexes. In conclusion, our sample preparation incorporated Th in the HAp matrix in the form of a Th phosphate disordered phase with the possible occurrence of carbonate anion(s) in its coordination sphere.

2.5. Thermodynamics of HAp–Th Chelate Systems: Dose–Response Curves

The capacity of the two chelating polymers to extract Th from the contaminated HAp was measured, at day 8, in TBS (50 mM Tris, 150 mM NaCl, pH 7.4) at different monomer concentrations and compared to DTPA as a reference by titrating the elementary Th (ICP-MS) present in the filtrate and, hence, complexed to the chelating agents.

The uptake curves of Th(IV) from contaminated HAp–Th with PEI-MP, PEI-MC, and DTPA are shown in Figure 5 and adjusted using a four-parameter logistic equation: response = min + (max − min)/(1 + (EC_{50}/dose)hill), where response is the Th% decontaminated, dose is the monomer concentration, min is the response in the absence of chelate (blank), max is the plateau, EC_{50} is the efficient concentration required to produce 50% response, and hill is the slope at the EC_{50}. Very importantly, for each experiment, the Th content remaining bound to HAp was also determined to ensure 100% recovery (see Supplementary Materials).

First of all, despite a better affinity for HAp (sixfold when compared to DTPA; see thermogravimetric results), PEI-MC showed unexpected outcomes. Indeed, the maximum Th content, 2.3%, recovered from the filtrate (green diamonds in Figure 5) was quite low, as with the blank (0.75%, at 0 mM). This could be due either to much more unfavorable desorption kinetics of the PEI-MC/Th complexes from the HAp or to a precipitation of these complexes. Secondly, despite similar values of EC_{50} and hill factors (i.e., the slope at the EC_{50} describing the transition rate from the concave to convex part of the sigmoid), the efficiency (maximum value at the plateau) strongly differed between the PEI-MP and the reference. Both of them undoubtedly showed decontamination abilities, but the phosphonate polymer was almost twice as effective: 29% versus 17% for DTPA (test for equal means provided a probability p (same mean) < 0.005). This result is undoubtedly significant. Indeed, DTPA is unlikely to reach the sequestering actinides site because of its low bioavailability, unless it is injected locally in the vicinity of the contaminated bone. However, even in this particular case, its activity would be limited due to its low affinity for the mineral part of the bone matrix, i.e., HAp. On the contrary, the thermodynamics of PEI-MP is more favorable, with higher affinity toward HAp and twofold better efficiency at extracting Th(IV). Furthermore, biodistribution, $t_{1/2}$, affinity, and uptake are all compatible with the envisioned objective [22].

The minimal monomer concentration needed for maximum decontamination under experimental conditions was determined and set at 6.3 mM, just after reaching the plateau and below the toxicity threshold of DTPA. This concentration, corresponding to a monomer/Th molar ratio of ~60, was used for studying the kinetics of decontamination with the chelates (see here after).

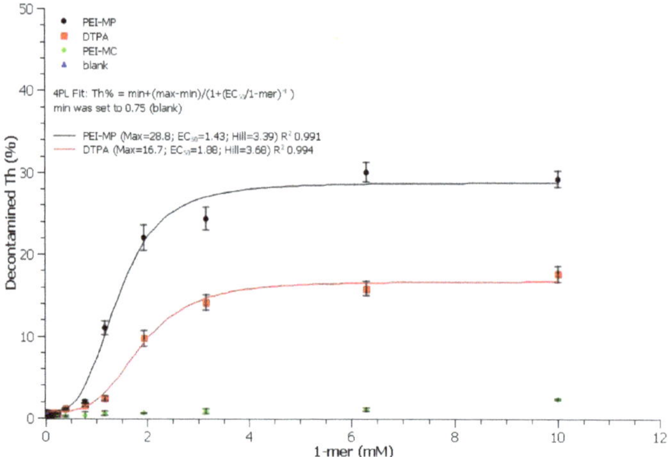

Figure 5. Dose–response curves of HAp–Th (5 mg, 0.7% Th) versus monomer chelate in 1.5 mL of TBS at day 8.

2.6. EXAFS of PEI-MP–Th and PEI-MP after Contact with HAp–Th

The fit of the EXAFS spectrum of PEI-MP–Th is shown in Figure 4, and best-fit parameters are provided in Table 1. Distances are characteristic of monodentate phosphonate coordination from a Th–phosphopeptide complex (2.35 Å for the Th–O(P) distance) that has already been described in a previous study [17]. It is quite remarkable that, using different sample preparations (present study from Th carbonate, previous study from Th NTA), very similar parameters were obtained. The presence of additional Cl anions in the second sphere is surprising but arose from the high chlorine content of the physiological medium (150 mM). This was also observed in Lahrouch et al. [17].

The EXAFS spectrum of HAp–Th in contact with PEI-MP was compared with the EXAFS spectra of HAp–Th and PEI-MP–Th (see Figure 4). Best-fit parameters were very similar to those obtained for PEI-MP–Th, meaning that Th incorporated into the HAp matrix changed speciation in the presence of PEI-MP. The resulting coordination sphere was similar to the environment of Th when complexed directly with PEI-MP alone and also similar to the previous preparation from Th-NTA. This confirmed the ability of PEI-MP to extract Th from the HAp matrix.

2.7. Kinetics of HAp–Th Chelate Systems

We then investigated kinetics, at 6.3 mM (minimal experimental monomer concentration to reach the highest Th extraction with both DTPA and PEI-MP) to see whether or not it would be possible to enhance the maximum efficiency obtained at day 8. The results are presented in Figure 6. Again, complementary Th content was checked to ensure 100% recovery (see Supplementary Materials). As for thermodynamics, curves were fitted using the same equation in Section 2.5 with the "min" parameter set to 3.4 (mean blank value across the whole time period).

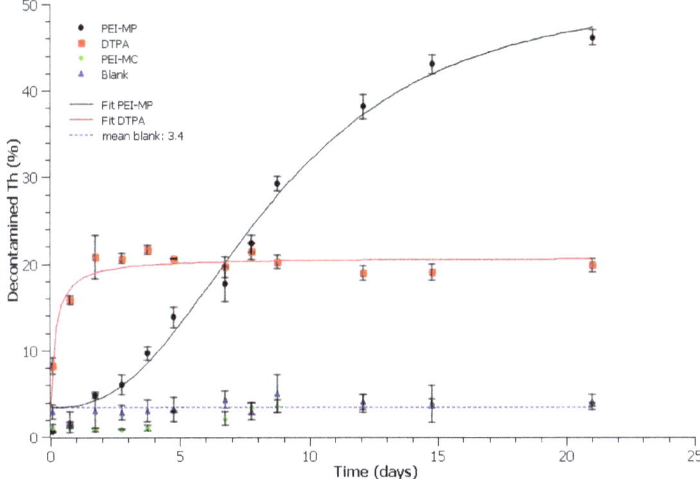

Figure 6. Decontamination kinetics of HAp–Th (5 mg, 0.7% Th) in the presence of 6.3 mM chelate in 1.5 mL of TBS over 21 days.

First of all, decontamination due to Th solubilization and/or release from HAp only was marginal (3.4% ± 1.3%), as seen with blank samples across the time period. This mean value could be attributed to the phosphate ions released from HAp at equilibrium in TBS. Secondly, the extraction of Th content with PEI-MC remained very inefficient with values of the same order of magnitude as the blanks. Most importantly, here again, huge behavior differences could be drawn between DTPA and PEI-MP. DTPA decontamination kinetics were very fast, with the highest value by day 2, but reached a plateau at 20% that could not be exceeded whatever the incubation time. It should be noted that this maximum efficiency value could not be enhanced even with a higher concentration of the chelate, as seen before with the dose–response curve. On the contrary, the kinetics with PEI-MP were much slower. Thus, the effectiveness of DTPA was matched by days 6–7. However, contrary to the DTPA trend, the Th decontamination increased across the used time period of 21 days. Strikingly, the maximum efficiency reached a top value of 46% at day 21 (p (same mean as DTPA) < 0.005). Furthermore, it should be noted that Th extraction at day 8, 29%, remained consistent with the value found during the dose–response experiment (30%). Additionally, when the filtrate was replaced at day 12 (35.8% Th extraction) with a fresh amount of 6.3 mM monomer PEI-MP, the cumulative Th decontamination reached 64.9% at day 21 (see Table 2). This indicates that additional treatment with PEI-MP over time could allow for almost a full recovery of all the mobilizable Th content.

Most importantly, these results also indicated that, unlike DTPA, the desorption of PEI-MP/Th complexes from the contaminated HAp was a slow time-dependent process that, once the effective dose was reached, continuously dragged the Th out of the HAp matrix. It should be noted that we set up here a particularly unfavorable situation where the HAp was highly contaminated (0.7% Th). Despite these conditions, PEI-MP showed an interesting ability to extract Th(IV) even though the kinetics were slow. In vivo, this could eventually be compensated for/modulated by its higher affinity for the HAp, by adjusting the polymer mass fraction and, hence, the desorption equilibrium from HAp and/or by using it in combination with the DTPA.

Table 2. Effect of cumulative dose administration.

	Th % Removal at d_{12}	Total Th % Removal at d_{21}
1 dose only from d_1	38.2 ± 0.85	46.1 [b] ± 0.52
2 doses [a]	35.8 ± 0.49	64.9 [b,c] ± 0.98

[a] Independent experiment: medium was withdrawn at d_{12} and replaced with a fresh amount of 6.3 mM monomer PEI-MP. [b] Probability p (same mean as d_{12}) < 0.005. [c] Probability p (same mean with one or two doses) < 0.005.

3. Materials and Methods

3.1. Reagents

Bromoacetic acid, sodium carbonate, acid chloride (37%), phosphorous acid H_3PO_4, branched polyethyleneimine (bPEI, 25 kDa MW), formaldehyde (37% solution), and hydroxyapatite nanopowder (<200 nm particle size (BET), ≥97%, synthetic) were purchased from Sigma-Aldrich, Saint-Louis, Mo, USA. This hydroxyapatite (HAp in the text) was provided with the following elemental analysis: Ca, 39.89%; H, 0.20%; O, 41.41%; P, 18.50%, corresponding to a formula of $Ca_5(OH)(PO_4)_3$, and it was kept under inert atmosphere. Nitric acid (67–70%, Plasmapure plus degree) and ICP-MS standard solution were purchased from SCP Science, Villebon sur Yvette, France and Honeywell Fluka, Guyancourt, France. A Th(IV) stock solution was prepared from thorium nitrate solution (5.8 mg $Th(NO_3)_4.5H_2O$ in 1 mL of HNO_3 0.1 M, (Th(IV)) = 0.01 M). Ultrafiltration was carried out with a stirred ultrafiltration cell (Millipore, 76 mm) equipped with ultrafiltration membrane disc filters Omega™ membrane, OM005076, 5 kDa MWCO (Pall corporation, port Washington, NY, USA) for the polymer purification. Ionic chromatography was performed with a Metrohm, Villebon sur Yvette, France 761 compact apparatus, equipped with a Metrosep Anion Dual 1 (3 × 150 mm) column, using 2.4 mmol/L $NaHCO_3$/2.5 mmol/L Na_2CO_3 + 2% acetone (conductivity after chemical suppression approximately 16 µS/cm) as eluent for the determination of chloride content. Na^+ counterions and phosphorus content were determined from ICP-AAS optima 8000 (Perkin Elmer, Villebon sur Yvette, France) with, respectively, Na standard solution and ICP P standard solution (Honeywell Fluka, Guyancourt, France) (see specific conditions below). Quantification of the thorium content was carried out by resorting to external calibration with standards of Th in HNO_3 1% prepared from ICP-MS Th standard solutions plasmaCAL. (SCP Science, Villebon sur Yvette, France). All ICP-MS experiments were performed with ELAN 9000 ((Perkin Elmer, Villebon sur Yvette, France).

3.2. Synthesis of PEI Chelates

3.2.1. Synthesis of PEI-MC

Branched PEI 25 kDa (10 g, 77.52 mmol of monomeric units, $C_6H_{15}N_3$) was dissolved in 500 mL of sodium carbonate solution (0.1 M). Then, bromoacetic acid (53.9 g, 387.6 mmol, 5 eq) diluted in water was added dropwise under constant stirring. The reaction mixture was stirred at room temperature overnight. The resulting solution was acidified with HCl to pH ≈ 7, and then purified by ultrafiltration (5 kDa MWCO) with a three-step sequence procedure. First, the neutral reaction mixture coming from the functionalization step was passed through the 5 kDa MWCO membrane (under adequate pressure); the resulting residue was then rinsed thoroughly with a saturated solution of NaCl (250 mL) and then with ultrapure water (250 mL × 2). Finally, the product was freeze-dried and conserved under inert (argon) to prevent moisture addition. This yielded 26 g of the water soluble polyethyleneimine methylcarboxylate sodium salt. Microanalysis revealed a C/N mass ratio of 3.43 (C/N molar ratio of 4) indicating that a full level of methylene carboxylation was achieved. Furthermore, the counterion amount, sodium for carboxylate and eventually chloride for tertiary amine, was determined with ICP-AAS and ionic chromatography, respectively (see Supplementary Materials). The following molecular formula per monomer was derived: $C_{12}H_{19}N_3O_6Na_2$ (MW 347.28 g/mol), suggesting that the dried polymer, as a sodium salt, dissociated into polyampholyte.

3.2.2. Synthesis of PEI-MP

PEI-MP was synthesized as described elsewhere [29]. Basically, phosphorous acid H_3PO_4 (19.1 g) was dissolved in concentrated HCl solution (50 mL) and heated at 80 °C. Then, formaldehyde 37% (37.8 mL) was added dropwise. Branched 25 kDa PEI (10.0 g) was dissolved in water (48 mL), and this solution was added dropwise to the reaction mixture. The reaction mixture was stirred at 90 °C for 2 h and then cooled slowly overnight. The product was separated as a viscous oil. After decanting, this viscous oil was washed with water to form a doughy substance. This decanting/washing procedure was repeated twice. The resulting oily residue was basified with Na_2CO_3 to pH $\approx$ 5, concentrated under vacuum, and then purified by ultrafiltration (5 kDa MWCO) as above, before being freeze-dried and conserved under inert (argon) to prevent moisture addition. This yielded 7.8 g of the water-soluble polyethyleneimine methylphosphonate sodium salt). Microanalysis revealed a C/N mass ratio of 2.57 (C/N molar ratio of 3) indicating that a full level of methylene phosphonylation was achieved. As for the PEI-MC, counterions were determined with ICP-AAS and ionic chromatography (see Supplementary Materials). The following molecular formula per monomer was derived: $C_9H_{21.5}N_3P_3O_9Na_{3.5}$ (MW 488.16 g/mol), suggesting that the dried polymer, as a sodium salt, dissociated into polyampholyte.

3.3. *Toxicity of PEI Chelates toward Bone Constitutive Cells*

The SAOS-2 cell line was purchased from the American Type Culture Collection. Briefly, SAOS-2 cells were maintained in McCoy's 5A medium without phenol red (HyClone, Thermo Fisher Scientific, Waltham, MA, USA) supplemented with 15% heat-inactivated fetal bovine serum (Biowest, Nuaille, France) and antibiotics (100 IU/mL penicillin, 100 µg/mL streptomycin, Sigma-Aldrich, Saint-Louis, MO, USA). Prior to assessment of toxicity, SAOS-2 cells were plated in 96-well plates (12,000 cells/well).

The MLO-A5 (murine-like osteocytes) cell line obtained from Linda Bonewald lab [30] was maintained at 37 °C, 5% CO_2 using rat tail collagen I-coated (Sigma-Aldrich, Saint-Louis, MO, USA) wells, in α-MEM with nucleotides and Ultraglutamine (BE02-002F, Lonza, Basel, Switzerland) supplemented with 1% penicillin/streptomycin (P/S, Sigma-Aldrich, Saint-Louis, MO, USA), 5% heat-inactivated fetal bovine serum (FBS: Hyclone SH30071.03 GE Healthcare, Chicago, IL, USA) and 5% heat-inactivated calf serum (CS: Hyclone SH30072.03, GE-Healthcare, Chicago, IL, USA).

For the assessment of toxicity, the medium was replaced by complete culture medium supplemented with DTPA, PEI-MP, or PEI-MC (0.1 µM, 1 µM, 10 µM, 100 µM, 1 mM, 10 mM, 50 mM, and 100 mM). Cells were further incubated for 1 h and washed three times with PBS buffer to remove the noninternalized polymers or DTPA. Next, 200 µL of culture medium was added, and cells were incubated for 24 or 48 h. At the end of the incubation period, cytotoxicity was assessed using the MTT assay. Briefly, culture medium was removed and replaced by 100 µL of EMEM containing MTT. After 1 h at 37 °C, EMEM containing MTT was removed and replaced by 150 µL of DMSO. After 15 min, absorbance was measured at 570 nm. The mean absorbance of nonexposed cells was taken as the reference value. The percentage of cell viability was calculated on the basis of the ratio between the absorbance of each sample compared to the average absorbance of the untreated cells. Results were expressed as the percentage mean (±SD) from two independent experiments performed in triplicate.

3.4. *Determination of the Affinity of the Chelating Agents toward HAp by Thermogravimetric Analysis*

HAp powder (10 mg) was dispersed in a 1.5 mL solution of DTPA, PEI-MC or PEI-MC (10 mM, based on monomer concentration), 1.5 µmol monomer/mg, and stirred overnight.

Samples were then centrifuged at 15,000 rpm during 15 min and washed (three times), to remove unbound DTPA, PEI-MC, or PEI-MP. The resulting HAp–chelate powder was freeze-dried to eliminate the water excess and kept under inert atmosphere. Thermogravimetric analyses (TGA) were performed on a Mettler Toledo, Columbus, OH, USA, TGA 851e using STAR© software (version 13.00) for data analysis. Freeze-dried samples (~10 mg) were placed in 70 µL alumina pans and heated at 10 °C·min^{-1} from 25 to 800 °C under N_2 flow (50 mL·min^{-1}). Calculation of the mass loss percentage, in the 200–400 °C range, allowed directly determining the affinity of each chelate toward HAp their comparison.

3.5. Contamination of HAp

To avoid hydrolysis at physiological pH, $Th(NO_3)_4$, was converted into $Th(CO_3)_4$. Briefly, a $Th(CO_3)_4$ (1.33 mM) solution was prepared by adding 400 µL of $Th(NO_3)_4$ (0.1M) in 29.6 mL of Na_2CO_3 (0,1M). The resulting Th(IV) solution, 30 mL (1.33 mM), was then incorporated into the hydroxyapatite (HAp) (500 mg). This suspension was stirred for 48 h. The contaminated HAp–Th powder was submitted to three cycles of centrifugation/washing steps with ultrapure water until no Th could be detected (ICP-MS) into the last filtrate (see Supplementary Materials). HAp–Th powder was then freeze-dried to yield 480 mg of a white powder. The incorporation level of Th(IV) into the HAp was ensured by quantifying the elementary Th via ICP-MS from different aliquots of the HAp–Th powder. It should be noted that the incorporation level of Th(IV) into the HAp matrix could be very precisely controlled using this procedure. This procedure was independently repeated onto different HAp samples, and contamination rates were found to be highly repeatable. Overall, the total Th content into the contaminated HAp samples was 0.7%.

3.6. Efficiency of Th Extraction: Dose–Response of HAp–Th Subjected to the Chelates

First, 5 mg of HAp–Th powder (0.7% Th content as determined by ICP-MS) were added to 1.5 mL of chelates (PEI-MC, PEI-MP, or DTPA), at different concentrations, in TBS buffer (50 mM Tris, 150 mM NaCl, pH 7.4) and mixed (orbitally) for 7 days. The corresponding chelate monomer molar concentrations were equal to 0 (blank), 0.010, 0.015, 0.039, 0.050, 0.077, 0.15, 0.23, 0.39, 0.77, 1.16, 1.93, 3.14, 6.29, and 10.0. Each of these samples was prepared in triplicate. After 8 days, a purification step (15,000 rpm, 15 min) was carried out. Then, 500 µL of supernatant was recovered from two centrifugations (250 µL 2×) for each sample and digested in 5 mL of 67–70% HNO_3 (Plasmapure plus degree, SCP Science, Villebon sur Yvette, France) at 120 °C during 2 h. The digested samples were then evaporated to dryness at 90 °C using a heating block. Finally, 5 mL of 1.5% HNO_3 was added to the tubes. Each sample was analyzed by ICP-MS (Perkin Elmer ELAN 9000). Operation conditions were daily optimized using a tuning solution. Determination of the Th concentrations was carried out by resorting to external calibration with standards of Th prepared from single-element ICP-MS standard solutions (SPEX CertiPrep, Inc., Vernon Hills, IL, USA) for thorium. An analytical blank consisting of HAp–Th without polymers or DTPA was prepared in the same conditions. Bismuth was added to each sample at a concentration of 10 ppb to correct for sample matrix effects. Dose–response curves represent the percentage of Th recovered from the filtrate versus the chelate concentration (monomer). Results were expressed as the mean (±SD) from triplicates.

3.7. Kinetics of Th Extraction from the HAp–Th Subjected to the Chelates

Firstly, 5 mg of HAp–Th powder (9.3 ppm Th content as determined by ICP-MS) was added to a 1.5 mL solution of PEI-MC, PEI-MP, or DTPA at a concentration of 6.3 mM (monomer) in TBS buffer (50 mM Tris, 150 mM NaCl, pH 7.4). Each sample was prepared in triplicate. A centrifugation step (15,000 rpm, 15 min) was carried out at different times (2, 18, 42, 66, 90, 114, 162, 186, 210, 290, 354, and 504 h). Blank samples (absence of chelates) were also evaluated in the same conditions. Next, 500 µL of supernatant was recovered from two centrifugations (250 µL 2×) for each sample and digested in 5 mL of 67–70% HNO_3 (Plasmapure plus degree, SCP Science) at 120 °C for 2 h. The digested samples were then evaporated to dryness at 90 °C using a heating block. Finally, 5 mL of 1.5% HNO_3 was added to each tube, and samples were analyzed by ICP-MS (Perkin Elmer ELAN 9000) as previously described. Kinetic curves represented the percentage of Th recovered from the filtrate with a 6.3 mM chelate concentration (monomer) at the specified time. Results were expressed as the mean (±SD) from triplicates.

3.8. EXAFS of HAp–Th, PEI-MP–Th, and PEI-MP–Th–HAp

3.8.1. Sample Preparation

Solid pellets were prepared by mixing HAp–Th (5mg) with polyethylene to obtain homogeneous solid pellets.

PEI-MP–Th was also prepared by using the same stock solution of Th(IV) as described above ($Th(NO_3)_4$ (0.1 M)). Then, 50 μL of $Th(NO_3)_4$), [Th] = 2.5×10^{-3} M pH = 1, was mixed with 250 μL of PEI-MP solution (5 mM of monomeric units) in Tris/NaCl buffer (50 mM, 150 mM). The pH was increased slowly to pH 7.0 by adding NaOH.

PEI-MP–Th–HAp was prepared by directly using the sample from the dose–response experiments of HAp–Th subjected to the chelates. Then, 1.5 mL of supernatant was recovered for each experiment. A centrifugation step (12,000× *g*, 5 min, 20 °C) was carried out on 10 kDa microcon® centrifugal filter, allowing us to concentrate the sample for suitable EXAFS measurements.

3.8.2. Data Recording and Processing

XAS data were recorded at the Th L_{III} edge (16,300 eV) on the MARS beamline at the SOLEIL synchrotron facility, which is dedicated to the study of radioactive materials. The optics of the beamline consisted of a water-cooled double-crystal monochromator for incident energy selection and horizontal focalization and two large water-cooled reflecting mirrors for high-energy rejection (harmonic part), vertical collimation, and focalization. All measurements were recorded in double-layered solution cells (200 μL) specifically designed for radioactive samples at room temperature. A 13-element Ge detector was used for data collection in the fluorescence mode.

Data treatment was carried out using ATHENA code of Demeter 0.9.26 package [31]. The E_o energy was identified at the maximum of the absorption edge. Background removal was performed using a pre-edge linear function. Atomic absorption was simulated with a cubic spline function.

3.8.3. Data Fitting

The extracted EXAFS signal was fitted in R space without any additional filtering after Fourier transformation with a Hanning window in k^2 and k^3 using the ARTEMIS code of Demeter 0.9.26 package [31]. Phases and amplitudes were calculated with Feff7 code embedded in ARTEMIS code. Only one global amplitude factor S_{02} fixed to 1 and one energy threshold correction factor Δe_0 were used for all paths. The agreement factor R (in %) and quality factor-reduced χ^2 were both provided as an indication of the fit quality.

HAp–Th (Hanning window = 2.7–11 Å^{-1}, fit range = 1–4 Å). The model used for phases and amplitude calculations was a $Th_4(PO_4)_4P_2O_7$ crystallographic structure [26] where the coordination sphere of Th was composed of six monodentate and one bidentate phosphate. The first coordination sphere was adjusted with three contributions of oxygen atoms sharing the same Debye–Waller factor. The total number of oxygen atoms was set to nine atoms as the average for Th coordination. Two contributions of one and six phosphorus atoms were included in the fitting procedure. The triple Th–O–P and quadruple Th–O–P–O paths for the monodentate phosphorus contribution were also considered. The multiple scattering paths shared the same Debye–Waller and distance correction factors.

PEI-MP–Th and PEI-MP–Th–HAp (Hanning window = 2.7–10 Å^{-1}, fit range = 1–4 Å). The model used for phases and amplitude calculations was described elsewhere [32]. The fitting procedure was the same as that used to fit previous EXAFS data of PEI-MP–Th (but synthesized differently, as described in Lahrouch et al. [17]). The first coordination sphere was fitted with a single scattering path of nine (fixed) oxygen atoms. A single scattering path of phosphorus atoms and the corresponding quadruple scattering path Th–O–P–O were also included in the fitting procedure. The addition of a single scattering path of chlorine atoms significantly improved the quality of the fit, as already observed elsewhere [17].

4. Conclusions

We showed herein, on the basis of strongly supported literature data and previous work we performed on actinide complexation with chelating polymers based on a PEI scaffold, that PEI-MP, i.e., the polyphosphonate analog of DTPA, can be considered as serious candidate for decontamination of Pu(IV) specifically targeted to the bones. Firstly, the PEI-MP, synthetized from a branched PEI (25 kDa), used without mass fractionation, did not show any toxicity toward bones constitutive cells, SAOS-2 and MLO-A5, unlike DTPA, which substantially decreased the viability at a concentration of 10 mM. Secondly, we demonstrated, using thermogravimetric analysis, that PEI-MP had a 15-fold higher affinity than DTPA toward hydroxyapatite, which constitutes the major part of the mineral bone matrix. We then successfully prepared hydroxyapatite contaminated with 0.7% Th(IV) (used as a Pu(IV) surrogate) and demonstrated through EXAFS experiments the full incorporation of this actinide into the bone mimicking matrix. Under the tested conditions, an optimum chelate concentration of 6.3 mM was sufficient to achieve maximum Th extraction with both compounds (DTPA and PEI-MP). However, the PEI-MP was able, in this case, to extract twice the amount of Th(IV), 29% versus 17%, than the gold standard DTPA. Additionally, when a kinetic study was performed over a 21 day period, this difference continued to increase, and the decontamination with the PEI-MP reached almost 50% at day 21, whereas it remained constant at 20% from day 1 with DTPA. Subsequent treatment with a fresh dose even increased the final level to almost 65%. In a more general context of actinide contamination, targeting the biological sites (lungs, liver, bones) that are involved in the sequestering of a substantial proportion of these radioelements could constitute a real advance in the field. This polymeric strategy, thus, provides a first, but nonetheless interesting complementary approach to the chelating therapy currently used based on Ca-/Zn-DTPA. Future studies aimed at tuning the kinetics and limiting the depletion of endogen cations are currently in progress.

Supplementary Materials: The following supporting information can be downloaded at: https://www.mdpi.com/article/10.3390/ijms23094732/s1.

Author Contributions: Conceptualization, C.D.G.; methodology, J.F. and C.D.G.; validation, J.F.; formal analysis, J.F., A.J., C.D.G., S.S.-D. and V.P.-C.; investigation, J.F., E.L., S.S.-D. and V.P.-C.; resources, S.S.-D., V.P.-C., G.F.C. and C.D.A.; data curation, A.J.; writing—original draft preparation, C.D.G.; writing—review and editing, G.F.C., C.D.A., A.J. and C.D.G.; visualization, J.F. and C.D.G.; supervision, C.D.A. and C.D.G.; project administration, C.D.G.; funding acquisition, C.D.A. and C.D.G. All authors have read and agreed to the published version of the manuscript.

Funding: This research was funded by DGA, grant number 2018025 and UCA (Université Côte d'Azur), more specifically the Academy Space, Environment, Risk, and Resilience of IDEX projet Investissements d'Avenir UCAJEDI, reference n° ANR-15-IDEX-01. XAS data were collected at the MARS beam line of SOLEIL synchrotron, Gif sur Yvette, France.

Institutional Review Board Statement: Not applicable.

Informed Consent Statement: Not applicable.

Data Availability Statement: The authors confirm that the data supporting the findings of this study are available within the article and its Supplementary Materials.

Acknowledgments: We would like to acknowledge Luc Vincent (MAPEC team at the ICN) for the ATG experiments and Nicolas Dacheux (ICSM Marcoule) for providing the phosphate diphosphate model sample for EXAFS data analysis.

Conflicts of Interest: The authors declare no conflict of interest.

References

1. Braakhuis, H.M.; Park, M.V.; Gosens, I.; De Jong, W.H.; Cassee, F.R. Physicochemical Characteristics of Nanomaterials That Affect Pulmonary Inflammation. *Part. Fibr. Toxicol.* **2014**, *11*, 18. [CrossRef] [PubMed]
2. *International Commission on Radiological Protection ICRP Publication*; Elsevier: Oxford, UK, 1994; Volume 66, pp. 1–3.
3. Kreyling, W.G.; Semmler, M.; Möller, W. Dosimetry and Toxicology of Ultrafine Particles. *J. Aerosol Med.* **2004**, *17*, 140–152. [CrossRef]
4. Bailey, M. The New ICRP Model for the Respiratory Tract. *Radiat. Protect Dosim.* **1994**, *53*, 107–114. [CrossRef]
5. Cassatt, D.R.; Kaminski, J.M.; Hatchett, R.J.; DiCarlo, A.L.; Benjamin, J.M.; Maidment, B.W. Medical Countermeasures against Nuclear Threats: Radionuclide Decorporation Agents. *Radiat. Res.* **2008**, *170*, 540–548. [CrossRef] [PubMed]
6. *ICRP Publication 48. Radiation Protection—The Metabolism of Plutonium and Related Elements*; Annals of the ICRP; Pergamon Press: Oxford, UK, 1986; Volume 16, pp. 2–3.
7. Stather, J.W.; Smith, H.; Bailey, M.R.; Birchall, A.; Bulman, R.A.; Crawley, F.E. The Retention of 14C-DTPA in Human Volunteers after Inhalation or Intravenous Injection. *Health Phys.* **1983**, *44*, 45–52. [CrossRef] [PubMed]
8. McAfee, J.G.; Gagne, G.; Atkins, H.L.; Kirchner, P.T.; Reba, R.C.; Blaufox, M.D.; Smith, E.M. Biological Distribution and Excretion of DTPA Labeled with Tc-99m and In-111. *J. Nucl. Med.* **1979**, *20*, 1273–1278.
9. Breustedt, B.; Blanchardon, E.; Berard, P.; Fritsch, P.; Giussani, A.; Lopez, M.A.; Luciani, A.; Nosske, D.; Piechowski, J.; Schimmelpfeng, J.; et al. Biokinetic Modelling of DTPA Decorporation Therapy: The CONRAD Approach. *Radiat. Prot. Dosim.* **2009**, *134*, 38–48. [CrossRef] [PubMed]
10. Ramounet-Le Gall, B.; Grillon, G.; Rateau, G.; Burgada, R.; Bailly, T.; Fritsch, P. Comparative Decorporation Efficacy of 3,4,3-LIHOPO, 4,4,4-LIHOPO and DTPA after Contamination of Rats with Soluble Forms of 238Pu and 233U. *Radiat. Prot. Dosim.* **2003**, *105*, 535–538. [CrossRef]
11. Durbin, P.W.; Lauriston, S. Taylor Lecture: The Quest for Therapeutic Actinide Chelators. *Health Phys.* **2008**, *95*, 465–492. [CrossRef]
12. Griffiths, N.M.; Van der Meeren, A.; Grémy, O. Comparison of Local and Systemic DTPA Treatment Efficacy According to Actinide Physicochemical Properties Following Lung or Wound Contamination in the Rat. *Front. Pharmacol.* **2021**, *12*, 289. [CrossRef] [PubMed]
13. Miccoli, L.; Ménétrier, F.; Laroche, P.; Grémy, O. Chelation Treatment by Early Inhalation of Liquid Aerosol DTPA for Removing Plutonium after Rat Lung Contamination. *Radiat. Res.* **2019**, *192*, 630. [CrossRef] [PubMed]
14. Phan, G.; Le Gall, B.; Grillon, G.; Rouit, E.; Fouillit, M.; Benech, H.; Fattal, E.; Deverre, J.-R. Enhanced Decorporation of Plutonium by DTPA Encapsulated in Small PEG-Coated Liposomes. *Biochimie* **2006**, *88*, 1843–1849. [CrossRef] [PubMed]
15. Wang, X.; Dai, X.; Shi, C.; Wan, J.; Silver, M.A.; Zhang, L.; Chen, L.; Yi, X.; Chen, B.; Zhang, D.; et al. A 3,2-Hydroxypyridinone-Based Decorporation Agent That Removes Uranium from Bones In Vivo. *Nat. Commun.* **2019**, *10*, 2570. [CrossRef] [PubMed]
16. Lahrouch, F.; Chamayou, A.C.; Creff, G.; Duvail, M.; Hennig, C.; Lozano Rodriguez, M.J.; Den Auwer, C.; Di Giorgio, C. A Combined Spectroscopic/Molecular Dynamic Study for Investigating a Methyl-Carboxylated PEI as a Potential Uranium Decorporation Agent. *Inorg. Chem.* **2017**, *56*, 1300–1308. [CrossRef]
17. Lahrouch, F.; Sofronov, O.; Creff, G.; Rossberg, A.; Hennig, C.; Auwer, C.D.; Giorgio, C.D. Polyethyleneimine Methylphosphonate: Towards the Design of a New Class of Macromolecular Actinide Chelating Agents in the Case of Human Exposition. *Dalton Trans.* **2017**, *46*, 13869–13877. [CrossRef] [PubMed]
18. Lahrouch, F.; Siberchicot, B.; Leost, L.; Aupiais, J.; Rossberg, A.; Hennig, C.; Den Auwer, C.; Di Giorgio, C. Polyethyleneimine Methylenecarboxylate: A Macromolecular DTPA Analogue to Chelate Plutonium(IV). *Chem. Commun* **2018**, *54*, 11705–11708. [CrossRef]
19. Lahrouch, F.; Siberchicot, B.; Fèvre, J.; Leost, L.; Aupiais, J.; Solari, P.L.; Den Auwer, C.; Di Giorgio, C. Carboxylate- and Phosphonate-Modified Polyethylenimine: Toward the Design of Actinide Decorporation Agents. *Inorg. Chem.* **2020**, *59*, 128–137. [CrossRef] [PubMed]
20. Serafini, A.N. Systemic Metabolic Radiotherapy with Samarium-153 EDTMP for the Treatment of Painful Bone Metastasis. *J. Nucl. Med.* **2001**, *45*, 91–99.
21. Jarvis, N.; Zeevaart, J.; Wagener, J.; Louw, W.; Dormehl, I.; Milner, R.; Killian, E. Metal-Ion Speciation in Blood Plasma Incorporating the Water-Soluble Polymer, Polyethyleneimine Functionalised with Methylenephosphonate Groups, in Therapeutic Radiopharmaceuticals. *Radiochim. Acta* **2002**, *90*, 237–246. [CrossRef]
22. Dormehl, I.C.; Louw, W.K.A.; Milner, R.J.; Kilian, E.; Schneeweiss, F.H.A. Biodistribution and Pharmacokinetics of Variously Sized Molecular Radiolabelled Polyethyleneiminomethyl Phosphonic Acid as a Selective Bone Seeker for Therapy in the Normal Primate Model. *Arzneimittelforschung* **2001**, *51*, 258–263. [CrossRef] [PubMed]
23. Romanov, S.A.; Efimov, A.V.; Aladova, ..; Suslova, .G.; Kuznetsova, I.S.; Sokolova, A.B.; Khokhryakov, V.V.; Sypko, S.A.; Ishunina, M.V.; Khokhryakov, V.F. Plutonium Production and Particles Incorporation into the Human Body. *J. Environ. Radioact.* **2020**, *211*, 106073. [CrossRef] [PubMed]
24. Lopez, C.; Deschanels, X.; Auwer, C.D.; Cachia, J.-N.; Peuget, S.; Bart, J.-M. X-Ray Absorption Studies of Borosilicate Glasses Containing Dissolved Actinides or Surrogates. *Phys. Scr.* **2005**, *2005*, 342. [CrossRef]

25. Tyagi, A.K.; Ambekar, B.R.; Mathews, M.D. Simulation of Lattice Thermal Expansion Behaviour of $Th_{1-x}Pu_xO_2$ ($0.0 \leq x \leq 1.0$) Using CeO_2 as a Surrogate Material for PuO_2. *J. Alloys Compd.* **2002**, *337*, 277–281. [CrossRef]
26. Bénard, P.; Brandel, V.; Dacheux, N.; Jaulmes, S.; Launay, S.; Lindecker, C.; Genet, M.; Louër, D.; Quarton, M. $Th_4(PO_4)_4P_2O_7$, a New Thorium Phosphate: Synthesis, Characterization, and Structure Determination. *Chem. Mater.* **1996**, *8*, 181–188. [CrossRef]
27. Siddiqi, S.A.; Azhar, U. Carbonate Substituted Hydroxyapatite. In *Handbook of Ionic Substituted Hydroxyapatites*, 1st ed.; Samad Khan, A., Anwar Chaudhry, A., Eds.; Elsevier: Amsterdam, The Netherlands, 2020; pp. 149–173. [CrossRef]
28. Zurita, C.; Tsushima, S.; Bresson, C.; Garcia Cortes, M.; Solari, P.L.; Jeanson, A.; Creff, G.; Den Auwer, C. How Does Iron Storage Protein Ferritin Interact with Plutonium (and Thorium)? *Chem. Eur. J.* **2021**, *27*, 2393–2401. [CrossRef]
29. Moedritzer, K.; Irani, R.R. The Direct Synthesis of α-Aminomethylphosphonic Acids. Mannich-Type Reactions with Orthophosphorous Acid. *J. Org. Chem.* **1966**, *31*, 1603–1607. [CrossRef]
30. Kato, Y.; Boskey, A.; Spevak, L.; Dallas, M.; Hori, M.; Bonewald, L.F. Establishment of an Osteoid Preosteocyte-like Cell MLO-A5 That Spontaneously Mineralizes in Culture. *J. Bone Miner. Res.* **2001**, *16*, 1622–1633. [CrossRef]
31. Ravel, B.; Newville, M. ATHENA, ARTEMIS, HEPHAESTUS: Data Analysis for X-ray Absorption Spectroscopy Using IFEFFIT. *J. Synchrotron Rad.* **2005**, *12*, 537–541. [CrossRef]
32. Creff, G.; Safi, S.; Roques, J.; Michel, H.; Jeanson, A.; Solari, P.L.; Basset, C.; Simoni, E.; Vidaud, C.; den Auwer, C. Actinide(IV) Deposits on Bone: Potential Role of the Osteopontin-Thorium Complex. *Inorg. Chem.* **2016**, *55*, 29–36. [CrossRef]

International Journal of
Molecular Sciences

Communication

Conductive Supramolecular Polymer Nanocomposites with Tunable Properties to Manipulate Cell Growth and Functions

Cheng-You Wu [1], Ashenafi Zeleke Melaku [1], Fasih Bintang Ilhami [1], Chih-Wei Chiu [2] and Chih-Chia Cheng [1,3,*]

1 Graduate Institute of Applied Science and Technology, National Taiwan University of Science and Technology, Taipei 10607, Taiwan; johnnywu2262756@gmail.com (C.-Y.W.); ashenafichem@gmail.com (A.Z.M.); fasihilhami17@gmail.com (F.B.I.)
2 Department of Materials Science and Engineering, National Taiwan University of Science and Technology, Taipei 10607, Taiwan; cwchiu@mail.ntust.edu.tw
3 Advanced Membrane Materials Research Center, National Taiwan University of Science and Technology, Taipei 10607, Taiwan
* Correspondence: cccheng@mail.ntust.edu.tw

Abstract: Synthetic bioactive nanocomposites show great promise in biomedicine for use in tissue growth, wound healing and the potential for bioengineered skin substitutes. Hydrogen-bonded supramolecular polymers (3A-PCL) can be combined with graphite crystals to form graphite/3A-PCL composites with tunable physical properties. When used as a bioactive substrate for cell culture, graphite/3A-PCL composites have an extremely low cytotoxic activity on normal cells and a high structural stability in a medium with red blood cells. A series of in vitro studies demonstrated that the resulting composite substrates can efficiently interact with cell surfaces to promote the adhesion, migration, and proliferation of adherent cells, as well as rapid wound healing ability at the damaged cellular surface. Importantly, placing these substrates under an indirect current electric field at only 0.1 V leads to a marked acceleration in cell growth, a significant increase in total cell numbers, and a remarkable alteration in cell morphology. These results reveal a newly created system with great potential to provide an efficient route for the development of multifunctional bioactive substrates with unique electro-responsiveness to manipulate cell growth and functions.

Keywords: bioactive supramolecular polymer; conductive graphene nanosheet; cell culture; hydrogen bonding; indirect electrical stimulation

Citation: Wu, C.-Y.; Melaku, A.Z.; Ilhami, F.B.; Chiu, C.-W.; Cheng, C.-C. Conductive Supramolecular Polymer Nanocomposites with Tunable Properties to Manipulate Cell Growth and Functions. *Int. J. Mol. Sci.* **2022**, 23, 4332. https://doi.org/10.3390/ijms23084332

Academic Editor: Raghvendra Singh Yadav

Received: 19 March 2022
Accepted: 12 April 2022
Published: 14 April 2022

Publisher's Note: MDPI stays neutral with regard to jurisdictional claims in published maps and institutional affiliations.

Copyright: © 2022 by the authors. Licensee MDPI, Basel, Switzerland. This article is an open access article distributed under the terms and conditions of the Creative Commons Attribution (CC BY) license (https://creativecommons.org/licenses/by/4.0/).

1. Introduction

Polymer-based bioengineering approaches have risen to prominence in the biomedical field as potential materials for use in fabrication with a wide range of tissue engineering applications [1–4]. Several studies have shown that synthetic bioactive polymers with a tailorable structural composition, surface microstructure and wettability can substantially affect cellular response and growth through the addition of specific functional monomers and control of the monomer addition sequence during the polymerization process [5–7]. However, the development of synthetic bioactive polymers is frequently constrained by our limited understanding of how cell adhesion and proliferation are regulated by the extracellular matrix (ECM) [8–11]. Recently, supramolecular polymers produced via reversible non-covalent interactions have attracted attention owing to the unique mechanical properties of their matrices, such as their environmental stimuli-responsiveness [12,13], self-healing [14,15] and shape-memory behavior [16,17]. Specifically, supramolecular hydrogen-bonding moieties (or synthons) within the polymer that drive self-assembly behavior might be exploited to manipulate supramolecular polymer–cell junctions [18–22]. For example, Dankers and coworkers synthesized novel synthetic supramolecular polymers with quadruple hydrogen-bonding ureido-pyrimidinone (UPy) moieties that could spontaneously self-assemble into a membrane-like structure and improve the bioactivity of cell growth and proliferation, thereby achieving a biocompatible polymer [23,24]. Therefore, functional polymeric materials with supramolecular moieties and strong hydrogen-bonding capability have critical factors required for the development of multifunctional tissue engineering scaffolds with tunable physical properties that can enhance the overall growth of cultured cells.

Graphene is a one-atom-thick two-dimensional material with a hexagonal structure that provides distinct features, including a large surface area, excellent thermo-electrical conductivity, high mechanical strength and light transmittance [25–27]. In particular, conductive materials such as graphene have the potential to provide an essential role in tissue regeneration that can accelerate cell adhesion and migration during wound healing by electrical stimulation. Cells have an innate self-electroactivity ability that enhances the overall efficiency of their cellular wound healing after electrical stimulation [28–32]. Nevertheless, full carbon-based graphene nanosheets have limitations. In particular, they are extremely hydrophobic, making it a challenge to interface them with biological systems and thus causing the significant inhibition of cell growth and functions [28–30]. Given the hydrophobic nature of graphene, we speculate that introducing hydrogen-bonded supramolecular polymers into the graphene matrix could significantly improve the hydrophilicity of the composites, thereby improving the affinity between the composite substrate and cells [18,19]. We further propose that electrical stimulation—as an exogenous physiological stimulus—could enhance the cellular affinity of adhesion, proliferation, and differentiation on materials substrates [33–35]. As a promising strategy with electrical stimuli, direct electrical stimulation (DES) implies the interaction of electron transport chain components with a working electrode surface into cells which may cause burn damage to living tissues treated with DES [36]. In contrast, indirect electrical stimulation (IES) involves the transfer of electrons from a working electrode to a microorganism without direct interaction into the cellular environment that allows safe stimulation to promote cellular growth [37]. Therefore, we speculated that a combination of hydrogen-bonded supramolecular polymers with graphene nanosheets for cell and tissue culture using IES may old great potential as a high-performance conductive bioactive substrate for manipulating cells in engineered tissues.

Recent studies in our laboratory demonstrated that supramolecular exfoliated graphene nanosheets with tunable physical properties could be obtained by controlling the amount of three-arm adenine-end-capped polycaprolactone polymer (3A-PCL) [38]. This results from the strong affinity of the self-assembled lamellar and spherical nanostructures of 3A-PCL to be strongly absorbed onto the surface of the graphite crystals, which subsequently lead to the formation of exfoliated graphene nanosheets. The tailorable graphene-exfoliation level and controlled conductive performance of these graphite/3A-PCL composites inspired us to explore their ability as a conductive bioactive substrate for cell culture in vitro (Scheme 1).

The objectives of this work were to achieve improved the surface bioactivity of conductive graphite/3A-PCL substrates to promote the adhesion, migration and proliferation of the cells cultured on the substrates via IES at very low voltage levels. In this paper, we show that graphite/3A-PCL composites not only exhibit extremely low cytotoxic activity against normal cells and high structural stability in a red blood cell-containing medium, but also significantly enhance wound-healing and cell-growth rates. In addition, we showed that graphite/3A-PCL substrates under an indirect current electric field at only 0.1 V can rapidly and efficiently produce cell adhesion, spreading and proliferation, resulting in a substantial increase in the total cell numbers and significant alteration in cell morphology. To the best of our knowledge, this is the first study demonstrating a conductive bioactive supramolecular substrate based on hydrogen-bonded adenine units and exfoliated graphene nanosheets that can efficiently control cell growth when exposed to IES. This newly created system has an advantageous combination of composite, amphipathic, conductive and bioactive characteristics with promise as a multifunctional, soft cell-culture scaffold for skin substitute bioconstructs and tissue-engineered regeneration.

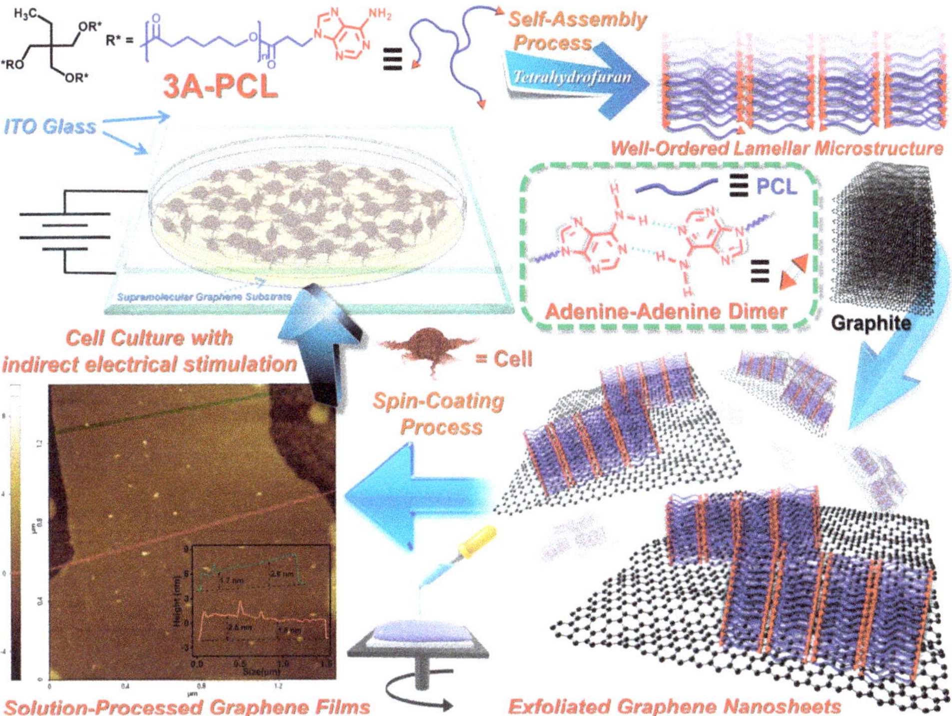

Scheme 1. Graphical illustration for the development of supramolecular composites: self-assembled lamellar nanostructures are formed by hydrogen-bonded adenine units within the 3A-PCL macromer, which subsequently adsorbs on the surface of exfoliated graphene nanosheets after blending with graphite in THF. The resulting composite is applicable as a conductive bioactive substrate for cell culture under indirect electrical stimulus.

2. Results and Discussion

The molecular structure of tri-adenine end-capped polycaprolactone macromer (3A-PCL) and its direct self-assembly into well-ordered lamellar structures in solid state via the noncovalent linkage of the self-complementary adenine-adenine (A-A) hydrogen-bonding interactions are shown in the upper part of Scheme 1. When 3A-PCL is directly blended with natural graphite in tetrahydrofuran (THF) using ultrasonic treatment, 3A-PCL efficiently undergoes self-assembly in THF to form a stable lamellar structure on the graphite surface. This promotes the exfoliation of the bulk graphite into variable layers of graphene nanosheets. The number of layers depends on the amount of 3A-PCL incorporated into the graphite crystals (Figure 1) due to the presence of the strong noncovalent intermolecular interactions between the exfoliated graphene surface and the self-organized lamellar 3A-PCL nanostructures. A well-tunable layer number and physical properties of the graphite/3A-PCL nanosheets in the solid state were analyzed and discussed in detail in our previous work [38]. Here, we expanded on our previous work, exploring the physical properties and potential applications for the cell culturing of graphite/3A-PCL composites.

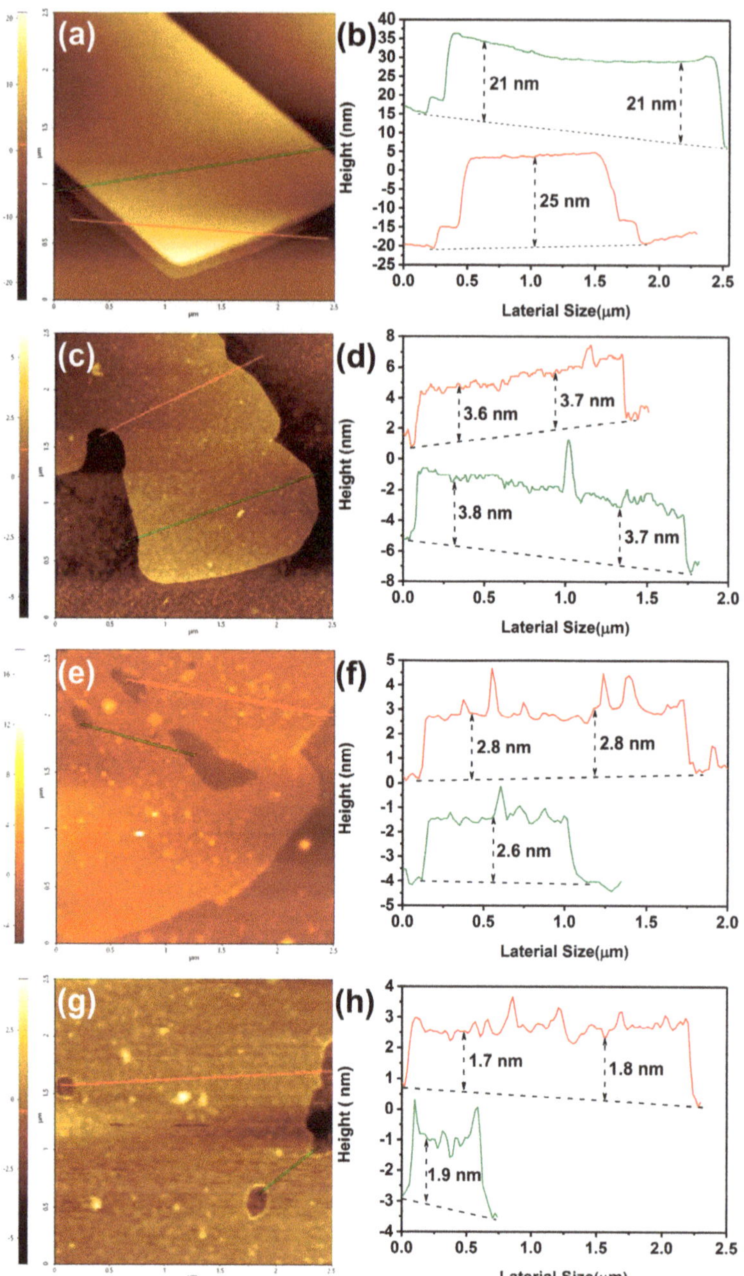

Figure 1. Atomic force microscopy (AFM) images (left side) and corresponding height profiles (right sides) of (**a**,**b**) pristine crystal graphite and the resulting graphite/3A-PCL composites at (**c**,**d**) 5/10, (**e**,**f**) 3/10 and (**g**,**h**) 1/10 blend ratios. The green and red profiles presented in (**b**,**d**,**f**) and (**h**) correspond with the green and red solid lines seen in the images of (**a**,**c**,**e**,**g**), respectively.

2.1. Physical Properties of Graphite/3A-PCL Composites

To further extend our previous findings, we explore here the effects of the self-assembled lamellar structures on the surface wettability of graphite/3A-PCL composites at 25 °C by measuring the water contact angle (WCA). Spin-coated commercial polycaprolactone (PCL; average molecular weight = 80,000 g/mol) and adenine-functionalized 3A-PCL thin-films had WCA values of approximately 80° and 49°, respectively. This confirms that introducing adenine moieties into the end groups of the PCL oligomer increases the surface hydrophilicity of 3A-PCL (Figure 2a). The WCA values of all spin-coated graphite/3A-PCL thin-films exhibited a gradual decrease from 74° to 54° with increasing weight fractions of 3A-PCL content, indicating that adjusting the content of 3A-PCL within composites not only significantly affected the surface hydrophilicity of composites, but also effectively regulated their level of surface wettability.

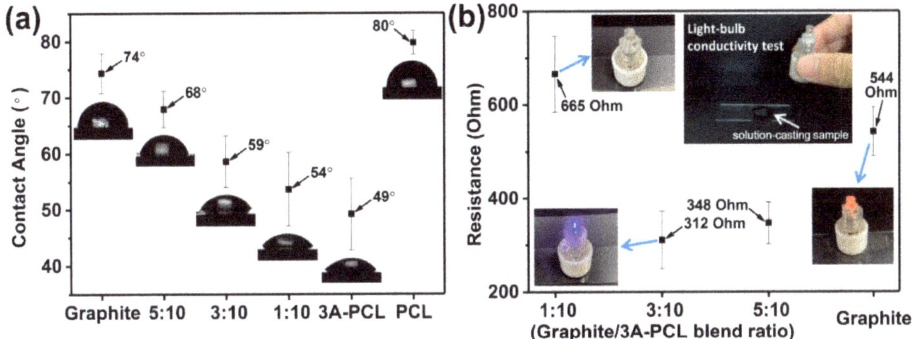

Figure 2. (**a**) WCA analysis of pristine graphite, graphite/3A-PCL composites, 3A-PCL and commercial PCL; (**b**) electrical resistance values for the graphite and graphite/3A-PCL composites. The photographs in the inset of (**b**) present the conductivity performance of graphite and graphite/3A-PCL composites through a simple light bulb conductivity apparatus.

An ideal exfoliated graphene-based composite for engineering applications must have high electrical conductivity in a thin-film state. We investigated the electrical resistance of spin-coated graphite/3A-PCL films using a light bulb conductivity apparatus at 25 °C. The light bulbs instantly lit up after being placed on the substrates of the 3/10 and 5/10 graphite/3A-PCL composites, but not on the 1/10 graphite/3A-PCL composite (Figure 2b). This suggests that an increased proportion of graphite forms enough exfoliated graphene nanosheets to enable overall electrical conductivity in composites. In addition, a four-point probing measurement of sensor resistance at 25 °C and relative humidity of approximately 35% showed similar trends for all composites (Figure 2b), further demonstrating that the 3A-PCL macromer promotes the efficient exfoliation process of graphite. The resulting graphene nanosheets have a substantial reduction in electrical resistance compared to pristine graphite. For example, the 3/10 and 5/10 graphite/3A-PCL composites had electric resistances of the 312 and 348 Ohm, respectively, which were approximately 1.5–2.0 times lower than pristine graphite (544 Ohm) and the 1/10 graphite/3A-PCL composite (665 Ohm). These results further indicate that the combination of surface hydrophilicity and electrical conductivity properties that can be tailored for suitability suggests that these composites have strong potential for electrical stimulation cell culture applications.

2.2. Graphite/3A-PCL Composites for Cell Culture Applications

We therefore further explored the structural stability and cytotoxic activity of the graphite/3A-PCL composites toward normal mouse embryonic fibroblasts (NIH/3T3 cells) and sheep red blood cells (SRBCs). As shown in Figure 3a, all sample solutions at concentrations ranging from 0.01 μg/mL to 100 μg/mL showed no significant effect on the viability of NIH/3T3 cells after 24 h treatment using MTT [3-(4,5-dimethylthiazol-2-yl)-2,5-diphenyltetrazolium bromide] assays. Thus, the graphite/3A-PCL composites exerted a low cytotoxic effect on normal NIH/3T3 cells. More surprisingly, the SRBC hemolysis assay clearly demonstrated that the graphite/3A-PCL composites appeared to exhibit greater structural stability and biocompatibility with SRBCs than pristine graphite does (see Supporting Information, Figure S1). This might be attributable to the reversible A-A hydrogen-bonding interactions within the composite matrix (see Figure S2 for the explanation and results), reducing the hemolytic effect of graphite and enabling its potential use for in vivo applications [39,40]. To further evaluate the effects of the hydrogen-bonded adenine moieties and self-assembled structures on the growth of cell numbers, NIH/3T3 cells were seeded on commercial PCL film as a control substrate and on spin-coated 3A-PCL and composite films, before being incubated at 37 °C for various periods (24, 48 and 72 h). After 72 h incubation, the average number of cells in pristine 3A-PCL (1.1×10^6) was 1.5 times higher than on commercial hydrophobic PCL film (7.2×10^5; Figure 3b), indicating that the adenine molecules on the end groups of PCL-enhanced surface hydrophilicity promotes accelerated cell growth in NIH/3T3 cells [19]. Interestingly, the spin-coated 3/10 graphite/3A-PCL substrate was similar in cell growth and number (1.0×10^6) to pristine 3A-PCL, suggesting that the self-assembled structures of 3A-PCL can play a major role in facilitating the proliferation and survival of cells even in the presence of exfoliated graphene nanosheets. However, the 1/10 and 5/10 graphite/3A-PCL substrates had significantly lower cell numbers compared to 3/10 graphite/3A-PCL substrate—cell numbers decreased to approximately 7.7×10^5 in both—suggesting that the exfoliated graphene surfaces are optimally covered by 3A-PCL to achieve a desired surface morphology when the blending ratio of 3A-PCL and graphite is 3:10 [38]. Thus, a further decrease or increase in the 3A-PCL content in the composites apparently alters the surface roughness and wettability in a significant way, leading to a decrease in the total number of produced NIH/3T3 cells. These results confirm that tuning the 3A-PCL content of composites can allow the efficient control of cellular growth and proliferation in adherent cells. This is perhaps due to the presence of strong, multiple high-affinity interactions between NIH/3T3 cells and the hydrogen-bonded adenine group of 3A-PCL, creating a cell culture platform with tailorable physical substrate properties [19].

To confirm the results of cell growth and assess the interaction and relationship at the interface between the adenine units, exfoliated nanosheets and cell surface, confocal laser scanning microscopy (CLSM) was employed to observe the cell cytoskeleton (F-actin, green) and nuclei (bright blue) using phalloidin and 4′,6-diamidino-2-phenylindole (DAPI) staining, respectively. The NIH/3T3 cells seeded and cultured for 24 h on pristine graphite or PCL films displayed an unhealthy cellular shape, insufficient cell density and lacked the characteristic features of the filamentous cytoskeletal network in cell morphology (Figure 3c). Thus, NIH/3T3 cells did not seem to attach to the hydrophobic graphite or PCL surfaces, leading to the inhibition of cell movement and little proliferation. In contrast, NIH/3T3 cells did adhere, spread, and grow well on 3A-PCL film and generated a large area of highly aligned fibroblast-like morphology with the elongation of dendritic filopodia, indicating that the introduction of the adenine moieties in the PCL structure effectively improved the binding affinity between the cell surface and the adenine-modified PCL substrate. This increased binding affinity in turn helped accelerate the cell alignment into clusters from adjacent cells via intercellular adhesions, further promoting cell migration and proliferation. Surprisingly, on the 3/10 graphite/3A-PCL film, NIH/3T3 cells exhibited a highly uniform cellular structure without the presence of a large area covered by filamentous cytoskeletal structures on the substrate. The restricted formation of cytoskeletal filamentous networks between neighboring cells was possibly due to the presence of exfoliated graphene nanosheets within the composites, altering the mechanism of cell proliferation and inducing a change in the cellular behavior or characteristics [41,42]. While 3A-PCL can help maintain the rate of cell growth, these changes in cell behavior led to differential morphological attributes in NIH/3T3 cells when compared to cells cultured in pristine 3A-PCL substrate. Thus, it appears that the combination of exfoliated graphene and bioactive 3A-PCL can create a surface with high cell affinity and tailorable effects on cell culture and that the resulting composite substrate can be efficiently controlled to regulate the cellular functions involved in cell growth, proliferation, and survival. This has the potential to play a vital role in cell and tissue cultures [43].

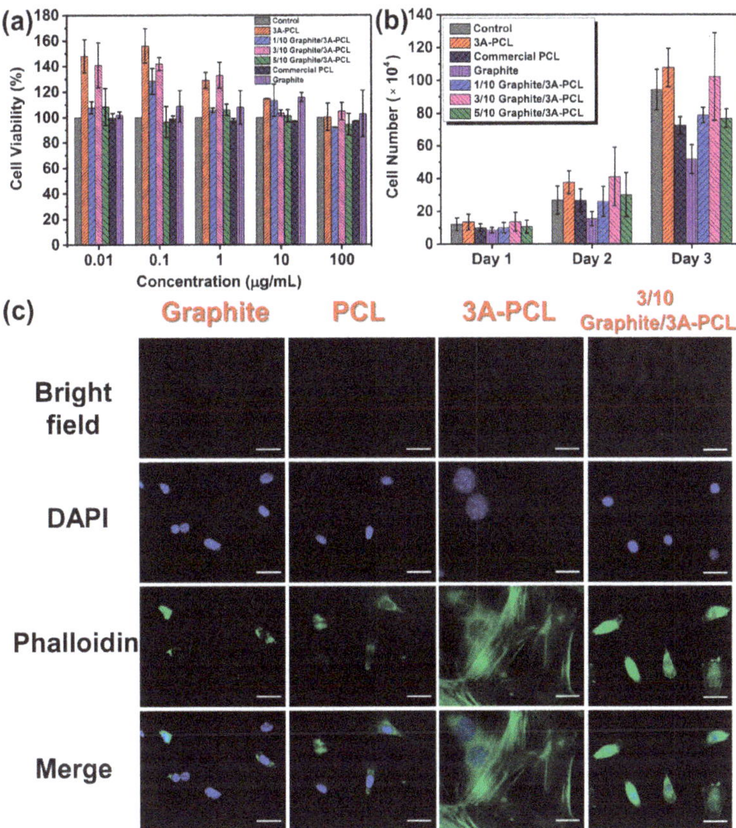

Figure 3. (**a**) The viability of NIH/3T3 cells incubated with pristine graphite, PCL, 3A-PCL, and graphite/3A-PCL composites for 24 h. These viability results were obtained by MTT assay. (**b**) The numbers of NIH/3T3 cells on the graphite, PCL, 3A-PCL, and graphite/3A-PCL substrates after 1–3 days of culture. (**c**) CLSM images of NIH/3T3 cells cultured on pristine graphite, PCL, 3A-PCL, and 3/10 graphite/3A-PCL substrates. After 24 h of culture, the resulting cells were stained with green-fluorescent phalloidin and blue-fluorescent DAPI. White scale bars are 20 μm in all images.

2.3. Evaluation of the In Vitro Wound-Healing Activity on Conductive Bioactive Substrates

To determine whether graphite/3A-PCL composites can significantly improve NIH/3T3 cell growth behavior, we evaluated the effects on cell attachment, migration, and proliferation, using an in vitro scratch wound healing assay [44,45]. As shown in Figures 4 and 5a, after 24 h of culture, the wound closure percentages of NIH/3T3 cells on pristine 3A-PCL can reach up to 64.5 ± 1.4%, whereas cells on control tissue culture polystyrene (TCPS) substrate had only 56.5 ± 2.1% wound closure. We thus conclude that the 3A-PCL substrate can promote cell adhesion and enhance cell spreading and migration through specific interactions between NIH/3T3 cells and hydrophilic adenine-functionalized 3A-PCL substrate, thereby accelerating cell proliferation and wound closure. After 48 h, NIH/3T3 cells on the 3A-PCL and TCPS substrates showed nearly complete wound closure, revealing that 3A-PCL can be used as a high-efficiency bioactive scaffold for cell and tissue culture applications. In contrast to pristine 3A-PCL, graphite substrates led to a delay in wound closure efficiency, with the scratch-damaged NIH/3T3 monolayer showing only 38.5 ± 2.0% wound closure after 48 h. This suggests that the hydrophobicity of the graphite surface suppresses cell growth, migration, and cycle progression. However, with the incorporation of 3A-PCL into the graphite matrix, the resulting composite substrates significantly enhance the wound closure rate in NIH/3T3 cells. For example, the wound closure efficiency of NIH/3T3 cells on the 3/10 and 5/10 graphite/3A-PCL substrates after 48 h of culture reached 83 ± 1.4% and 74 ± 2.8%, respectively. Overall, these findings demonstrate that adjusting the content of 3A-PCL is critical to controlling the physical properties of exfoliated graphene nanosheets [38] and the resulting composites can be used as a bioactive substrate to efficiently manipulate and regulate the cell growth and wound healing ability, eventually achieving desirable cell performance in cell culture.

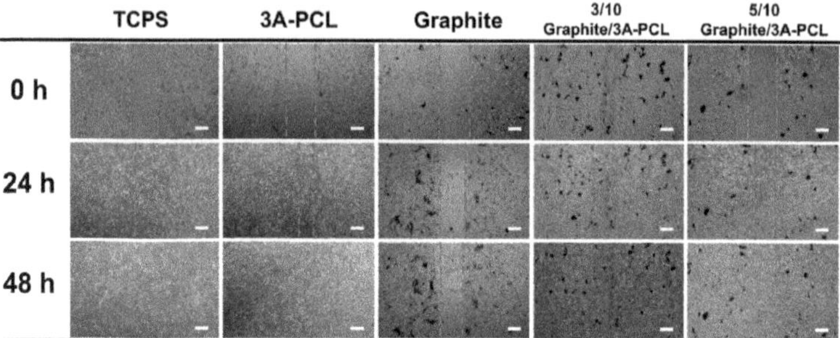

Figure 4. Results of in vitro scratch wound healing assay. Linear scratches were created in the monolayer of NIH/3T3 cells cultured on TCPS, 3A-PCL, graphite, 3/10 and 5/10 graphite/3A-PCL substrates. Images of the scratched areas were monitored through a phase-contrast microscope at 0, 24 and 48 h after wounding. The dashed red lines denote wound edges. All white scale bars represent 100 μm.

2.4. Assessment of Cell Growth on Conductive Bioactive Substrates with IES

Given that graphite/3A-PCL composites have excellent electrical conductivity, we decided to further explore whether these newly developed substrates could enhance cell growth and manipulate the cell morphology and shape through the use of an indirect-current electric field [46–48]. The cell culture experiment using IES was performed under low electric voltage at 0.1 V/mm (approximately 1~2 mA current intensity); this level does not affect cell growth or have negative effects during cell culture [49,50]. After 24 h of culture under IES, the number of NIH/3T3 cells on the 3:10 graphite/3A-PCL substrate increased from 11.4×10^4 to 15.7×10^4 while little change in cell number was seen on control TCPS or pristine 3A-PCL substrate with IES treatment (Figure 5b). These results suggest that exfoliated graphite nanosheets present in composites create an electric field-sensing medium that effectively stimulates the structural motion of the composite under low-voltage electric fields. This probably promotes an increased interaction of composite and cells at the interface, leading to a significant increase in total cell number. To explore how the graphite/3A-PCL substrates under IES treatment influence cell morphology, NIH/3T3 cells were cultured with pristine 3A-PCL or graphite/3A-PCL substrates for 24 h under 0.1 V, and then observed under CLSM to detect changes in cell shape and morphology. The CLSM images showed no significant differences in cell morphology on the pristine 3A-PCL substrate before and after IES treatments (Figure 5c). In contrast, cells on the 3/10 graphite/3A-PCL substrate cell morphology changed remarkably after IES treatment, with cells displaying interlocking and close-packing bundles of spindle-shaped cells of increased overall cell area and pseudopod numbers after IES treatment (Figure 5c, far right). The increased coalescence and proliferation of cells with neighboring cells can be attributed to the fact that exfoliated graphene nanosheets in graphite/3A-PCL composites not only act as an electrical stimulation unit, like an "active trigger", to facilitate the segmental motion of the 3A-PCL polymer chains, but also to efficiently facilitate interaction between the cells and substrate through the effect of an external electric field. This facilitated interaction accelerates the formation of a closely connected cell morphology via the promotion of intercellular adhesion, resulting in a significant increase in total cell numbers on the substrate. Overall, we concluded that the introduction of the adenine moieties in the PCL matrix substantially enhances the binding affinity with cells, which promotes the attachment of cells to the substrate and regulates cellular characteristics, i.e., adhesion, proliferation, and migration. Incorporating graphite into the 3A-PCL substrates enabled the effective tuning of the wettability and conductivity of the substrate surface and thus altered the growth behavior and characteristics of the cells. Importantly, under an indirect-current electric field of only 0.1 V, the graphite/3A-PCL substrates rapidly stimulated the spread and proliferation of cells and significantly increased the total cell numbers. Thus, the adenine and exfoliated graphene-containing bioactive substrates exhibit unique physical and biological properties that efficiently enhance cell growth and manipulate cell morphology and function through IES-responsive characteristics. These important features suggest great potential for a wide variety of biomedical applications, especially as a highly effective scaffold for tissue and cell cultures [43].

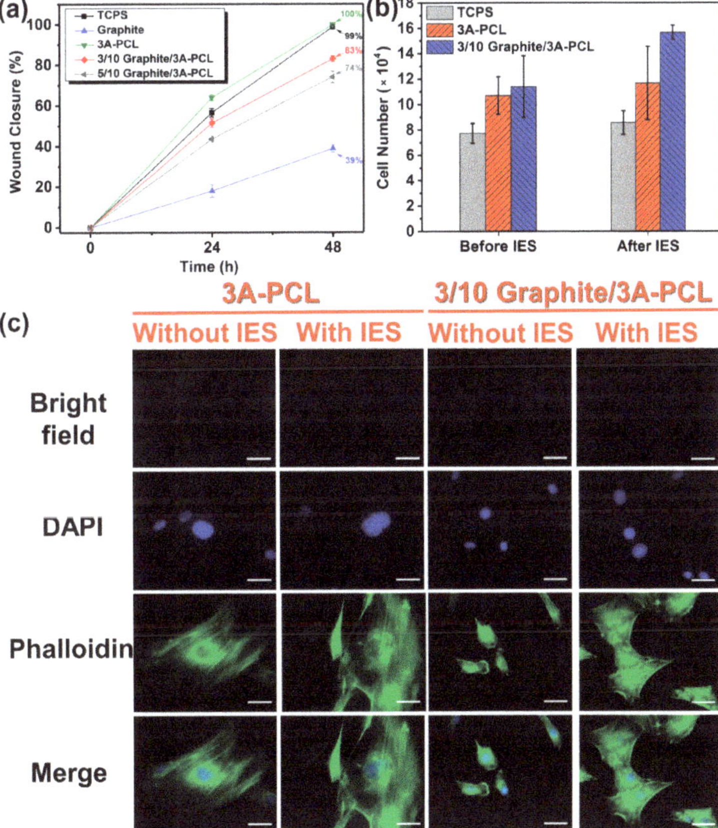

Figure 5. (**a**) The healing rate (percentage) of NIH/3T3 cells by scratch wound healing assay cultured on TCPS, 3A-PCL, 3/10, and 5/10 graphite/3A-PCL substrates at 0, 24, and 48 h after scratching. The (**b**) numbers and (**c**) CLSM images of NIH/3T3 cells harvested on TCPS, 3A-PCL and graphite/3A-PCL substrates treated with or without IES at 0.1 V for 24 h. White scale bars in (**c**) are 20 μm in all images.

3. Materials and Methods

Details regarding the synthetic procedures, the cell experiments and instrumentation used in this study are given in the Supplementary Information.

4. Conclusions

In summary, we successfully created a high-performance conductive supramolecular nanocomposite containing a hydrogen-bonded adenine-functionalized PCL and exfoliated graphene nanosheets that serve as a highly efficient bioactive substrate for the cell culture and manipulation of cell biophysical properties. The exfoliation of graphite within the 3A-PCL matrix promotes the formation of well-dispersed graphene nanosheets with unique structural and physical properties due to the presence of strong interaction between the exfoliated graphene nanosheets and the self-assembled nanostructures of 3A-PCL. The resulting spin-coated composite films can be easily tuned by altering the blending ratio of the graphite and 3A-PCL to obtain the required level of surface roughness and achieve the desired surface wettability and electrical conductivity. The combination of a wide range of tunable physical properties and stable thermo-reversible behavior of graphite/3A-PCL composites is rare and has strong potential for use as cell culture substrates or tissue culture scaffolds. When these newly developed composites were evaluated under in vitro environmental conditions, they exhibited extremely low cytotoxic activity against NIH/3T3 normal cells, high structural stability, and biocompatibility in the SRBC-containing medium. Cell culture, scratch experiments, and fluorescence images confirmed that pristine 3A-PCL substrates can efficiently interact with cell surfaces to enhance cell attachment, spreading, migration, and proliferation. With the incorporation of graphite into the 3A-PCL matrix, the resulting composite substrates can efficiently regulate cellular functions involved in cellular morphological features, without affecting wound healing abilities. More importantly, placing the graphite/3A-PCL substrates under an indirect current electric field of only 0.1 V rapidly stimulated cell responses in terms of adhesion, spreading, viability, and proliferation, leading to a substantial increase in the total cell numbers and a significant alteration in cell morphology, especially a gradual increase in cell size distribution. The presence of both adenine moieties and exfoliated graphene nanosheets within the composite substrates are crucial for the manipulation of cell growth, morphology, and functions by IES-responsive characteristics. This newly created strategy provides a simple, rapid, and efficient path to produce biocompatible and biodegradable conductive supramolecular nanocomposites for the development of multifunctional bioactive substrates that can substantially improve the cell culturing process.

Supplementary Materials: The following supporting information can be downloaded at: https://www.mdpi.com/article/10.3390/ijms23084332/s1.

Author Contributions: C.-Y.W.: data curation, methodology, writing—original draft. A.Z.M.: investigation, validation, writing—original draft. F.B.I.: investigation, writing—original draft. C.-W.C.: investigation, resources, C.-C.C.: conceptualization, funding acquisition, investigation, methodology, resources, supervision, visualization, writing—review and editing. All authors have read and agreed to the published version of the manuscript.

Funding: Ministry of Science and Technology, Taiwan (contract No. MOST 110-2221-E-011-003-MY3).

Institutional Review Board Statement: Not applicable.

Informed Consent Statement: Not applicable.

Data Availability Statement: Not applicable.

Acknowledgments: This study was financially supported by the Ministry of Science and Technology, Taiwan (contract no. MOST 110-2221-E-011-003-MY3).

Conflicts of Interest: The authors declare no competing financial interest.

References

1. Liang, Y.; Kiick, K.L. Heparin-functionalized polymeric biomaterials in tissue engineering and drug delivery applications. *Acta Biomater.* **2014**, *10*, 1588–1600. [CrossRef] [PubMed]
2. Zhang, J.; Liu, K.; Müllen, K.; Yin, M. Self-assemblies of amphiphilic homopolymers: Synthesis, morphology studies and biomedical applications. *Chem. Commun.* **2015**, *51*, 11541–11555. [CrossRef] [PubMed]
3. Suh, J.K.F.; Matthew, H.W.T. Application of chitosan-based polysaccharide biomaterials in cartilage tissue engineering: A review. *Biomaterials* **2000**, *21*, 2589–2598. [PubMed]
4. Cao, L.; Cao, B.; Lu, C.; Wang, G.; Yu, L.; Ding, J. An injectable hydrogel formed by in situ cross-linking of glycol chitosan and multi-benzaldehyde functionalized PEG analogues for cartilage tissue engineering. *J. Mater. Chem. B* **2015**, *3*, 1268–1280. [CrossRef]
5. Aizawa, Y.; Owen, S.C.; Shoichet, M.S. Polymers used to influence cell fate in 3D geometry: New trends. *Prog. Polym. Sci.* **2012**, *37*, 645–658. [CrossRef]
6. Seo, J.-H.; Yui, N. The effect of molecular mobility of supramolecular polymer surfaces on fibroblast adhesion. *Biomaterials* **2013**, *34*, 55–63. [CrossRef]
7. Dou, X.-Q.; Zhang, D.; Feng, C.; Jiang, L. Bioinspired hierarchical surface structures with tunable wettability for regulating bacteria adhesion. *ACS Nano* **2015**, *9*, 10664–10672. [CrossRef]
8. Galbraith, C.G.; Sheetz, M.P. Forces on adhesive contacts affect cell function. *Curr. Opin. Cell Biol.* **1998**, *10*, 566–571. [CrossRef]
9. Gumbiner, B.M. Cell adhesion: The molecular basis of tissue architecture and morphogenesis. *Cell* **1996**, *84*, 345–357. [CrossRef]
10. Zhao, K.; Deng, Y.; Chen, J.C.; Chen, G.-Q. Polyhydroxyalkanoate (PHA) scaffolds with good mechanical properties and biocompatibility. *Biomaterials* **2003**, *24*, 1041–1045. [CrossRef]
11. Qi, W.; Xue, Z.; Yuan, W.; Wang, H. Layer-by-layer assembled graphene oxide composite films for enhanced mechanical properties and fibroblast cell affinity. *J. Mater. Chem. B* **2014**, *2*, 325–331. [CrossRef]
12. Yu, G.; Zhao, X.; Zhou, J.; Mao, Z.; Huang, X.; Wang, Z.; Hua, B.; Liu, Y.; Zhang, F.; He, Z.; et al. Supramolecular polymer-based nanomedicine: High therapeutic performance and negligible long-term immunotoxicity. *J. Am. Chem. Soc.* **2018**, *140*, 8005–8019. [CrossRef]
13. Torchilin, V.P. Multifunctional, stimuli-sensitive nanoparticulate systems for drug delivery. Nat. Rev. *Drug Discov.* **2014**, *13*, 813–827. [CrossRef]
14. Hillewaere, X.K.D.; Du Prez, F.E. Fifteen chemistries for autonomous external self-healing polymers and composites. *Prog. Polym. Sci.* **2015**, *49–50*, 121–153. [CrossRef]
15. Roy, N.; Bruchmann, B.; Lehn, J.-M. Dynamers: Dynamic polymers as self-healing materials. *Chem. Soc. Rev.* **2015**, *44*, 3786–3807. [CrossRef]
16. Li, Z.-Y.; Zhang, Y.; Zhang, C.-W.; Chen, L.-J.; Wang, C.; Tan, H.; Yu, Y.; Li, X.; Yang, H.-B. Cross-linked supramolecular polymer gels constructed from discrete multi-pillar[5]arene metallacycles and their multiple stimuli-responsive behavior. *J. Am. Chem. Soc.* **2014**, *136*, 8577–8589. [CrossRef]
17. Liu, J.; Tan, C.S.Y.; Yu, Z.; Li, N.; Abell, C.; Scherman, O.A. Tough supramolecular polymer networks with extreme stretchability and fast room-temperature self-healing. *Adv. Mater.* **2017**, *29*, 1605325. [CrossRef]
18. Cheng, C.C.; Lee, D.J.; Chen, J.K. Self-assembled supramolecular polymers with tailorable properties that enhance cell attachment and proliferation. *Acta Biomater.* **2017**, *50*, 476–483. [CrossRef]
19. Cheng, C.C.; Yang, X.J.; Fan, W.L.; Lee, A.W.; Lai, J.Y. Cytosine-functionalized supramolecular polymer-mediated cellular behavior and wound healing. *Biomacromolecules* **2020**, *21*, 3857–3866. [CrossRef]
20. Zalatan, J.G.; Lee, M.E.; Almeida, R.; Gilbert, L.A.; Whitehead, E.H.; La Russa, M.; Tsai, J.C.; Weissman, J.S.; Dueber, J.E.; Qi, L.S.; et al. Engineering complex synthetic transcriptional programs with CRISPR RNA scaffolds. *Cell* **2015**, *160*, 339–350. [CrossRef]
21. Huo, Y.; He, Z.; Wang, C.; Zhang, L.; Xuan, Q.; Wei, S.; Wang, Y.; Pan, D.; Dong, B.; Wei, R.; et al. The recent progress of synergistic supramolecular polymers: Preparation, properties and applications. *Chem. Commun.* **2021**, *57*, 1413–1429. [CrossRef] [PubMed]
22. Cordier, P.; Tournilhac, F.; Soulié-Ziakovic, C.; Leibler, L. Self-healing and thermoreversible rubber from supramolecular assembly. *Nature* **2008**, *451*, 977–980. [CrossRef] [PubMed]
23. Mollet, B.B.; Bogaerts, I.L.J.; van Almen, G.C.; Dankers, P.Y.W. A bioartificial environment for kidney epithelial cells based on a supramolecular polymer basement membrane mimic and an organotypical culture system. *J. Tissue Eng. Regen. Med.* **2017**, *11*, 1820–1834. [CrossRef] [PubMed]
24. Mollet, B.B.; Comellas-Aragonès, M.; Spiering, A.J.H.; Söntjens, S.H.M.; Meijer, E.W.; Dankers, P.Y.W. A modular approach to easily processable supramolecular bilayered scaffolds with tailorable properties. *J. Mater. Chem. B* **2014**, *2*, 2483–2493. [CrossRef]
25. Liao, G.; Hu, J.; Chen, Z.; Zhang, R.; Wang, G.; Kuang, T. Preparation, properties, and applications of graphene-based hydrogels. *Front. Chem.* **2018**, *6*, 450. [CrossRef]
26. Li, L.; Zhou, M.; Jin, L.; Liu, L.; Mo, Y.; Li, X.; Mo, Z.; Liu, Z.; You, S.; Zhu, H. Research progress of the liquid-phase exfoliation and stable dispersion mechanism and method of grapheme. *Front. Mater.* **2019**, *6*, 325. [CrossRef]
27. Marinho, B.; Ghislandi, M.; Tkalya, E.; Koning, C.E.; de With, G. Electrical conductivity of compacts of graphene, multi-wall carbon nanotubes, carbon black, and graphite powder. *Powder Technol.* **2012**, *221*, 351–358. [CrossRef]

28. Rajendran, S.B.; Challen, K.; Wright, K.L.; Hardy, J.G. Electrical stimulation to enhance wound healing. *J. Funct. Biomater.* **2021**, *12*, 40. [CrossRef]
29. Huang, Z.; Guo, Z.; Sun, M.; Fang, S.; Li, H. A study on graphene composites for peripheral nerve injury repair under electrical stimulation. *RSC Adv.* **2019**, *9*, 28627–28635. [CrossRef]
30. Shin, S.R.; Li, Y.-C.; Jang, H.L.; Khoshakhlagh, P.; Akbari, M.; Nasajpour, A.; Zhang, Y.S.; Tamayol, A.; Khademhosseini, A. Graphene-based materials for tissue engineering. *Adv. Drug Deliv. Rev.* **2016**, *105*, 255–274. [CrossRef]
31. Dybowska-Sarapuk, L.; Sosnowicz, W.; Krzeminski, J.; Grzeczkowicz, A.; Granicka, L.H.; Kotela, A.; Jakubowska, M. Printed graphene layer as a base for cell electrostimulation—preliminary results. *Int. J. Mol. Sci.* **2020**, *21*, 7865. [CrossRef] [PubMed]
32. Aydin, T.; Gurcan, C.; Taheri, H.; Yilmazer, A. Graphene based materials in neural tissue regeneration. *Cell Biol. Transl. Med.* **2018**, *3*, 129–142.
33. Wang, W.; Hou, Y.; Martinez, D.; Kurniawan, D.; Chiang, W.-H.; Bartolo, P. Carbon nanomaterials for electro-active structures: A review. *Polymers* **2020**, *12*, 2946. [CrossRef] [PubMed]
34. Xue, J.; Gao, Z.; Xiao, L. The application of stimuli-sensitive actuators based on graphene materials. *Front Chem.* **2019**, *7*, 803. [CrossRef]
35. Lovley, D.R. Bug juice: Harvesting electricity with microorganisms. *Nat. Rev. Microbiol.* **2006**, *4*, 497–508. [CrossRef]
36. Cheah, Y.J.; Buyong, M.R.; Mohd Yunus, M.H. Wound healing with electrical stimulation technologies: A review. *Polymers* **2021**, *13*, 3790. [CrossRef]
37. Thrash, J.C.; Coates, J.D. Review: Direct and indirect electrical stimulation of microbial metabolism. *Environ. Sci. Technol.* **2008**, *42*, 3921–3931. [CrossRef]
38. Wu, C.Y.; Melaku, A.Z.; Chuang, W.T.; Cheng, C.C. Manipulating the self-assembly behavior of graphene nanosheets via adenine-functionalized biodegradable polymers. *Appl. Surf. Sci.* **2022**, *572*, 151437. [CrossRef]
39. Choi, J.; Reipa, V.; Hitchins, V.M.; Goering, P.L.; Malinauskas, R.A. Physicochemical characterization and *in vitro* hemolysis evaluation of silver nanoparticles. *Toxicol. Sci.* **2011**, *123*, 133–143. [CrossRef]
40. Hisey, B.; Ragogna, P.J.; Gillies, E.R. Phosphonium-functionalized polymer micelles with intrinsic antibacterial activity. *Biomacromolecules* **2017**, *18*, 914–923. [CrossRef]
41. Lee, W.C.; Lim, C.H.Y.X.; Shi, H.; Tang, L.A.L.; Wang, Y.; Lim, C.T.; Loh, K.P. Origin of enhanced stem cell growth and differentiation on graphene and graphene oxide. *ACS Nano* **2011**, *5*, 7334–7341. [CrossRef] [PubMed]
42. Kostarelos, K.; Novoselov, K.S. Exploring the interface of graphene and biology. *Science* **2014**, *344*, 261–263. [CrossRef] [PubMed]
43. Chen, L.; Yan, C.; Zheng, Z. Functional polymer surfaces for controlling cell behaviors. *Mater. Today* **2018**, *21*, 38–59. [CrossRef]
44. Jonkman, J.E.N.; Cathcart, J.A.; Xu, F.; Bartolini, M.E.; Amon, J.E.; Stevens, K.M.; Colarusso, P. An introduction to the wound healing assay Using live-cell microscopy. *Cell Adhes. Migr.* **2014**, *8*, 440–451. [CrossRef]
45. Martinotti, S.; Ranzato, E. Scratch wound healing assay. *Methods Mol. Biol.* **2020**, *2109*, 225–229.
46. Tehovnik, E.J.; Tolias, A.S.; Sultan, F.; Slocum, W.M.; Logothetis, N.K. Direct and indirect activation of cortical neurons by electrical microstimulation. *J. Neurophysiol.* **2006**, *96*, 512–521. [CrossRef]
47. Park, H.-H.; Jo, S.; Seo, C.H.; Jeong, J.H.; Yoo, Y.-E.; Lee, D.H. An indirect electric field-induced control in directional migration of rat mesenchymal stem cells. *Appl. Phys. Lett.* **2014**, *105*, 244109. [CrossRef]
48. Ning, C.; Zhou, Z.; Tan, G.; Zhu, Y.; Mao, C. Electroactive polymers for tissue regeneration: Developments andperspectives. *Prog. Polym. Sci.* **2018**, *81*, 144–162. [CrossRef]
49. Onuma, E.K.; Hui, S.W. Electric field-directed cell shape changes, displacement, and cytoskeletal reorganization are calcium dependent. *J. Cell Biol.* **1988**, *106*, 2067–2075. [CrossRef]
50. Balint, R.; Cassidy, N.J.; Cartmell, S.H. Electrical stimulation: A novel tool for tissue engineering. *Tissue Eng. Part B Rev.* **2013**, *19*, 48–57. [CrossRef]

International Journal of *Molecular Sciences*

MDPI

Article

$Cu_xCo_{1-x}Fe_2O_4$ (x = 0.33, 0.67, 1) Spinel Ferrite Nanoparticles Based Thermoplastic Polyurethane Nanocomposites with Reduced Graphene Oxide for Highly Efficient Electromagnetic Interference Shielding

Anju [1], Raghvendra Singh Yadav [1,*], Petra Pötschke [2], Jürgen Pionteck [2], Beate Krause [2], Ivo Kuřitka [1], Jarmila Vilčáková [1], David Škoda [1], Pavel Urbánek [1], Michal Machovský [1], Milan Masař [1] and Michal Urbánek [1]

1 Centre of Polymer Systems, University Institute, Tomas Bata University in Zlín, Trida Tomase Bati 5678, 760 01 Zlín, Czech Republic; deswal@utb.cz (A.); kuritka@utb.cz (I.K.); vilcakova@utb.cz (J.V.); dskoda@utb.cz (D.Š.); urbanek@utb.cz (P.U.); machovsky@utb.cz (M.M.); masar@utb.cz (M.M.); murbanek@utb.cz (M.U.)

2 Leibniz Institute of Polymer Research Dresden (IPF Dresden), 01069 Dresden, Germany; poe@ipfdd.de (P.P.); pionteck@ipfdd.de (J.P.); krause-beate@ipfdd.de (B.K.)

* Correspondence: yadav@utb.cz or raghvendra.nac@gmail.com; Tel.: +42-057-603-1725

Abstract: $Cu_xCo_{1-x}Fe_2O_4$ (x = 0.33, 0.67, 1)-reduced graphene oxide (rGO)-thermoplastic polyurethane (TPU) nanocomposites exhibiting highly efficient electromagnetic interference (EMI) shielding were prepared by a melt-mixing approach using a microcompounder. Spinel ferrite $Cu_{0.33}Co_{0.67}Fe_2O_4$ (CuCoF1), $Cu_{0.67}Co_{0.33}Fe_2O_4$ (CuCoF2) and $CuFe_2O_4$ (CuF3) nanoparticles were synthesized using the sonochemical method. The CuCoF1 and CuCoF2 exhibited typical ferromagnetic features, whereas CuF3 displayed superparamagnetic characteristics. The maximum value of EMI total shielding effectiveness (SE_T) was noticed to be 42.9 dB, 46.2 dB, and 58.8 dB for CuCoF1-rGO-TPU, CuCoF2-rGO-TPU, and CuF3-rGO-TPU nanocomposites, respectively, at a thickness of 1 mm. The highly efficient EMI shielding performance was attributed to the good impedance matching, conductive, dielectric, and magnetic loss. The demonstrated nanocomposites are promising candidates for a lightweight, flexible, and highly efficient EMI shielding material.

Keywords: electromagnetic interference shielding; magnetic nanoparticles; reduced graphene oxide; nanocomposites; spinel ferrite

Citation: Anju; Yadav, R.S.; Pötschke, P.; Pionteck, J.; Krause, B.; Kuřitka, I.; Vilčáková, J.; Škoda, D.; Urbánek, P.; Machovský, M.; et al. $Cu_xCo_{1-x}Fe_2O_4$ (x = 0.33, 0.67, 1) Spinel Ferrite Nanoparticles Based Thermoplastic Polyurethane Nanocomposites with Reduced Graphene Oxide for Highly Efficient Electromagnetic Interference Shielding. *Int. J. Mol. Sci.* **2022**, *23*, 2610. https://doi.org/10.3390/ijms23052610

Academic Editor: Daniel Arcos

Received: 26 January 2022
Accepted: 24 February 2022
Published: 26 February 2022

Publisher's Note: MDPI stays neutral with regard to jurisdictional claims in published maps and institutional affiliations.

Copyright: © 2022 by the authors. Licensee MDPI, Basel, Switzerland. This article is an open access article distributed under the terms and conditions of the Creative Commons Attribution (CC BY) license (https://creativecommons.org/licenses/by/4.0/).

1. Introduction

Electromagnetic interference, which is generated by the rapid procreation of electronic and communication technology devices, has become a serious concern in everyday life [1,2]. The transmission of electromagnetic waves from potential sources, such as mobile phones, radar systems, and different electronic appliances, are cause for interfering with electronic devices, which influences the lifetime and functionality of such electronic instruments [3]. Therefore, to resolve this problem, the designing of a shielding material endowed with efficient shielding characteristics has become a considerable research interest. Traditional metal-based composites present with many demerits, such as high density, poor corrosion resistance, high processing cost, etc. Moreover, in the case of conventional metals, attenuation of incident electromagnetic (EM) wave occurs by reflection, which has very little contribution towards the reduction of EM pollutions. The material should contain electrical and magnetic dipoles to be an EM wave absorber [4]. In this context, polymer nanocomposites along with magnetic and dielectric nanofillers opened a new pathway due to their lightweight, flexibility, good absorption, low cost, and resistance to corrosion [5]. Epoxy resin, thermoplastic polyurethane (TPU), polyvinylidene fluoride (PVDF), and polydimethylsiloxane are some of the most commonly used polymer matrices [6,7]. Among them, TPU based composites have attained an incredible research interest due to their flexibility, stretchability, superior mechanical properties, and wearable resistance [8]. For example, Valentini et al. used exfoliated graphite-TPU nanocomposite and recorded remarkable shielding efficiency in the microwave region [9]. Similarly, Zahid et al. [10] fabricated nanocomposite based on reduced graphene oxide (rGO) and TPU matrix and reported a high shielding performance of 53 dB. Moreover, Sobha et al. [11] recorded the EMI shielding efficiency of 31.5 dB by utilizing multi-walled carbon nanotube-based TPU composites.

The magnetic properties of spinel ferrite nanoparticles make them an ideal filler for the development of robust electromagnetic shielding polymer nanocomposite material. Moreover, the incorporation of rGO as a second filler along with magnetic spinel ferrites can help in the enhancement of interfacial polarization, high electrical conductivity, and good impedance matching. In our recent work, we demonstrated excellent EMI shielding with $MnFe_2O_4$ and rGO in a polypropylene matrix [12]. Further, the research group of Kumar et al. [13] reported total shielding effectiveness of $\approx$ 38.2 dB for $NiFe_2O_4$ and rGO nanocomposite in the X-band frequency range. Among all spinel ferrites, there are several studies on $CoFe_2O_4$ along with rGO nanocomposites in a polymer matrix for applications in EMI shielding, which proves $CoFe_2O_4$ as a potential candidate. Dey et al. [14] investigated the EMI shielding efficiency of $Co_{0.5}Zn_{0.4}Cu_{0.1}Fe_2O_4$-GO/paraffin wax hybrid nanocomposite and reported a shielding efficiency of 53.2 dB in the X-band frequency region. Gulzar et al. [8] reported a shielding efficiency of 35 dB in the frequency range from 0.1 to 8 GHz for cobalt ferrites along with coal-fly in the TPU matrix. In addition, Ismail et al. [15] investigated EM shielding and microwave absorption properties of $CoFe_2O_4$ and polyaniline doped with para toluene sulfonic acid and reported a maximum return loss of -28.4 dB at 8.1 GHz.

In this contribution, we utilized hybrid filler systems of $Cu_xCo_{1-x}Fe_2O_4$ (x = 0.33, 0.67, 1) spinel ferrite nanoparticles and rGO inside a TPU matrix for developing nanocomposites with light weight, good flexibility, and highly efficient electromagnetic shielding performance. For the development of thermoplastic polyurethane nanocomposites with reduced graphene oxide and spinel ferrite nanoparticles, we aimed to utilize three samples of spinel ferrite nanoparticles, one with a low content of Cu^{2+}, another with a high content of Cu^{2+}, and a last one of pure $CuFe_2O_4$ nanoparticles. Therefore, x = 0.33, 0.67, and 1 was selected for $Cu_xCo_{1-x}Fe_2O_4$ spinel ferrite system. To the best of our knowledge, this is the first report on $Cu_xCo_{1-x}Fe_2O_4$ (x = 0.33, 0.67, 1) spinel ferrite nanoparticles along with rGO in the TPU matrix for this purpose.

2. Results and Discussion

2.1. XRD Study

Figure 1a depicts the X-ray diffraction (XRD) pattern of CuCoF1, CuCoF2, and CuF3 nanoparticles. The distinctive XRD peaks at 2θ = 18.2°, 30.2°, 35.5°, 43.2°, 53.4°, 57°, 62.6°, 66.2°, and 68.3° can be seen, which are assigned to crystalline planes (111), (220), (311), (400), (422), (511), (440), (532) and (442), respectively [16]. The diffraction pattern revealed the monophasic formation of the inverse spinel ferrite crystal structure with space group Fd-$\bar{3}$m in all the samples [17]. However, in the case of the CuF3 sample, some additional peaks of CuO indexed to (111), (−202), and (−113) diffraction planes were also noticed [18]. In the sonochemical synthesis of CuF3 samples, there is an excess concentration of Cu^{2+} in reaction solution, which agglomerates with NaOH under ultrasonic waves and formation of CuO occurs with rising the temperature of the reaction mixture up to 85 °C in sonochemical approach. M.A. Shilpa Amulya et al. [19] also noticed the formation of CuO phases with $CuFe_2O_4$ during the calcination process. Further, the presence of CuO can improve the EMI performance by improving the impedance matching of CuF3-rGO-TPU nanocomposites. Shangyu Gao et al. [20] also noticed the role of the secondary phase on the EMI performance of magnesium alloy. In addition, Lulu Song et al. [21] adjusted the electromagnetic wave absorption characteristics with control of multiple phases.

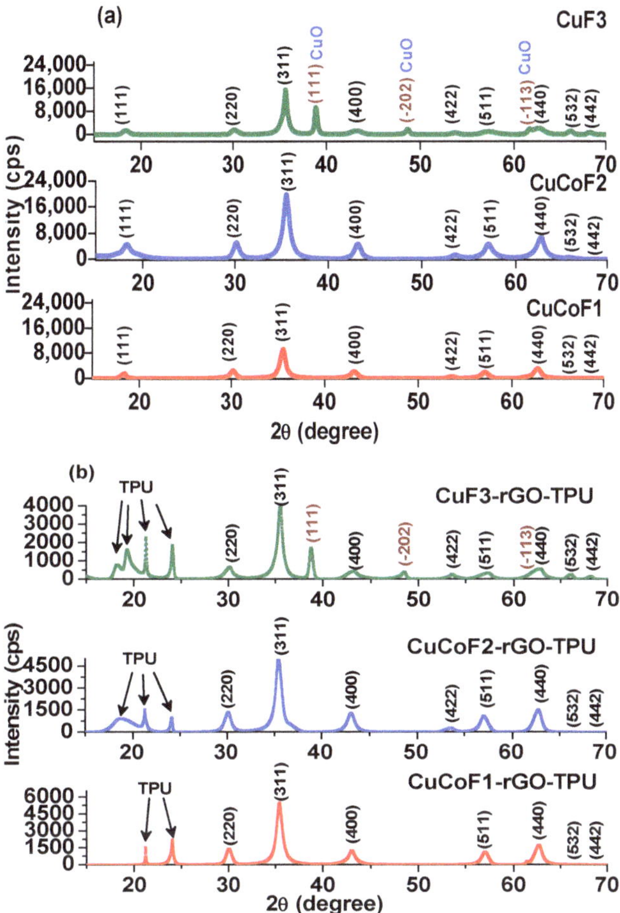

Figure 1. XRD pattern of (**a**) CuCoF1, CuCoF2 and CuF3 nanoparticles (**b**) nanocomposites CuCoF1, CuCoF2 and CuF3 nanoparticles along with rGO in TPU matrix.

The crystallite size of the synthesized spinel ferrite nanoparticles was evaluated using Scherrer's formula:

$$D = \frac{0.94\,\lambda}{\beta Cos\theta} \quad (1)$$

where D is the crystallite size (nm), β is the full width of diffraction line at half maxima (in radians), λ is the wavelength of the source (Cu-Kα radiation) and θ signifies diffraction angle. The value of crystallite size was calculated for the most prominent peak corresponding to the d_{311} diffraction plane and was found to be 5.51 nm, 4.96 nm, and 3.94 nm for CuCoF1, CuCoF2, and CuF3 nanoparticles, respectively [22,23]. Jnaneshwara et al. [24] also reported a similar trend in crystallite size with the substitution of Cu^{2+} ions in the $CoFe_2O_4$ lattice.

The XRD diffraction pattern for TPU polymer nanocomposites based on CuCoF1, CuCoF2, and CuF3 nanoparticles along with rGO is shown in Figure 1b. As can be seen, the diffraction planes of the face-centered cubic structure of spinel ferrites were present in all the prepared nanocomposite samples [25]. Further, the signature peak for rGO was not present due to its fine dispersion and small size [26]. TPU exhibits broad diffraction peaks ranging from 18° to 24° associated with a mixture of the ordered structure of the hard phase and disordered structure of the amorphous phase [27]. In the case of CuCoF1-rGO-TPU nanocomposite, a distinct peak at $2\theta = 21.08°$ and another peak at 23° remarked the presence of TPU [28]. After the addition of more crystalline CuCoF2 in TPU, the ordering is further improved with appearance of additional peak at 18.7°. In addition, ordering was further improved with addition of CuF3 consisted of CuO phase, and additional peak at 19.2° was noticed. S. Kumar et al. [29] also noticed the appearance of additional peaks with an improvement in the crystallinity of TPU with the addition of MWCNT. The pure TPU exhibited a broad amorphous diffraction peak centered at about $2\theta = 19.7°$ of the (110) reflection plane with the interchain spacing of 4.44 Å [30,31].

2.2. Raman Study

Raman spectroscopy was employed for further investigation of the structural characteristics of prepared nanoparticles and nanocomposites. Figure 2a displays the Raman spectra for CuCoF1, CuCoF2, and CuF3 spinel ferrite nanoparticles. As can be seen, spinel ferrite nanoparticles exhibited Raman bands around 276 cm^{-1}, 371 cm^{-1}, 463 cm^{-1}, 561 cm^{-1}, 604 cm^{-1}, and 674 cm^{-1} attributed to $T_{1g}(3)$, E_g, $T_{1g}(2)$, $T_{1g}(1)$, $A_{1g}(2)$, and $A_{1g}(1)$ Raman modes [32,33]. The lower frequency modes are a consequence of vibration at the tetrahedral site while higher frequency modes reflect vibrations at the octahedral site of spinel ferrite lattice [32,34]. Further, Figure 2b represents the Raman spectra for GO and rGO. Two distinct peaks attributed to the D band at 1353 cm^{-1} and G-band at 1596 cm^{-1} can be noticed for GO. The D vibration band arises from the imperfection edges, also known as the breathing mode of j-point photons of A_{1g} symmetry [35]. However, the peak assigned to the G band is a consequence of the first-order scattering of E_{2g} phonons by sp^2 carbon at the Brillouin zone center [36]. Similarly, the Raman peaks for D and G bands in Raman spectra were obtained at 1342 cm^{-1} and 1573 cm^{-1} for rGO [37]. Moreover, the intensity ratio (I_D/I_G) which is a measure of the extent of the disorder, was noted to be increased after GO was reduced to rGO. The intensity ratio of I_D/I_G was found to be 0.95 and 1.26 for GO and rGO, respectively. In addition, after the reduction of GO, the highest intensity was observed in the case of the D-band [38,39]. Besides the D and G vibration band, other Raman bands at 2674 cm^{-1} and 2911 cm^{-1} assigned to 2D and D+G band were also observed for rGO [40]. The 2D band was related to the inelastic scattering of phonons whereas D+G is the consequence of the summation of both D and G-bands [41].

Figure 2c depicts the Raman spectrum of spinel ferrite nanocomposites with rGO in the TPU matrix. As can be seen, the signature peaks at 2932 cm^{-1}, 1533 cm^{-1}, 1730 cm^{-1}, and 1445 cm^{-1} attributed to stretching vibration of $-CH_2$, amide (II), C=O stretching, and bending vibration of $-CH_2$ confirmed the existence of polyurethane in nanocomposites [42]. Further, two characteristic peaks at 1342 cm^{-1} and 1573 cm^{-1} attributed to the D and G band of rGO confirmed the presence of rGO in all the composite samples [43].

A slight peak at 606 cm^{-1} assigned to the $A_{1g}(2)$ mode of spinel ferrite was noticed [32]. Another Raman band observed at 467 cm^{-1} is assigned with the T_{1g} (2) mode of stretching. The intensity of the peaks is reduced due to the interaction of nanofillers in polyurethane. These modes associated with space group Fd-$\bar{3}$m are the signature of a cubic inverse spinel structure, thus confirming the presence of a spinel ferrite structure inside the nanocomposites [32]. Raman spectroscopy provides confirmation of the presence of spinel ferrite nanoparticles and reduced graphene oxide (rGO) in the thermoplastic polyurethane (TPU) matrix.

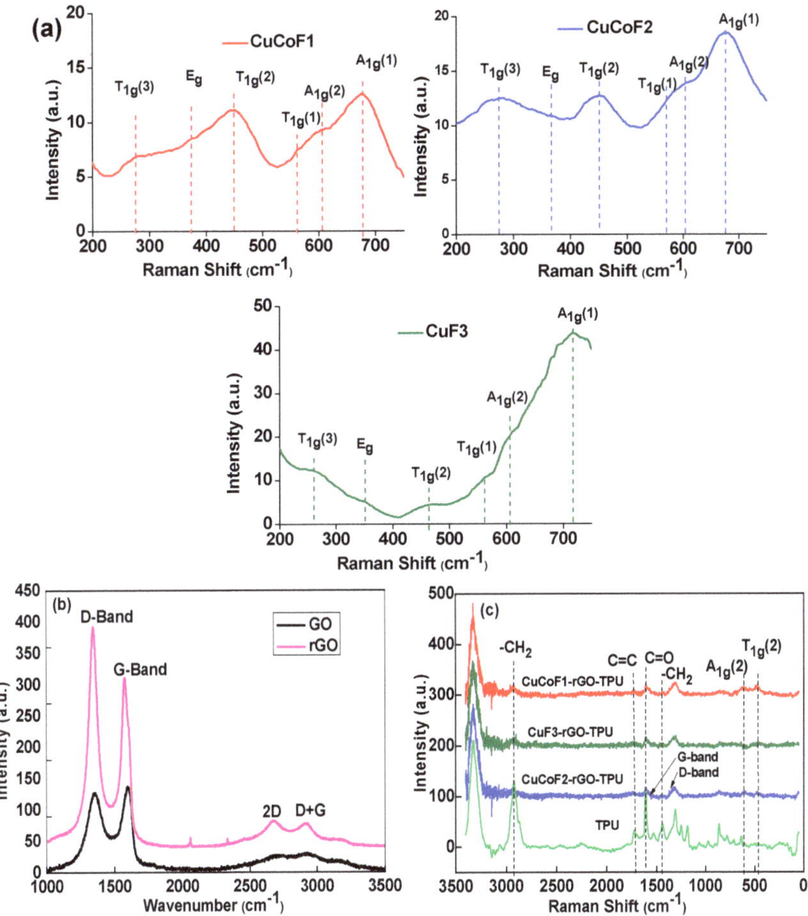

Figure 2. Raman Spectra of (**a**) CuCoF1, CuCoF2 and CuF3 nanoparticles (**b**) GO and rGO (**c**) CuCoF1, CuCoF2, and CuF3 nanoparticles based nanocomposites along with rGO in TPU matrix.

2.3. Fourier Transform Infra-Red Spectroscopy (FTIR) Analysis

FTIR spectroscopy was employed to investigate the structural characteristics of spinel ferrite nanoparticles, RGO, and its TPU nanocomposites. Figure 3a displays the information on vibrational spectra of CuCoF1, CuCoF2, and CuF3 nanoparticles. One absorption band v_1 at around 550 cm^{-1} is associated with the intrinsic symmetric vibration due to the metal-oxygen bond at the tetrahedral site and the other absorption band v_2 around 302–322 cm^{-1} is relevant to octahedral metal stretching [44,45]. The presence of the two major absorption bands v_1 and v_2 in FTIR spectra, as shown in Figure 3a, confirmed the cubic spinel structure of all the nanoparticles samples [46,47]. The obtained results are in good agreement with the reported literature [48].

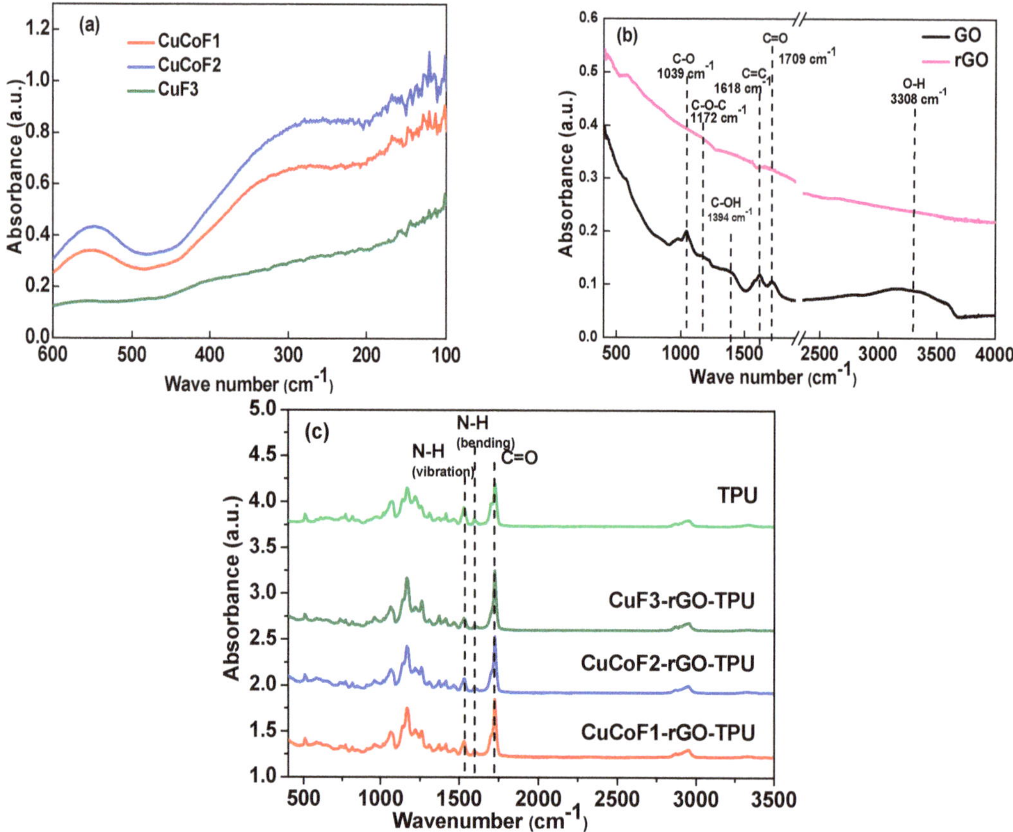

Figure 3. FTIR spectra of (**a**) CuCoF1, CuCoF2 and CuF3 nanoparticles (**b**) GO and rGO (**c**) CuCoF1, CuCoF2 and CuF3 nanoparticles based nanocomposites along with rGO in TPU matrix.

The presence of oxygen-containing functional groups was confirmed by FTIR spectra of GO, represented in Figure 3b. As can be observed, the absorption bands at 1709 cm^{-1} and 1618 cm^{-1} represent the carbonyl stretching (C=O) and skeletal stretching (C=C) of the alkene group [49,50]. Moreover, the high intensity of prominent peaks in GO divulges the presence of a large amount of oxygen-containing functional groups. Furthermore, another absorption band centered around 1000–1100 cm^{-1} assigned to the C-O epoxide group was also noticed [51]. A broad absorption band between 2500 and 3500 cm^{-1} was attributed to the carboxyl (–COOH) groups [52]. It can be observed in Figure 3b that, after the reduction of GO, there was a significant decrease in the intensity of absorption spectra of alkoxy and hydroxyl groups [53]. Indeed, all attributed peaks possess weaker intensity as compared to the intensity of the FTIR spectrum of GO, which proves the successful reduction of GO.

Figure 3c depicts the FTIR spectra of CuCoF1, CuCoF2, and CuF3 nanocomposites along with rGO in the TPU matrix. As can be seen in the graph, a slight absorption peak at 3329 cm^{-1} for all the samples is attributed to the stretching vibration of -NH in the urethane group [42]. Moreover, the characteristic bands of TPU regardless of synthesis route were also noticed at 1725 cm^{-1} and 1527 cm^{-1} ascribed to C=O stretching and N-H bending vibration in polyurethane, thus confirming the existence of TPU in nanocomposites [54]. In consolidation with Raman and FTIR spectra analysis, the existence of spinel ferrite nanoparticles and rGO in TPU was confirmed.

2.4. TEM and HRTEM Analysis of Nanoparticles

Figure 4 depicts the TEM and HRTEM images of CuCoF1, CuCoF2, and CuF3 nanoparticles. The spherical nanoparticles of 3–5 nm with slight agglomeration can be noticed from Figure 4a for the CuCoF1 sample. The HRTEM image of the CuCoF1 sample as shown in Figure 4b, displays lattice of (220) planes (d spacing 0.29 nm) and (400) planes (d spacing 0.21 nm) of spinel ferrite [55]. Further, the TEM image (Figure 4c) of the CuCoF2 sample shows spherical particles of 2–4 nm with moderate agglomeration. The lattice of (311) planes (d spacing 0.25 nm), (220) planes (d spacing 0.29 nm), and (400) planes (d spacing 0.21 nm) of spinel ferrite can be observed in the HRTEM image (Figure 4d) for the CuCoF2 sample. Furthermore, spherical nanoparticles of 2–3 nm can be observed in Figure 4e for the CuF3 sample. The HRTEM image of CuF3 as displayed in Figure 4f, reveals the lattice of (400) planes (d spacing 0.21 nm) and (111) planes (d spacing 0.46 nm) of spinel ferrite.

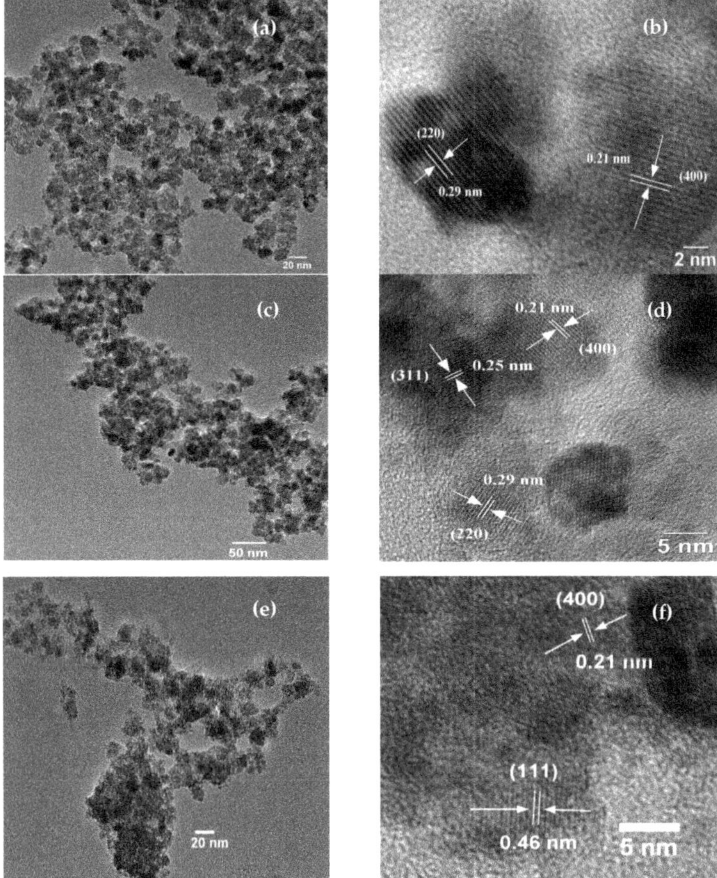

Figure 4. (**a**) TEM and (**b**) HRTEM of CuCoF1; (**c**) TEM, and (**d**) HRTEM of CuCoF2; (**e**) TEM and (**f**) HRTEM image of CuF3.

2.5. FE-SEM and EDX Study of Nanocomposites

To investigate the nanofillers in the TPU matrix, FE-SEM images were studied. Figure 5a depicts the FE-SEM image of the cryo-fractured surface of CuCoF1-rGO-TPU nanocomposite which depicts the dispersion of CuCoF1 and rGO nanofillers in the TPU matrix. Figure 5b shows the EDX spectrum of CuCoF1-rGO-TPU nanocomposite, which confirm the presence of Cu, Co, Fe, C, and O. Further, the good dispersion of spinel ferrite nanoparticles and rGO in TPU matrix of other prepared nanocomposites CuCoF2-rGO-TPU, and CuF3-rGO-TPU can be seen in FE-SEM image, as shown in Figure 5c,d, respectively.

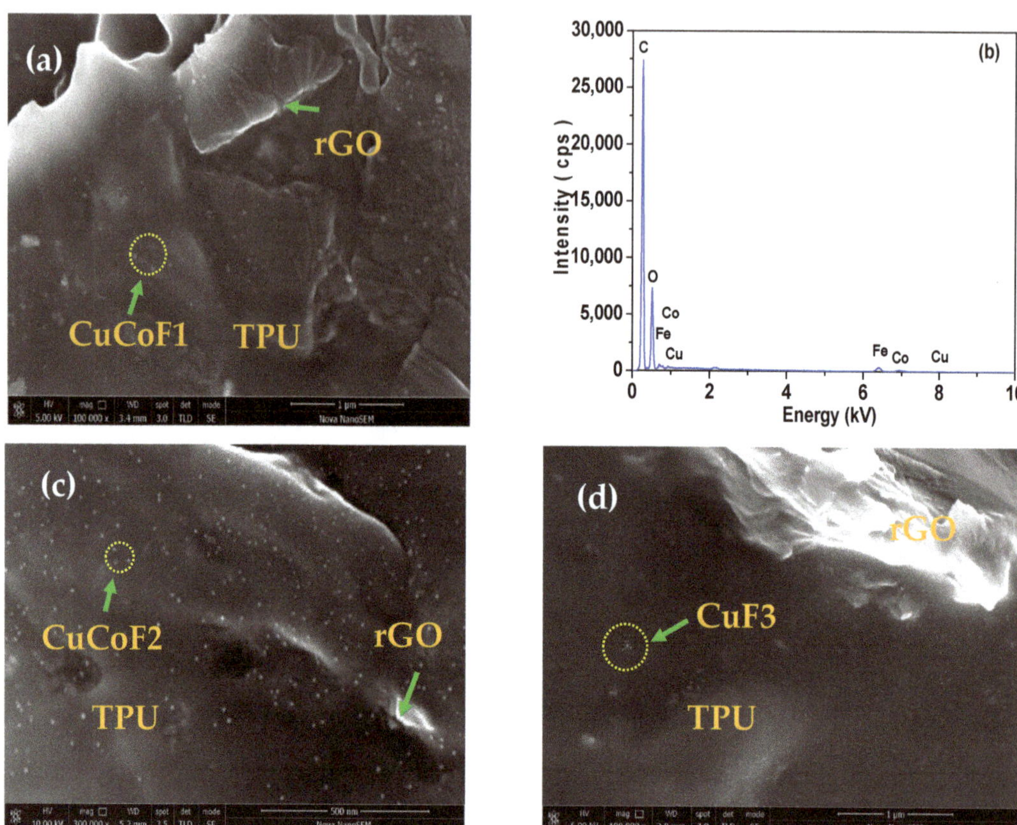

Figure 5. (**a**) FE-SEM image of CuCoF1-rGO-TPU, (**b**) EDX pattern of CuCoF1-rGO-TPU, (**c**) FE-SEM image of CuCoF2-rGO-TPU, and (**d**) FE-SEM image of CuF3-rGO-TPU.

2.6. Magnetic Property

Figure 6 depicts the magnetic hysteresis curves of prepared CuCoF1, CuCoF2, and CuF3 spinel ferrite nanoparticles. The magnetic hysteresis curves of CuCoF1, and CuCoF2 show typical ferromagnetic features. The value of saturation magnetization (M_s) of the CuCoF1, CuCoF2, and CuF3 samples are 37.2 emu/g, 31.5 emu/g, and 15.6 emu/g, respectively. The decrease in the value of M_s with a decrease of grain size is associated with an increase of the surface spin canting and dead magnetic layer [56]. The decrease in saturation magnetization (M_s) with an increase of Cu^{2+} content can also be explained by the probable replacement of Co^{2+} by Cu^{2+} at the tetrahedral sites of the spinel ferrite lattice systems [57]. The observed variation in M_s can be explained by the help of Neel's two sub-lattice magnetization model [58]. According to this model, magnetization is obtained with the help of $M(\mu_B) = M_B - M_A$, where M_A and M_B are the net magnetic moment of tetrahedral (A) and octahedral (B) sites, respectively. The decrease in the saturation magnetization can be attributed to the lower magnetic moment of Cu^{2+} ($1\mu_B$) than Co^{2+} ($3\mu_B$). Thus, the magnetic moment in the B-sublattice was sequentially decreased with the increase in copper substitution, which results in lower magnetic moment of copper substituted spinel ferrite nanoparticles [59]. Similar results were noticed by other researchers [57–59]. The remanent (M_r) magnetization values were 3.2 emu/g, 0.48 emu/g, and 0 emu/g for CuCoF1, CuCoF2, and CuF3 sample, respectively. The coercivity (H_c) value was 50.4 Oe, 9.5 Oe, and 0 Oe for CuCoF1, CuCoF2, and CuF3 samples, respectively. In nanosized particles, a single magnetic domain with no residual magnetism are known as superparamagnetic characteristics [60]. Below a critical size of a ferromagnetic material, the anisotropy energy is lower than the thermal energy, which leads to superparamagnetic characteristics [61]. The magnetic hysteresis curve of CuF3 display an S-shape with no coercivity and remanence, which points to the superparamagnetic characteristic, and can be beneficial for utilization as high-performance electromagnetic interference shielding material at high frequency [62].

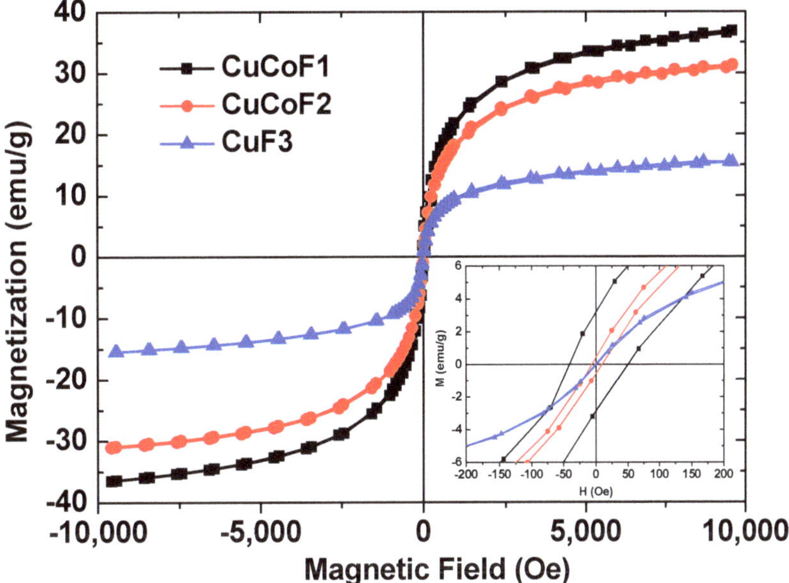

Figure 6. Magnetic hysteresis curves of prepared CuCoF1, CuCoF2, and CuF3 spinel ferrite nanoparticles.

2.7. Electromagnetic Interference Shielding

Electromagnetic interference shielding is a mechanism of reflection and absorption of electromagnetic radiation by a material that prevents the penetration of harmful electromagnetic waves. It is defined as the ratio of incident power (P_i) to the outgoing power (P_t) of electromagnetic waves and is represented in the decibel (dB) unit. The mechanisms which contribute to total shielding effectiveness, SE_T, are shielding due to reflection SE_R, absorption SE_A, and multiple internal reflections SE_M. The total shielding effectiveness SE_T can be mathematically written as follows [63]:

$$SE_T = -10\log(P_t/P_i) = SE_R + SE_A + SE_M$$

The reflection of EM wave is due to the impedance mismatch between the air and the absorber (shielding material) by mobile charge carriers such as electrons and holes whereas in the case of multiple reflections, the scattering effect is responsible due to inhomogeneity and large interfacial area inside the matrix [64]. Absorption is associated with the dissipation of EM waves in form of heat energy through shielding material. Moreover, several other factors such as the thickness of the shielding material, ohmic loss, polarization loss, and magnetic loss have their role. Ohmic losses in a shielding material are related to the dissipation of energy due to charge hopping, tunneling, and conducting mechanisms. Polarization losses originate due to the dissipation of energy required for overcoming the state of reorientation of dipoles in every half-cycle of EM radiations caused by functional groups, interfaces, and defects in material [64]. The main magnetic losses are associated with the eddy current loss, natural resonance, hysteresis loss, and exchange resonance [65]. Although according to Schelkunoff's theory, if the shielding exceeds 10 dB and the shielding material has thickness higher than the material skin depth (δ), then the role of multiple reflections (SE_M) can be disregarded in practical EMI applications and only SE_A and SE_R are taken into consideration for the total shielding effectiveness SE_T [64]. Therefore, the higher is the value of SE_T, the lower the energy transmission through a shielding material will be.

In the present study, we investigated the electromagnetic interference shielding effectiveness of CuCoF1-rGO-TPU, CuCoF2-rGO-TPU, and CuF3-rGO-TPU nanocomposites in the X-band frequency (8.2–12.4 GHz), as depicted in Figure 7a. As can be seen, the maximum value of SE_T was found to be 42.9 dB, 46.2 dB, and 58.8 dB for CuCoF1-rGO-TPU, CuCoF2-rGO-TPU, and CuF3-rGO-TPU nanocomposites, respectively. Next, Figure 7b elaborates effective shielding performance more in terms of another electromagnetic parameter which is absorption, SE_A. As observed, the maximum value of SE_A was 23.1 dB, 25.1 dB, and 35.0 dB for developed CuCoF1-rGO-TPU, CuCoF2-rGO-TPU, and CuF3-rGO-TPU nanocomposites, respectively. For further assessment, the value of SE_R was also evaluated. Figure 7c represents the values of SE_R, which were found to be 19.8 dB, 21.1 dB, and 23.8 dB for CuCoF1-rGO-TPU, CuCoF2-rGO-TPU, and CuF3-rGO-TPU nanocomposites, respectively. Notably, the high value of SE_A in these nanocomposites indicates that the absorption of EM waves is the dominant mechanism. The plot in Figure 7d displays an evaluation and comparison of electromagnetic shielding parameters. As can be seen, the highest values of SE_T, SE_A, and SE_R for the CuF3-rGO-TPU nanocomposite signify that this composition possesses good electromagnetic shielding properties. In addition, EMI shielding efficiency (%), which signifies the capability of shielding material to block EM waves in terms of percentage, can be evaluated by the following relation with the EMI shielding effectiveness (dB) [66]:

$$\text{Shielding Efficiency } (\%) = 100 - \left(\frac{1}{10^{\frac{SE}{10}}}\right) \times 100$$

The EMI shielding efficiency (%) was 99.9948%, 99.9976%, and 99.9998% for CuCoF1-rGO-TPU, CuCoF2-rGO-TPU, and CuF3-rGO-TPU nanocomposites, respectively. It signifies that CuF3-rGO-TPU nanocomposite has a blockage of 99.9998% of the incident EM waves with only 0.0002% transmission. In addition, the shielding ability, as well as the lightweight property of material, can be expressed by specific shielding effectiveness, SSE (=EMISE/density) and the absolute shielding effectiveness, SSE/t (=SSE/thickness) [67]. The evaluated value of SSE was 90.232 dB cm^3 g^{-1} and SSE/t was 902.320 dB cm^2 g^{-1} for CuF3-rGO-TPU nanocomposite.

A research group, Ali et al. [68] investigated microwave absorption characteristics of Polyaniline (PANI)/NiZn ferrite nanocomposites by varying ferrite percentage and reported a maximum reflection loss of −39.56 dB with 2.5 mm sample thickness. Similarly, another research group Kumar et al. [13] studied the electromagnetic shielding behavior of $NiFe_2O_4$/rGO possessing a thickness of 2.0 mm and observed SE_T of 38.2 dB in the X-band frequency range. Further, a maximum value of SE_T of 53 dB by fabrication of rGO-TPU nanocomposite has been reported by Zahid et al. [10]. Furthermore, Gunasekaran et al. [69] observed an optimum reflection loss RL of −31.89 dB with rGO/zirconium substituted cobalt ferrite ($Co_{0.5}Zr_{0.5}$ Fe_2O_4) nanocomposites having 5 mm sample thickness. Moreover, Gahlout et al. [70] studied electromagnetic shielding response for polypyrrole-MWCNT/polyurethane composites and reported a SE_T of 48 dB with a sample thickness of 3 mm in the X-band frequency range. Furthermore, Sulaiman et al. [71] reported excellent absorption characteristics of −24.86 dB for $Co_{0.5}Zn_{0.5}$ Fe_2O_4/PANI-PTSA nanocomposites (weight ratio 1:1) with an optimal matching thickness of 3 mm. Total shielding effectiveness, SE_T of 35 dB for Polystyrene/PANI/Nickel spinel ferrite composite, is noticed in a broad frequency range of 0.1 to 20 GHz by Shakir et al. [72]. Moreover, Li et al. [73] reported total shielding effectiveness SE_T of 48.4 dB with nitrogen-doped reduced graphene oxide/$CoFe_2O_4$ hybrid nanocomposites in X and Ku bands. The recent advances in the electromagnetic shielding performance of some nanocomposites are displayed in Table 1.

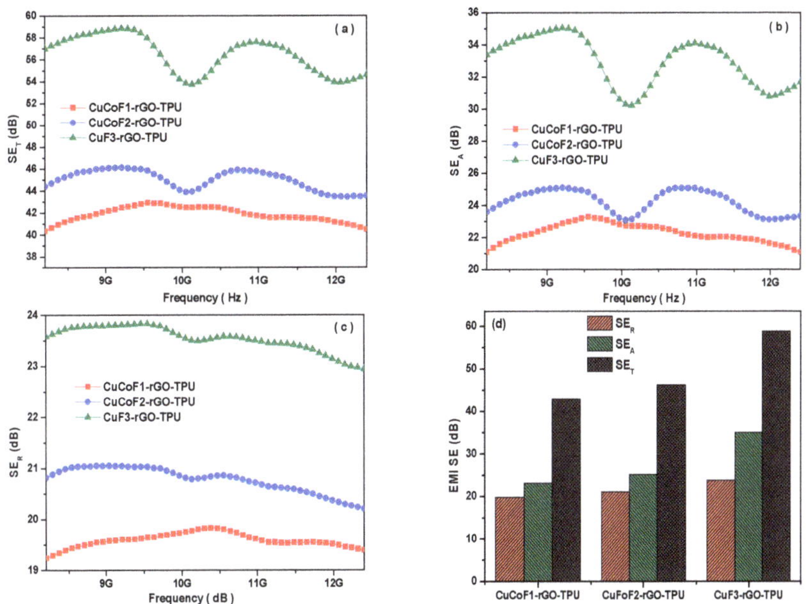

Figure 7. Electromagnetic interference shielding effectiveness (**a**) due to SE_T (**b**) SE_R (**c**) SE_A (**d**) comparison chart of the value of SE_R, SE_A, and SE_T for CuCoF1, CuCoF2, and CuF3 nanoparticles based nanocomposites along with rGO in TPU matrix.

Table 1. EMI shielding characteristics of nanocomposites reported in some previous literatures.

Shielding Material	Frequency Band	Sample Thickness	SE_T	Ref.
$NiFe_2O_4$/rGO	8.2–12.4 GHz	2 mm	38.2 dB	[13]
rGO-TPU nanocomposite	0.1–20 GHz	250 µm	53 dB	[10]
TPU-PCNT composites	8.2–12.4 GHz	3.0 mm	48 dB	[70]
Polystyrene/PANI/Nickel spinel ferrite composite	0.1 20 GHz	0.25 mm	35 dB	[72]
Nitrogen-doped reduced graphene oxide/$CoFe_2O_4$ hybrid nanocomposites	8–18 GHz	1.9 mm	48.4 dB	[73]
$CuFe_2O_4$-rGO-TPU	8.2–12.4 GHz	1 mm	58.8 dB	This work

2.8. Electromagnetic Parameters

For further evaluation and better comparison with different compositions of polymer nanocomposites, we investigated intrinsic parameters such as complex permittivity and complex permeability. In both complex parameters, the real term attributes to energy storage, while the imaginary term attributes to loss or energy dissipation within the material contributed from conduction, resonance, and relaxation mechanisms [74]. The graph in Figure 8a demonstrates the variation of the real part of permittivity, ε' with a change in the frequency from 8.2 to 12.4 GHz for CuCoF1-rGO-TPU, CuCoF2-rGO-TPU, and CuF3-rGO-TPU nanocomposites. As can be seen, the value of ε' was found to vary from 3.67 to 3.95, 3.56 to 3.85, and 2.89 to 3.16 for CuCoF1-rGO-TPU, CuCoF2-rGO-TPU, and CuF3-rGO-TPU nanocomposites, respectively. Various kinds of polarization such as interfacial polarization, dipolar polarization, electronic polarization, and space charge polarization contribute to the value of ε' [75,76]. For studying the dielectric loss of the material, the imaginary part of permittivity (ε'') is significant. The graph in Figure 8b depicts the variation of imaginary permittivity (ε'') with change in frequency for $Cu_xCo_{1-x}Fe_2O_4$ (x = 0.33, 0.67, and 1.00)-rGO TPU nanocomposites. As can be seen, the value of ε'' was found to be altering from 0.23 to 0.44, 0.21 to 0.36, and 0.14 to 0.41 for CuCoF1-rGO-TPU, CuCoF2-rGO-TPU, and CuF3-rGO-TPU nanocomposites, respectively. The imaginary part of permittivity provides information about energy dissipation. In the ε'' curves, the resonance peaks can be attributed to the leakage conductance and lags in polarization [77,78].

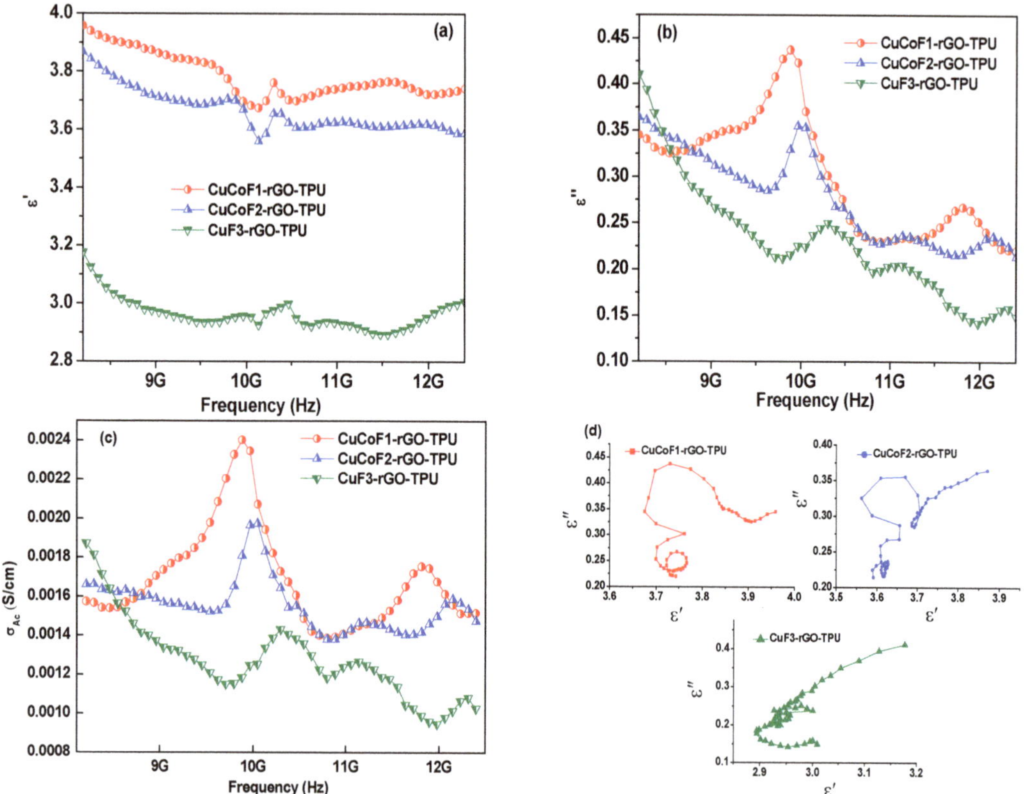

Figure 8. (**a**) the real part (ε') (**b**) the imaginary part (ε'') of permittivity (**c**) ac conductivity (σ_{ac}) versus frequency, and (**d**) Cole-cole plots for CuCoF1, CuCoF2 and CuF3 nanoparticles based nanocomposites along with rGO in TPU matrix.

It is worth noting that ε'' and electrical conductivity are dependent on each other with the following expression [79]:

$$\sigma_{ac} = \varepsilon_o \varepsilon'' 2\pi f$$

where σ_{ac} denotes electrical conductivity, ε_o stands for the absolute permittivity of vacuum, f represents the frequency of electromagnetic waves. The ac conductivity (σ_{ac}) with change in frequency of CuCoF1-rGO-TPU, CuCoF2-rGO-TPU, and CuF3-rGO-TPU nanocomposites at room temperature is displayed in Figure 8c. As can be seen, the value σ_{ac} is in the range 1.40×10^{-3}–2.41×10^{-3} S/cm, 1.36×10^{-3}–1.96×10^{-3} S/cm, and 9.44×10^{-4}–1.87×10^{-3} S/cm for CuCoF1-rGO-TPU, CuCoF2-rGO-TPU, and CuF3-rGO-TPU nanocomposites, respectively. The Debye theory of relaxation is employed for an explanation of the mechanism involved in the electromagnetic shielding material. According to the Debye theory, the value of ε' and ε'' can be expressed as follows [80]:

$$\varepsilon' = \varepsilon_\infty + \frac{\varepsilon_s - \varepsilon_\infty}{1 + (\omega\tau)^2}$$

$$\varepsilon'' = \frac{\varepsilon_s - \varepsilon_\infty}{1 + (\omega\tau)^2}\omega\tau + \frac{\sigma}{\omega\varepsilon_o}$$

where ε_∞ represents the relative dielectric permittivity at an infinite frequency, ε_s signifies the static dielectric permittivity, ω denotes angular frequency, and τ is the polarization relaxation time, respectively. On neglecting the role of σ to ε'' and removal of $\omega\tau$, the equation between ε' and ε'' can be deduced as follows:

$$\left(\varepsilon' - \frac{\varepsilon_s + \varepsilon_\infty}{2}\right)^2 + (\varepsilon'')^2 = \left(\frac{\varepsilon_s - \varepsilon_\infty}{2}\right)^2$$

According to the above expression, the Cole–Cole plot which is ε'' versus ε' should be a semicircle. Each Debye relaxation process possesses a semi-circle that can be upgraded through the interface, subsequently improving the tendency of EM absorption [81]. Figure 8d displays the Cole–Cole plots for CuCoF1-rGO-TPU, CuCoF2-rGO-TPU, and CuF3-rGO-TPU nanocomposites. As can be seen, at least three Cole–Cole semicircles are present which marks the role of multiple relaxation mechanisms involved in synthesized nanocomposites. Moreover, the semi-circles were found to be distorted, which suggests that apart from Debye relaxation, there could be involvement of other mechanisms in the nanocomposites, such as the presence of interfaces leading to interfacial polarization or the Maxwell Wagner effect in the developed samples [82].

Further, for the sake of better evaluation of the EMI shielding mechanism, the real part of permeability (μ'), which signifies the storage capacity of magnetic energy was also studied. Figure 9a displays the plot of μ' with a change of frequency from 8.2 to 12.4 GHz for $Cu_xCo_{1\text{-}x}Fe_2O_4$ (x = 0.33, 0.67, and 1.00)-rGO-TPU nanocomposites. As can be seen, the value of μ' is varying within the range of 0.77 to 1.04, 0.71 to 1.11, and 0.64 to 1.18 for CuCoF1-rGO-TPU, CuCoF2-rGO-TPU, and CuF3-rGO-TPU nanocomposites, respectively. The imaginary part of permeability (μ'') represents the energy loss of the magnetic field. Figure 9b displays the imaginary part (μ'') of permeability with change in the frequency of $Cu_xCo_{1\text{-}x}Fe_2O_4$ (x = 0.33, 0.67, and 1.00)-rGO-TPU nanocomposites. The value of μ'' is found to be varied between 0.00 to 0.15, 0.00 to 0.12, and 0.00 to 0.19 for CuCoF1-rGO-TPU, CuCoF2-rGO-TPU, and CuF3-rGO-TPU nanocomposites, respectively. Three peaks were observed for the value of μ'' in the frequency region of 8.2 GHz to 12.4 GHz for all samples, which correspond to strong natural resonance [81,83].

Dielectric loss plays a vital role in the attenuation of EM waves. To compare dielectric loss capabilities of prepared nanocomposites, it was calculated using the following expression:

$$\tan \delta_\varepsilon = \varepsilon'' / \varepsilon'$$

Figure 9c displays the dielectric loss ($\tan\delta\varepsilon$) of $Cu_xCo_{1\text{-}x}Fe_2O_4$ (x = 0.33, 0.67, and 1.00)-rGO-TPU nanocomposites. As can be seen, the value of $\tan\delta_\varepsilon$ is in the range of 0.06–0.12, 0.06 to 0.10, and 0.05 to 0.13 for CuCoF1-rGO-TPU, CuCoF2-rGO-TPU, and CuF3-rGO-TPU nanocomposites, respectively. It is worth noting that values of ε'' and $\tan\delta_\varepsilon$ follow a similar trend with a change in frequency for all the nanocomposites. The residual groups and defects in rGO contribute to dielectric loss in nanocomposites. In addition, as a result of polarization and associated relaxation, dielectric loss is developed in nanocomposites [84,85]. For further analysis, the magnetic loss was calculated for nanocomposites, which is given by the following expression

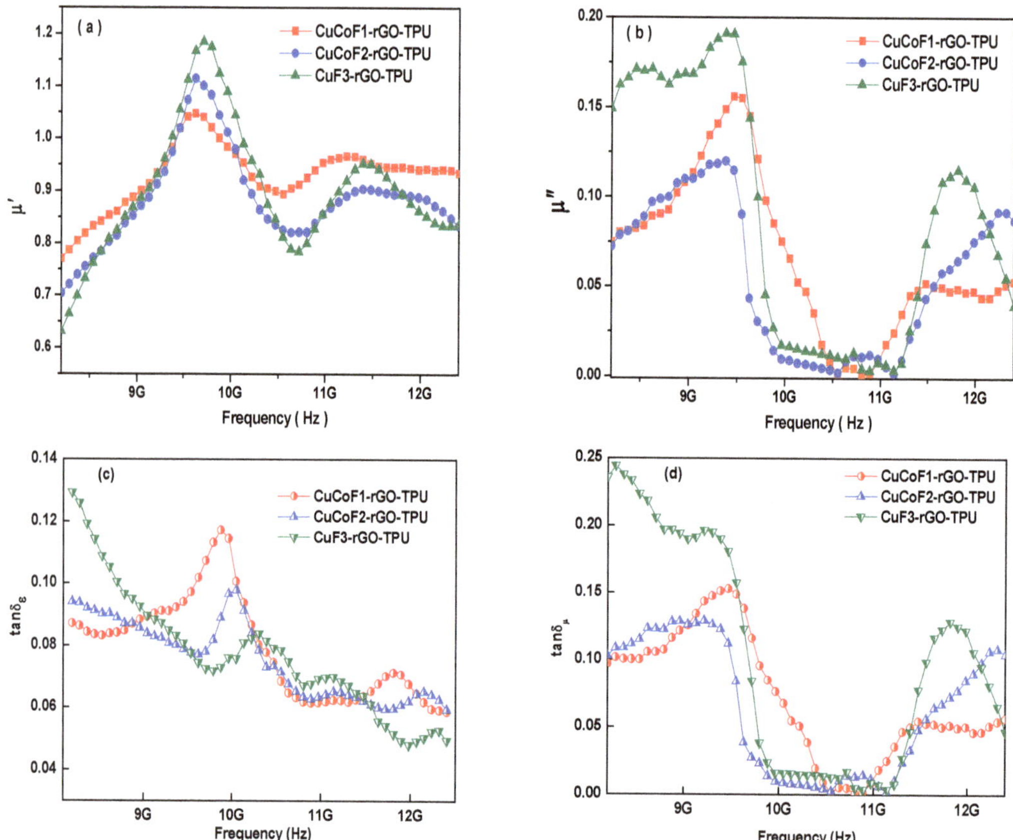

Figure 9. Variation of (**a**) the real part (μ′) of permeability, (**b**) the imaginary part (μ″) of permeability, (**c**) dielectric loss, $\tan\delta_\varepsilon$ (**d**) magnetic loss, $\tan\delta_\mu$ with change in frequency for CuCoF1, CuCoF2, and CuF3 nanoparticles based nanocomposites along with rGO in TPU matrix.

$$\tan \delta_\mu = \mu'' / \mu'$$

Figure 9d below displays the variation of magnetic loss $\tan\delta_\mu$ with change in frequency. As can be seen, the value of $\tan\delta_\mu$ varies from 0.00 to 0.15, 0.00 to 0.13, and 0.00 to 0.24 for CuCoF1-rGO-TPU, CuCoF2-rGO-TPU, and CuF3-rGO-TPU nanocomposites, respectively. The high value of $\tan\delta_\mu$ compared with $\tan\delta_\varepsilon$ reveals that magnetic loss is dominant in endowing for electromagnetic characteristics. It is well known that the magnetic loss, $\tan\delta_\mu$ arises due to eddy current, natural resonance, and anisotropic energy present inside the polymer nanocomposite [86]. The magnetic loss induced by eddy current loss can be evaluated using the following equation [87]:

$$C_o = \mu'' \left(\mu'\right)^{-2} f^{-1}$$

In general, if the value of C_o is not changing with the change in the frequency, then it is obvious that magnetic loss is induced by the eddy current. Figure 10a depicts the plot of C_o with change in frequency over X-band region for CuCoF1-rGO-TPU, CuCoF2-rGO-TPU, and CuF3-rGO-TPU nanocomposites. As can be seen, the value of C_o with the variation of frequency remains nearly constant with change in frequency from 9.7 GHz to 11.2 GHz for CuCoF2-rGO-TPU and CuF3-rGO-TPU nanocomposites. It shows that eddy currents are a contributor to the magnetic loss for CuCoF2-rGO-TPU and CuF3-rGO-TPU nanocomposites. However, in the case of CuCoF1-rGO-TPU nanocomposite, the value of C_o

was found constant in the frequency range 8.3–8.8 GHz, 10.4–10.9 GHz, and 11.4–12.3 GHz, which signifies that eddy current is contributing to the magnetic loss in this frequency range [88].

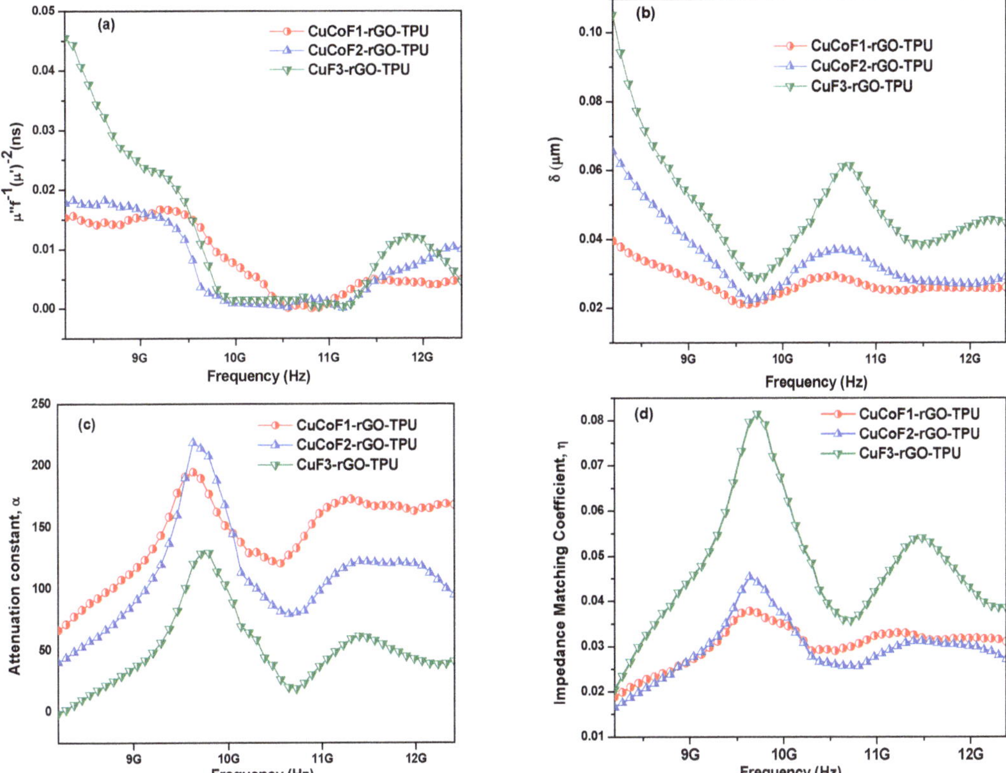

Figure 10. Variation of (**a**) eddy current loss, C_o (**b**) skin depth, δ (**c**) attenuation constant, α and (**d**) impedance matching, Z with change in frequency for CuCoF1, CuCoF2, and CuF3 nanoparticles based nanocomposites along with rGO in TPU matrix.

The EM wave enters only near the surface of the shielding material at higher frequencies and the strength of the EM field suffers exponential decay with the thickness. The skin depth (δ) of shielding material refers to the certain distance up to which the strength of the electric field suffers attenuation and drops to 1/e of its original incident EM wave [89]. Theoretically, it can be represented as follows [90]:

$$\delta = \frac{1}{\sqrt{\pi \mu_r \sigma f}}$$

where μ_r refers to the magnetic permeability of the material, σ is electrical conductivity and f is the frequency. The given plot in Figure 10b represents the variation of skin depth (δ) as a function of frequency for CuCoF1-rGO-TPU, CuCoF2-rGO-TPU, and CuF3-rGO-TPU nanocomposites. The value of skin depth was found to be fluctuating from 0.02 μm to 0.04 μm, 0.022 μm to 0.065 μm, and 0.03 μm to 0.11 μm for CuCoF1-rGO-TPU, CuCoF2-rGO-TPU, and CuF3-rGO-TPU nanocomposites, respectively.

To understand the underlying EMI shielding mechanism in the nanocomposites, attenuation constant (α) and impedance matching conditions were studied. The attenuation constant (α) was determined for calculating the degree of energy attenuated when an EM wave was incident on the surface of the shielding material [91]. The attenuation constant (α) can be expressed by the following equation [92]:

$$\alpha = \frac{\sqrt{2}\pi f}{c}\sqrt{(\mu''\varepsilon'' - \mu'\varepsilon') + \sqrt{(\mu''\varepsilon'' - \mu'\varepsilon')^2 + (\varepsilon'\mu'' + \varepsilon''\mu')^2}}$$

where c stands for speed for light in vacuum. Figure 10c represents the attenuation constant (α) of the developed for CuCoF1-rGO-TPU, CuCoF2-rGO-TPU, and CuF3-rGO-TPU nanocomposites with variation in frequency. It was observed that CuCoF1-rGO-TPU possesses maximum attenuation constant (α) value as compared with other developed nanocomposites. However, only attaining a high value of attenuation constant doesn't mark for good absorption characteristics. A good and balanced impedance match should be achieved between the permittivity and permeability of the shielding material [93]. Keeping this in mind, the impedance matching (Z) was also evaluated for different nanocomposites using the following relation [94]:

$$Z = \sqrt{\mu_r/\varepsilon_r} = \sqrt{\sqrt{\left(\mu'^2 + \mu''^2\right)}/\sqrt{\left(\varepsilon'^2 + \varepsilon''^2\right)}}$$

The impedance matching (Z) with change in frequency is displayed in Figure 10d. Notably, the highest value of impedance matching (Z) for CuF3-rGO-TPU nanocomposite as compared with other developed nanocomposites was noticed. Owing to the balanced impedance matching and moderate attenuation capacity of CuF3-rGO-TPU nanocomposite, it comes up with good shielding performance [95]. Referring to the above discussion, a possible EMI shielding mechanism based on developed $Cu_xCo_{1-x}Fe_2O_4$ (x = 0.33, 0.67, and 1.00)-rGO-TPU nanocomposites has been proposed and illustrated in Figure 11. When EM waves interact with developed nanocomposite, some EM waves are immediately reflected due to the presence of abundant free electrons on the surface [96]. The remaining EM waves move through the shielding material due to a good impedance matching condition. In this process, the EM waves are attenuated because of the dielectric loss, conduction loss, magnetic loss, and multiple scattering/reflection. In this developed nanocomposite, under the interaction of EM waves, the dielectric loss is mainly associated with interfacial polarization and dipole polarization. The interfacial polarization arises from interfaces between $Cu_xCo_{1-x}Fe_2O_4$ spinel ferrite nanoparticles and rGO. The dipole polarization derives from defects and residual functional groups of rGO [97]. Magnetic $Cu_xCo_{1-x}Fe_2O_4$ spinel ferrite nanoparticles induce magnetic loss characteristics and contribute to the absorption component of EMI shielding. Magnetic loss is mainly associated with eddy current loss and natural resonance [98]. The high conductivity of rGO sheets in favor of intrinsic or hopping conduction causes the enhancement of conduction loss. The synergistic effect between $Cu_xCo_{1-x}Fe_2O_4$ spinel ferrite nanoparticles and rGO with an efficient complementarity between the complex permeability and permittivity provides a good impedance matching condition [99]. Hence, the synergy of multiple loss mechanisms associated with good impedance matching provides CuF3-rGO-TPU nanocomposite with high EMI shielding performance.

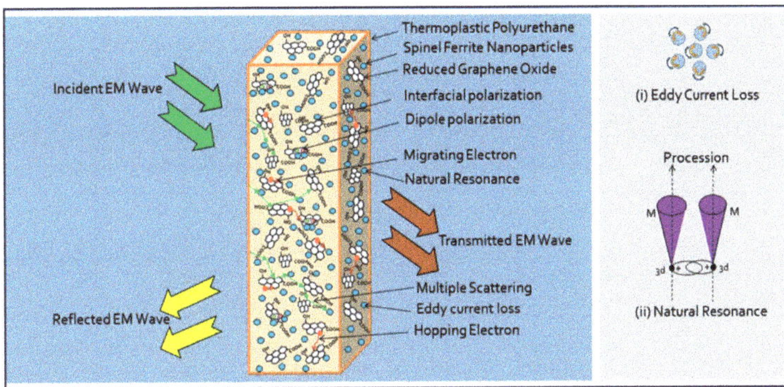

Figure 11. Schematic illustration of EMI shielding mechanism based on developed $Cu_xCo_{1-x}Fe_2O_4$ (x = 0.33, 0.67, and 1.00)-rGO-TPU nanocomposites.

3. Materials and Methods

3.1. Chemicals

Cobalt nitrate $Co(NO_3){\cdot}6H_2O$, copper nitrate $Cu(NO_3){\cdot}6H_2O$, and iron nitrate $Fe(NO_3)_3{\cdot}9H_2O$ were purchased from Alfa Aesar GmbH & Co. KG (Karlsruhe, Germany). Sodium nitrate ($NaNO_3$) was obtained from Lach-Ner, Czech Republic. Potassium permanganate ($KMnO_4$) powder and graphite flakes were sourced from Sigma-Aldrich, Munich, Germany. Vitamin C (Livsane) was a product of Dr. Kleine Pharma GmbH, Bielefeld, Germany.

3.2. Synthesis of $Cu_xCo_{1-x}Fe_2O_4$ (x = 0.33. 0.67, 1) Spinel Ferrite Nanoparticles

$Cu_xCo_{1-x}Fe_2O_4$ (x = 0.33, 0.67, 1) spinel ferrite nanoparticles with different compositions, such as $Cu_{0.33}Co_{0.67}Fe_2O_4$, $Cu_{0.67}Co_{0.33}Fe_2O_4$, and $CuFe_2O_4$, labelled as CuCoF1, CuCoF2, and CuF3, were synthesized using the sonochemical method. For $Cu_xCo_{1-x}Fe_2O_4$ (x = 0.33. 0.67, 1) spinel ferrite nanoparticle formation, appropriate stoichiometric amounts of cobalt nitrate, copper nitrate, and iron nitrate were taken and mixed with 60 mL of deionized water in a 100 mL beaker and stirred for 15 min at room temperature. A solution of NaOH was prepared and added to the above mixture slowly accompanying stirring for a further 2–3 min. The solution turned into a thick precipitate with the addition of a base solution. Further, it was exposed to ultrasonic irradiation (ultrasonic homogenizer UZ SONOPULS HD 2070) (frequency: 20 kHz and power: 70 W) for 60 min. Afterward, the precipitate was cooled down, washed with deionized water, and further centrifuged at 7000 rpm for 15 min. This process was repeated several times to remove any remaining impurities. The acquired product was further dried at 60 °C in an oven for 24 h.

3.3. Synthesis of Graphene Oxide

Graphene oxide (GO) was synthesized following the modified Hummer's method [100] using graphite flakes as a raw material. For this purpose, 3 g of graphite and 1.5 g of $NaNO_3$ were mixed with 75 mL of H_2SO_4 (98%) in a 1000 mL flask in an ice bath (0 °C) and was stirred for 15 min. Next, 9 g of $KMnO_4$ was added slowly and carefully during the vigorous magnetic stirring for 30 min. After that, the prepared mixture was again subject to magnetic stirring for 30 min. Further, the reaction temperature was fixed to room temperature and the mixture was stirred for an additional 48 h. Next, 138 mL of deionized water was added to the mixture, which was followed by stirring for 10 min with heating at 100 °C. Then, another 420 mL of warm deionized water and 30 mL of H_2O_2 were added to the mixture. Afterward, the synthesized yellow suspension was washed with an aqueous solution of H_2SO_4 (6 wt%) and H_2O_2 (1 wt%). Next, the prepared suspension was subjected to washing with deionized water until the pH turned neutral, and then the mixture was three times centrifuged at 6000 rpm for 10 min. The precipitate was then dried in the oven at 50 °C for

10–12 h to obtain graphite oxide powder. To utilize GO as a filler in TPU nanocomposite with spinel ferrite nanoparticles, the enhancement of the electrical conductivity of GO is required. The electrical conductivity of GO can be improved by doping or reduction of oxygen functional groups [101]. In the present work, the improvement of the electrical conductivity of GO via reduction of the oxygen functional group was achieved by a chemical approach with vitamin C as reducing agent.

3.4. Synthesis of Reduced Graphene Oxide

Reduced graphene oxide (rGO) was obtained by chemical reduction of synthesized graphene oxide (GO). Firstly, 3 g of graphene oxide (GO) was dissolved in 200 mL of deionized water in a 500 mL flask and the solution was stirred for 15 min. Next, 10 g of vitamin C was added, and the solution was stirred for 3 h to a fixed temperature of 100 °C. The mixture was cooled down and then washed with deionized water and ethanol. After that, it was centrifuged at 8000 rpm for 20 min. The product was then dried in a vacuum oven at 60 °C for 14 h.

3.5. Preparation of Nanocomposites

For the preparation of thermoplastic polyurethane (TPU) based nanocomposites of 20 wt% nanofiller (in which spinel ferrites nanoparticles and rGO was in 9:1 wt% ratio) were mixed with TPU (Elastollan® C80A10) in A microcompounder (Xplore Instruments B.V., Sittard, The Netherlands) with a capacity of 5 cm^3. TPU and fillers were dried at 90 °C for 12 h in a vacuum oven before mixing. The samples were melt-mixed in a microcompounder at 200 °C for 7 min at 150 rpm. The force was varied from 2405 Newton to 1680 Newton during preparation of nanocomposites. Further, the nanocomposite characteristics, such as electrical, mechanical, thermal, and optical, etc., have dependence on interfacial physical and chemical interactions [102]. The strong interfacial interactions between nanofillers (spinel ferrite and rGO) and polymer matrix, may provide high nanocomposites properties [103]. Three nanocomposite systems, namely CuCoF1-rGO-TPU, CuCoF2-rGO-TPU, and CuF3-rGO-TPU, were prepared. The rectangular-shaped sheets of 22.9 mm × 10.2 mm × 1.0 mm were developed by compression molding. Further, the schematic representation of the preparation of $Cu_xCo_{1-x}Fe_2O_4$-rGO-TPU nanocomposite is shown in Figure 12. Furthermore, a digital photograph for the demonstration of the dimension, lightweight, and flexibility of a prepared nanocomposite is shown in Figure 13.

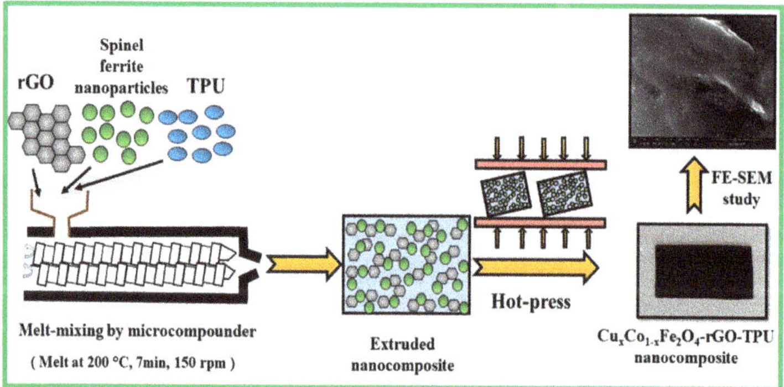

Figure 12. Schematic representation of the preparation of $Cu_xCo_{1-x}Fe_2O_4$-rGO-TPU nanocomposite.

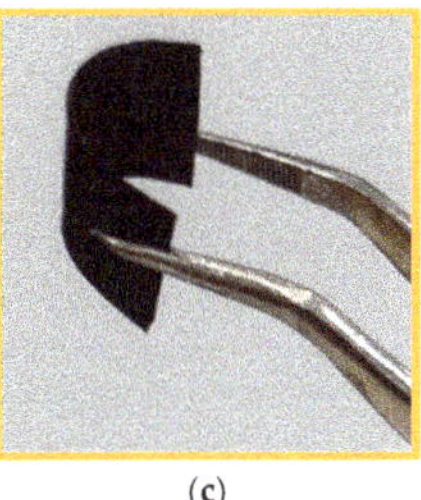

(**a**) (**b**) (**c**)

Figure 13. (**a–c**) Digital photograph for the demonstration of dimension, lightweight, and flexibility of prepared CuF3 and rGO based TPU nanocomposite.

4. Conclusions

In this work, lightweight and flexible TPU polymer matrix-based nanocomposites embedded with $Cu_xCo_{1-x}Fe_2O_4$ (x = 0.33, 0.67, and 1.00) spinel ferrite nanoparticles and rGO as nanofillers were fabricated by a melt-mixing approach using a microcompounder, which demonstrated remarkable shielding effectiveness values. Spinel ferrite $Cu_{0.33}Co_{0.67}Fe_2O_4$ (CuCoF1), $Cu_{0.67}Co_{0.33}Fe_2O_4$ (CuCoF2), and $CuFe_2O_4$ (CuF3) nanoparticles were developed by a facile sonochemical synthesis method. The developed nanocomposites with a thickness of 1 mm exhibited maximum total shielding effectiveness (SE_T) of 42.9 dB, 46.2 dB, and 58.8 dB for CuCoF1-rGO-TPU, CuCoF2-rGO-TPU, and CuF3-rGO-TPU nanocomposites, respectively. It was found that 99.9998% of electromagnetic waves can be shielded in the investigated frequency range, which indicates that the CuF3-rGO-TPU nanocomposite can be considered an effective electromagnetic shielding material. With abundant interfacial polarization, dipole relaxation, better impedance matching condition, and eddy current effect, the natural resonance of nanocomposites plays a key role in this outstanding EMI shielding performance. It is believed that the current investigation could be beneficial in the design and development of lightweight and flexible shielding materials with outstanding EMI shielding performance.

Author Contributions: A.: formal analysis, visualization, investigation, writing—original draft; R.S.Y.: conceptualization, methodology, visualization, investigation, formal analysis, writing—review & editing, supervision, project administration; P.P.: resources, writing—review & editing; J.P.: methodology, writing—review & editing; B.K.: writing—review & editing; I.K.: resources, writing—review & editing, funding acquisition; J.V.: formal analysis, data curation, validation; D.Š.: formal analysis, data curation; P.U.: formal analysis, data curation; M.M. (Michal Machovský): formal analysis, data curation; M.M. (Milan Masař): formal analysis, data curation; M.U.: formal analysis, data curation. All authors have read and agreed to the published version of the manuscript.

Funding: This research was funded by the financial support of the Czech Science Foundation (GA19-23647S) project at the Centre of Polymer Systems, Tomas Bata University in Zlin, Czech Republic. We also thank the Ministry of Education, Youth, and Sports of the Czech Republic—DKRVO (RP/CPS/2022/007).

Institutional Review Board Statement: Not applicable.

Informed Consent Statement: Not applicable.

Data Availability Statement: The data will be available from the corresponding author following reasonable request.

Acknowledgments: R.S.Y. also thanks to the Leibniz Institute of Polymer Research Dresden, Germany, for hosting a stay in its laboratories.

Conflicts of Interest: The authors declare no conflict of interest.

References

1. Yadav, R.S.; Kuřitka, I.; Vilčáková, J. *Advanced Spinel Ferrite Nanocomposites for Electromagnetic Interference Shielding Applications*; Elsevier: Cambridge, MA, USA, 2020; ISBN 978-0-12-821291-2.
2. Bhawal, P.; Ganguly, S.; Das, T.K.; Mondal, S.; Choudhury, S.; Das, N.C. Superior Electromagnetic Interference Shielding Effectiveness and Electro-Mechanical Properties of EMA-IRGO Nanocomposites through the in-Situ Reduction of GO from Melt Blended EMA-GO Composites. *Compos. Part B Eng.* **2018**, *134*, 46–60. [CrossRef]
3. Kumaran, R.; kumar, S.D.; Balasubramanian, N.; Alagar, M.; Subramanian, V.; Dinakaran, K. Enhanced Electromagnetic Interference Shielding in a Au–MWCNT Composite Nanostructure Dispersed PVDF Thin Films. *J. Phys. Chem. C* **2016**, *120*, 13771–13778. [CrossRef]
4. Ravindren, R.; Mondal, S.; Nath, K.; Das, N.C. Prediction of Electrical Conductivity, Double Percolation Limit and Electromagnetic Interference Shielding Effectiveness of Copper Nanowire Filled Flexible Polymer Blend Nanocomposites. *Compos. Part B Eng.* **2019**, *164*, 559–569. [CrossRef]
5. Duan, H.; He, P.; Zhu, H.; Yang, Y.; Zhao, G.; Liu, Y. Constructing 3D Carbon-Metal Hybrid Conductive Network in Polymer for Ultra-Efficient Electromagnetic Interference Shielding. *Compos. Part B Eng.* **2021**, *212*, 108690. [CrossRef]
6. Paun, C.; Obreja, C.; Comanescu, F.; Tucureanu, V.; Tutunaru, O.; Romanitan, C.; Ionescu, O.; Gavrila, D.E.; Paltanea, V.M.; Stoica, V.; et al. Studies on Structural MWCNT/Epoxy Nanocomposites for EMI Shielding Applications. *IOP Conf. Ser. Mater. Sci. Eng.* **2021**, *1009*, 012046. [CrossRef]
7. Rani, P.; Ahamed, B.; Deshmukh, K. Dielectric and Electromagnetic Interference Shielding Properties of Zeolite 13X and Carbon Black Nanoparticles Based PVDF Nanocomposites. *J. Appl. Polym. Sci.* **2021**, *138*, 50107. [CrossRef]
8. Gulzar, N.; Zubair, K.; Shakir, M.F.; Zahid, M.; Nawab, Y.; Rehan, Z.A. Effect on the EMI Shielding Properties of Cobalt Ferrites and Coal-Fly-Ash Based Polymer Nanocomposites. *J. Supercond. Nov. Magn.* **2020**, *33*, 3519–3524. [CrossRef]
9. Valentini, M.; Piana, F.; Pionteck, J.; Lamastra, F.R.; Nanni, F. Electromagnetic Properties and Performance of Exfoliated Graphite (EG)—Thermoplastic Polyurethane (TPU) Nanocomposites at Microwaves. *Compos. Sci. Technol.* **2015**, *114*, 26–33. [CrossRef]
10. Zahid, M.; Nawab, Y.; Gulzar, N.; Rehan, Z.A.; Shakir, M.F.; Afzal, A.; Abdul Rashid, I.; Tariq, A. Fabrication of reduced graphene oxide (RGO) and nanocomposite with thermoplastic polyurethane (TPU) for EMI shielding application. *J. Mater. Sci. Mater. Electron.* **2020**, *31*, 967–974. [CrossRef]
11. Sobha, A.P.; Sreekala, P.S.; Narayanankutty, S.K. Electrical, Thermal, Mechanical and Electromagnetic Interference Shielding Properties of PANI/FMWCNT/TPU Composites. *Prog. Org. Coat.* **2017**, *113*, 168–174. [CrossRef]
12. Yadav, R.S.; Anju; Jamatia, T.; Kuřitka, I.; Vilčáková, J.; Škoda, D.; Urbánek, P.; Machovský, M.; Masař, M.; Urbánek, M.; et al. Excellent, Lightweight and Flexible Electromagnetic Interference Shielding Nanocomposites Based on Polypropylene with $MnFe_2O_4$ Spinel Ferrite Nanoparticles and Reduced Graphene Oxide. *Nanomaterials* **2020**, *10*, 2481. [CrossRef]
13. Kumar, A.; Singh, A.K.; Tomar, M.; Gupta, V.; Kumar, P.; Singh, K. Electromagnetic Interference Shielding Performance of Lightweight $NiFe_2O_4$/RGO Nanocomposite in X- Band Frequency Range. *Ceram. Int.* **2020**, *46*, 15473–15481. [CrossRef]
14. Dey, C.C.; Mahapatra, A.S.; Sadhukhan, S.; Chakrabarti, P.K. Electromagnetic Shielding Performance of $Co_{0.5}Zn_{0.4}Cu_{0.1}Fe_2O_4$-GO/Paraffin Wax Hybrid Nanocomposite through Magnetic Energy Morphing Prepared by Facile Synthesis Method. *Mater. Today Commun.* **2021**, *27*, 102190. [CrossRef]
15. Ismail, M.M.; Rafeeq, S.N.; Sulaiman, J.M.A.; Mandal, A. Electromagnetic Interference Shielding and Microwave Absorption Properties of Cobalt Ferrite $CoFe_2O_4$/Polyaniline Composite. *Appl. Phys. A* **2018**, *124*, 380. [CrossRef]
16. Chitra, P.; Kumar, E.R.; Pushpagiri, T.; Steephen, A. Size and Phase Purity–Dependent Microstructural and Magnetic Properties of Spinel Ferrite Nanoparticles. *J. Supercond. Nov. Magn.* **2021**, *34*, 1239–1244. [CrossRef]
17. Thanh, N.K.; Loan, T.T.; Anh, L.N.; Duong, N.P.; Soontaranon, S.; Thammajak, N.; Hien, T.D. Cation Distribution in $CuFe_2O_4$ Nanoparticles: Effects of Ni Doping on Magnetic Properties. *J. Appl. Phys.* **2016**, *120*, 142115. [CrossRef]
18. Zhu, G.; Xu, H.; Xiao, Y.; Liu, Y.; Yuan, A.; Shen, X. Facile Fabrication and Enhanced Sensing Properties of Hierarchically Porous CuO Architectures. *ACS Appl. Mater. Interfaces* **2012**, *4*, 744–751. [CrossRef]
19. Amulya, M.A.S.; Nagaswarupa, H.P.; Kumar, M.R.A.; Ravikumar, C.R.; Kusuma, K.B.; Prashantha, S.C. Evaluation of bifunctional applications of $CuFe_2O_4$ nanoparticles synthesized by a sonochemical method. *J. Phys. Chem. Solids* **2021**, *148*, 109756. [CrossRef]
20. Gao, S.; Chen, X.; Pan, F.; Song, K.; Zhao, C.; Liu, L.; Liu, X.; Zhao, D. Efect of secondary phase on the electromagnetic shielding efectiveness of magnesium alloy. *Sci. Rep.* **2018**, *8*, 1625. [CrossRef] [PubMed]
21. Song, L.; Duan, Y.; Cui, Y.; Huang, Z. Fe-Doped MnO_2 Nanostructures for Attenuation−Impedance Balance-Boosted Microwave Absorption. *ACS Appl. Nano Mater.* **2022**, *5*, 2738. [CrossRef]
22. Tedjieukeng, H.M.K.; Tsobnang, P.K.; Fomekong, R.L.; Etape, E.P.; Joy, P.A.; Delcorte, A.; Lambi, J.N. Structural Characterization and Magnetic Properties of Undoped and Copper-Doped Cobalt Ferrite Nanoparticles Prepared by the Octanoate Coprecipitation Route at Very Low Dopant Concentrations. *RSC Adv.* **2018**, *8*, 38621–38630. [CrossRef]
23. Anugraha, A.; Lakshmi, V.K.; Kumar, G.S.; Raguram, T.; Rajni, K.S. Synthesis and Characterisation of Copper Substituted Cobalt Ferrite Nanoparticles by Sol-Gel Auto Combustion Route. *IOP Conf. Ser. Mater. Sci. Eng.* **2019**, *577*, 012059. [CrossRef]
24. Jnaneshwara, D.M.; Avadhani, D.N.; Daruka Prasad, B.; Nagabhushana, H.; Nagabhushana, B.M.; Sharma, S.C.; Prashantha, S.C.; Shivakumara, C. Role of Cu^{2+} Ions Substitution in Magnetic and Conductivity Behavior of Nano-$CoFe_2O_4$. *Spectrochim. Acta Part A Mol. Biomol. Spectrosc.* **2014**, *132*, 256–262. [CrossRef] [PubMed]

25. Mahalakshmi, S.; Jayasri, R.; Nithiyanatham, S.; Swetha, S.; Santhi, K. Magnetic Interactions and Dielectric Behaviour of Cobalt Ferrite and Barium Titanate Multiferroics Nanocomposites. *Appl. Surf. Sci.* **2019**, *494*, 51–56. [CrossRef]
26. Chen, Y.; Pötschke, P.; Pionteck, J.; Voit, B.; Qi, H. Multifunctional Cellulose/RGO/Fe_3O_4 Composite Aerogels for Electromagnetic Interference Shielding. *ACS Appl. Mater. Interfaces* **2020**, *12*, 22088–22098. [CrossRef] [PubMed]
27. Jing, X.; Mi, H.-Y.; Huang, H.-X.; Turng, L.-S. Shape Memory Thermoplastic Polyurethane (TPU)/Poly(ε-Caprolactone) (PCL) Blends as Self-Knotting Sutures. *J. Mech. Behav. Biomed. Mater.* **2016**, *64*, 94–103. [CrossRef]
28. Beniwal, A. Sunny Novel TPU/Fe_2O_3 and TPU/Fe_2O_3/PPy Nanocomposites Synthesized Using Electrospun Nanofibers Investigated for Analyte Sensing Applications at Room Temperature. *Sens. Actuators B Chem.* **2020**, *304*, 127384. [CrossRef]
29. Kumar, S.; Gupta, T.K.; Varadarajan, K.M. Strong, Stretchable and Ultrasensitive MWCNT/TPU Nanocomposites for Piezoresistive Strain Sensing. *Compos. Part B Eng.* **2019**, *177*, 107285. [CrossRef]
30. Anju; Yadav, R.S.; Pötschke, P.; Pionteck, J.; Krause, B.; Kuřitka, I.; Vilcakova, J.; Skoda, D.; Urbánek, P.; Machovsky, M.; et al. High-Performance, Lightweight, and Flexible Thermoplastic Polyurethane Nanocomposites with Zn^{2+}-Substituted $CoFe_2O_4$ Nanoparticles and Reduced Graphene Oxide as Shielding Materials against Electromagnetic Pollution. *ACS Omega* **2021**, *6*, 28098–28118. [CrossRef]
31. Barick, A.K.; Tripathy, D.K. Preparation, characterization and properties of acid functionalized multi-walled carbon nanotube reinforced thermoplastic polyurethane nanocomposites. *Mater. Sci. Eng. B* **2011**, *176*, 1435–1447. [CrossRef]
32. Chandramohan, P.; Srinivasan, M.P.; Velmurugan, S.; Narasimhan, S.V. Cation distribution and particle size effect on Raman spectrum of $CoFe_2O_4$. *J. Solid State Chem.* **2011**, *184*, 89–96. [CrossRef]
33. Nandan, B.; Bhatnagar, M.C.; Kashyap, S.C. Cation Distribution in Nanocrystalline Cobalt Substituted Nickel Ferrites: X-ray Diffraction and Raman Spectroscopic Investigations. *J. Phys. Chem. Solids* **2019**, *129*, 298–306. [CrossRef]
34. Sumalatha, M.; Shravan kumar Reddy, S.; Reddy, M.S.; Sripada, S.; Raja, M.M.; Reddy, C.G.; Reddy, P.Y.; Reddy, V.R. Raman and In-Field 57Fe Mössbauer Study of Cation Distribution in Ga Substituted Cobalt Ferrite ($CoFe_{2-x}Ga_xO_4$). *J. Alloy. Compd.* **2020**, *837*, 155478. [CrossRef]
35. Hidayah, N.M.S.; Liu, W.-W.; Lai, C.-W.; Noriman, N.Z.; Khe, C.-S.; Hashim, U.; Lee, H.C. Comparison on Graphite, Graphene Oxide and Reduced Graphene Oxide: Synthesis and Characterization. *AIP Conf. Proc.* **2017**, *1892*, 150002. [CrossRef]
36. Sadhukhan, S.; Ghosh, T.K.; Roy, I.; Rana, D.; Bhattacharyya, A.; Saha, R.; Chattopadhyay, S.; Khatua, S.; Acharya, K.; Chattopadhyay, D. Green Synthesis of Cadmium Oxide Decorated Reduced Graphene Oxide Nanocomposites and Its Electrical and Antibacterial Properties. *Mater. Sci. Eng. C* **2019**, *99*, 696–709. [CrossRef] [PubMed]
37. Konios, D.; Stylianakis, M.M.; Stratakis, E.; Kymakis, E. Dispersion Behaviour of Graphene Oxide and Reduced Graphene Oxide. *J. Colloid Interface Sci.* **2014**, *430*, 108–112. [CrossRef] [PubMed]
38. Gnanaseelan, M.; Samanta, S.; Pionteck, J.; Jehnichen, D.; Simon, F.; Pötschke, P.; Voit, B. Vanadium salt assisted solvothermal reduction of graphene oxide and the thermoelectric characterisation of the reduced graphene oxide in bulk and as composite. *Mater. Chem. Phys.* **2019**, *229*, 319–329. [CrossRef]
39. Aradhana, R.; Mohanty, S.; Nayak, S.K. Comparison of Mechanical, Electrical and Thermal Properties in Graphene Oxide and Reduced Graphene Oxide Filled Epoxy Nanocomposite Adhesives. *Polymer* **2018**, *141*, 109–123. [CrossRef]
40. Muzyka, R.; Drewniak, S.; Pustelny, T.; Chrubasik, M.; Gryglewicz, G. Characterization of Graphite Oxide and Reduced Graphene Oxide Obtained from Different Graphite Precursors and Oxidized by Different Methods Using Raman Spectroscopy. *Materials* **2018**, *11*, 1050. [CrossRef] [PubMed]
41. Khan, Q.A.; Shaur, A.; Khan, T.A.; Joya, Y.F.; Awan, M.S. Characterization of Reduced Graphene Oxide Produced through a Modified Hoffman Method. *Cogent Chem.* **2017**, *3*, 1298980. [CrossRef]
42. de León, A.S.; Domínguez-Calvo, A.; Molina, S.I. Materials with Enhanced Adhesive Properties Based on Acrylonitrile-Butadiene-Styrene (ABS)/Thermoplastic Polyurethane (TPU) Blends for Fused Filament Fabrication (FFF). *Mater. Des.* **2019**, *182*, 108044 [CrossRef]
43. Galindo, B.; Alcolea, S.G.; Gómez, J.; Navas, A.; Murguialday, A.O.; Fernandez, M.P.; Puelles, R.C. Effect of the Number of Layers of Graphene on the Electrical Properties of TPU Polymers. *IOP Conf. Ser. Mater. Sci. Eng.* **2014**, *64*, 012008. [CrossRef]
44. Ati, A.A.; Othaman, Z.; Samavati, A. Influence of Cobalt on Structural and Magnetic Properties of Nickel Ferrite Nanoparticles. *J. Mol. Struct.* **2013**, *1052*, 177–182. [CrossRef]
45. Samoila, P.; Cojocaru, C.; Cretescu, I.; Stan, C.D.; Nica, V.; Sacarescu, L.; Harabagiu, V. Nanosized Spinel Ferrites Synthesized by Sol-Gel Autocombustion for Optimized Removal of Azo Dye from Aqueous Solution. *J. Nanomater.* **2015**, *2015*, e713802. [CrossRef]
46. Karimi, Z.; Mohammadifar, Y.; Shokrollahi, H.; Asl, S.K.; Yousefi, G.; Karimi, L. Magnetic and Structural Properties of Nano Sized Dy-Doped Cobalt Ferrite Synthesized by Co-Precipitation. *J. Magn. Magn. Mater.* **2014**, *361*, 150–156. [CrossRef]
47. Patange, S.M.; Shirsath, S.E.; Toksha, B.G.; Jadhav, S.S.; Shukla, S.J.; Jadhav, K.M. Cation Distribution by Rietveld, Spectral and Magnetic Studies of Chromium-Substituted Nickel Ferrites. *Appl. Phys. A* **2009**, *95*, 429–434. [CrossRef]
48. Avazpour, L.; Zandi khajeh, M.A.; Toroghinejad, M.R.; Shokrollahi, H. Synthesis of Single-Phase Cobalt Ferrite Nanoparticles via a Novel EDTA/EG Precursor-Based Route and Their Magnetic Properties. *J. Alloy. Compd.* **2015**, *637*, 497–503. [CrossRef]
49. Sharma, N.; Sharma, V.; Jain, Y.; Kumari, M.; Gupta, R.; Sharma, S.K.; Sachdev, K. Synthesis and Characterization of Graphene Oxide (GO) and Reduced Graphene Oxide (RGO) for Gas Sensing Application. *Macromol. Symp.* **2017**, *376*, 1700006. [CrossRef]

50. Zhang, C.; Dabbs, D.M.; Liu, L.-M.; Aksay, I.A.; Car, R.; Selloni, A. Combined Effects of Functional Groups, Lattice Defects, and Edges in the Infrared Spectra of Graphene Oxide. *J. Phys. Chem. C* **2015**, *119*, 18167–18176. [CrossRef]
51. Al-Gaashani, R.; Najjar, A.; Zakaria, Y.; Mansour, S.; Atieh, M.A. XPS and Structural Studies of High Quality Graphene Oxide and Reduced Graphene Oxide Prepared by Different Chemical Oxidation Methods. *Ceram. Int.* **2019**, *45*, 14439–14448. [CrossRef]
52. Strankowski, M.; Włodarczyk, D.; Piszczyk, Ł.; Strankowska, J. Polyurethane Nanocomposites Containing Reduced Graphene Oxide, FTIR, Raman, and XRD Studies. *J. Spectrosc.* **2016**, *2016*, e7520741. [CrossRef]
53. Saleem, H.; Haneef, M.; Abbasi, H.Y. Synthesis Route of Reduced Graphene Oxide via Thermal Reduction of Chemically Exfoliated Graphene Oxide. *Mater. Chem. Phys.* **2018**, *204*, 1–7. [CrossRef]
54. Pielichowski, K.; Leszczyńska, A. TG-FTIR Study of the Thermal Degradation of Polyoxymethylene (POM)/Thermoplastic Polyurethane (TPU) Blends. *J. Therm. Anal. Calorim.* **2004**, *78*, 631–637. [CrossRef]
55. Zhou, X.; Li, X.; Sun, H.; Sun, P.; Liang, X.; Liu, F.; Hu, X.; Lu, G. Nanosheet-Assembled $ZnFe_2O_4$ Hollow Microspheres for High-Sensitive Acetone Sensor. *ACS Appl. Mater. Interfaces* **2015**, *7*, 15414–15421. [CrossRef]
56. Zheng, X.; Feng, J.; Zong, Y.; Miao, H.; Hu, X.; Bai, J.; Li, X. Hydrophobic Graphene Nanosheets Decorated by Monodispersed Superparamagnetic Fe_3O_4 Nanocrystals as Synergistic Electromagnetic Wave Absorbers. *J. Mater. Chem. C* **2015**, *3*, 4452–4463. [CrossRef]
57. Hammad, T.M.; Kuhn, S.; Amsha, A.A.; Hempelmann, R. Investigation of structural, optical, and magnetic properties of Co^{2+} ions substituted $CuFe_2O_4$ spinel ferrite nanoparticles prepared via precipitation approach. *J. Aust. Ceram. Soc.* **2021**, *57*, 543–553. [CrossRef]
58. Ahmad, I.; Abbas, T.; Islam, M.U.; Maqsood, A. Study of cation distribution for Cu–Co nanoferrites synthesized by the sol–gel method. *Ceram. Int.* **2013**, *39*, 6735–6741. [CrossRef]
59. Singh, C.; Bansal, S.; Kumar, V.; Tikoo, K.B.; Singhal, S. Encrustation of cobalt doped copper ferrite nanoparticles on solid scaffold CNTs and their comparison with corresponding ferrite nanoparticles: A study of structural, optical, magnetic and photo catalytic properties. *RSC Adv.* **2015**, *5*, 39052–39061. [CrossRef]
60. Chen, G.-H.; Chen, H.-S. Nanometer-Thick Sol–Gel Silica–Titania Film Infused with Superparamagnetic Fe_3O_4 Nanoparticles for Electromagnetic Interference Shielding. *ACS Appl. Nano Mater.* **2020**, *3*, 8858–8865. [CrossRef]
61. Liu, T.; Pang, Y.; Kikuchi, H.; Kamada, Y.; Takahashi, S. Superparamagnetic Property and High Microwave Absorption Performance of FeAl@(Al, Fe)$_2$O$_3$ Nanoparticles Induced by Surface Oxidation. *J. Mater. Chem. C* **2015**, *3*, 6232–6239. [CrossRef]
62. Li, X.; Feng, J.; Zhu, H.; Qu, C.; Bai, J.; Zheng, X. Sandwich-like Graphene Nanosheets Decorated with Superparamagnetic $CoFe_2O_4$ Nanocrystals and Their Application as an Enhanced Electromagnetic Wave Absorber. *RSC Adv.* **2014**, *4*, 33619–33625. [CrossRef]
63. Pawar, S.P.; Stephen, S.; Bose, S.; Mittal, V. Tailored Electrical Conductivity, Electromagnetic Shielding and Thermal Transport in Polymeric Blends with Graphene Sheets Decorated with Nickel Nanoparticles. *Phys. Chem. Chem. Phys.* **2015**, *17*, 14922–14930. [CrossRef]
64. Mondal, S.; Ganguly, S.; Das, P.; Khastgir, D.; Das, N.C. Low Percolation Threshold and Electromagnetic Shielding Effectiveness of Nano-Structured Carbon Based Ethylene Methyl Acrylate Nanocomposites. *Compos. Part B Eng.* **2017**, *119*, 41–56. [CrossRef]
65. Wei, H.; Zhang, Z.; Hussain, G.; Zhou, L.; Li, Q.; Ostrikov, K.K. Techniques to Enhance Magnetic Permeability in Microwave Absorbing Materials. *Appl. Mater. Today* **2020**, *19*, 100596. [CrossRef]
66. Jin, X.; Wang, J.; Dai, L.; Liu, X.; Li, L.; Yang, Y.; Cao, Y.; Wang, W.; Wu, H.; Guo, S. Flame-Retardant Poly(Vinyl Alcohol)/MXene Multilayered Films with Outstanding Electromagnetic Interference Shielding and Thermal Conductive Performances. *Chem. Eng. J.* **2020**, *380*, 122475. [CrossRef]
67. Hu, L.; Kang, Z. Enhanced flexible polypropylene fabric with silver/magnetic carbon nanotubes coatings for electromagnetic interference shielding. *Appl. Surf. Sci.* **2021**, *568*, 150845. [CrossRef]
68. Ali, N.N.; Atassi, Y.; Salloum, A.; Charba, A.; Malki, A.; Jafarian, M. Comparative Study of Microwave Absorption Characteristics of (Polyaniline/NiZn Ferrite) Nanocomposites with Different Ferrite Percentages. *Mater. Chem. Phys.* **2018**, *211*, 79–87. [CrossRef]
69. Gunasekaran, S.; Thanrasu, K.; Manikandan, A.; Durka, M.; Dinesh, A.; Anand, S.; Shankar, S.; Slimani, Y.; Almessiere, M.A.; Baykal, A. Structural, Fabrication and Enhanced Electromagnetic Wave Absorption Properties of Reduced Graphene Oxide (RGO)/Zirconium Substituted Cobalt Ferrite ($Co_{0.5}Zr_{0.5}Fe_2O_4$) Nanocomposites. *Phys. B Condens. Matter* **2021**, *605*, 412784. [CrossRef]
70. Gahlout, P.; Choudhary, V. EMI Shielding Response of Polypyrrole-MWCNT/Polyurethane Composites. *Synth. Met.* **2020**, *266*, 116414. [CrossRef]
71. Sulaiman, J.M.A.; Ismail, M.M.; Rafeeq, S.N.; Mandal, A. Enhancement of Electromagnetic Interference Shielding Based on $Co_{0.5}Zn_{0.5}Fe_2O_4$/PANI-PTSA Nanocomposites. *Appl. Phys. A* **2020**, *126*, 236. [CrossRef]
72. Shakir, M.F.; Tariq, A.; Rehan, Z.A.; Nawab, Y.; Abdul Rashid, I.; Afzal, A.; Hamid, U.; Raza, F.; Zubair, K.; Rizwan, M.S.; et al. Effect of Nickel-Spinal-Ferrites on EMI Shielding Properties of Polystyrene/Polyaniline Blend. *SN Appl. Sci.* **2020**, *2*, 706. [CrossRef]
73. Li, X.; Shu, R.; Wu, Y.; Zhang, J.; Wan, Z. Fabrication of Nitrogen-Doped Reduced Graphene Oxide/Cobalt Ferrite Hybrid Nanocomposites as Broadband Electromagnetic Wave Absorbers in Both X and Ku Bands. *Synth. Met.* **2021**, *271*, 116621. [CrossRef]

74. Arief, I.; Biswas, S.; Bose, S. Wool-Ball-Type Core-Dual-Shell FeCo@SiO_2@MWCNTs Microcubes for Screening Electromagnetic Interference. *ACS Appl. Nano Mater.* **2018**, *1*, 2261–2271. [CrossRef]
75. Luo, J.; Zuo, Y.; Shen, P.; Yan, Z.; Zhang, K. Excellent Microwave Absorption Properties by Tuned Electromagnetic Parameters in Polyaniline-Coated $Ba_{0.9}La_{0.1}Fe_{11.9}Ni_{0.1}O_{19}$/Reduced Graphene Oxide Nanocomposites. *RSC Adv.* **2017**, *7*, 36433–36443. [CrossRef]
76. Verma, M.; Singh, A.P.; Sambyal, P.; Singh, B.P.; Dhawan, S.K.; Choudhary, V. Barium Ferrite Decorated Reduced Graphene Oxide Nanocomposite for Effective Electromagnetic Interference Shielding. *Phys. Chem. Chem. Phys.* **2014**, *17*, 1610–1618. [CrossRef] [PubMed]
77. Yin, Y.; Zeng, M.; Liu, J.; Tang, W.; Dong, H.; Xia, R.; Yu, R. Enhanced High-Frequency Absorption of Anisotropic Fe_3O_4/Graphene Nanocomposites. *Sci. Rep.* **2016**, *6*, 25075. [CrossRef] [PubMed]
78. Tang, X.T.; Wei, G.T.; Zhu, T.X.; Sheng, L.M.; An, K.; Yu, L.M.; Liu, Y.; Zhao, X.L. Microwave Absorption Performance Enhanced by High-Crystalline Graphene and $BaFe_{12}O_{19}$ Nanocomposites. *J. Appl. Phys.* **2016**, *119*, 204301. [CrossRef]
79. Sharma, A.L.; Thakur, A.K. AC Conductivity and Relaxation Behavior in Ion Conducting Polymer Nanocomposite. *Ionics* **2011**, *17*, 135–143. [CrossRef]
80. Yu, H.; Wang, T.; Wen, B.; Lu, M.; Xu, Z.; Zhu, C.; Chen, Y.; Xue, X.; Sun, C.; Cao, M. Graphene/Polyaniline Nanorod Arrays: Synthesis and Excellent Electromagnetic Absorption Properties. *J. Mater. Chem.* **2012**, *22*, 21679–21685. [CrossRef]
81. Zhou, X.; Chuai, D.; Zhu, D. Electrospun Synthesis of Reduced Graphene Oxide (RGO)/NiZn Ferrite Nanocomposites for Excellent Microwave Absorption Properties. *J. Supercond. Nov. Magn.* **2019**, *32*, 2687–2697. [CrossRef]
82. Wang, H.; Zhang, Z.; Dong, C.; Chen, G.; Wang, Y.; Guan, H. Carbon Spheres@MnO_2 Core-Shell Nanocomposites with Enhanced Dielectric Properties for Electromagnetic Shielding. *Sci. Rep.* **2017**, *7*, 15841. [CrossRef]
83. Wang, F.; Li, X.; Chen, Z.; Yu, W.; Loh, K.P.; Zhong, B.; Shi, Y.; Xu, Q.H. Efficient low-frequency microwave absorption and solar evaporation properties of γ-Fe_2O_3 nanocubes/graphene composites. *Chem. Eng. J.* **2021**, *405*, 126676. [CrossRef]
84. Rostami, M.; Majles Ara, M.H. The Dielectric, Magnetic and Microwave Absorption Properties of Cu-Substituted Mg-Ni Spinel Ferrite-MWCNT Nanocomposites. *Ceram. Int.* **2019**, *45*, 7606–7613. [CrossRef]
85. Cao, M.; Wang, X.; Cao, W.; Fang, X.; Wen, B.; Yuan, J. Thermally Driven Transport and Relaxation Switching Self-Powered Electromagnetic Energy Conversion. *Small* **2018**, *14*, 1800987. [CrossRef] [PubMed]
86. Liu, P.; Yao, Z.; Zhou, J.; Yang, Z.; Kong, L.B. Small Magnetic Co-Doped NiZn Ferrite/Graphene Nanocomposites and Their Dual-Region Microwave Absorption Performance. *J. Mater. Chem. C* **2016**, *4*, 9738–9749. [CrossRef]
87. Guo, J.; Song, H.; Liu, H.; Luo, C.; Ren, Y.; Ding, T.; Khan, M.A.; Young, D.P.; Liu, X.; Zhang, X.; et al. Polypyrrole-Interface-Functionalized Nano-Magnetite Epoxy Nanocomposites as Electromagnetic Wave Absorbers with Enhanced Flame Retardancy. *J. Mater. Chem. C* **2017**, *5*, 5334–5344. [CrossRef]
88. Raju, P.; Shankar, J.; Anjaiah, J.; Kalyani, C.; Rani, G.N. Complex Permittivity and Permeability Properties Analysis of NiCuZn Ferrite-Polymer Nanocomposites for EMI Suppressor Applications. *J. Phys. Conf. Ser.* **2020**, *1495*, 012001. [CrossRef]
89. Mondal, S.; Ghosh, S.; Ganguly, S.; Das, P.; Ravindren, R.; Sit, S.; Chakraborty, G.; Das, N.C. Highly Conductive and Flexible Nano-Structured Carbon-Based Polymer Nanocomposites with Improved Electromagnetic-Interference-Shielding Performance. *Mater. Res. Express* **2017**, *4*, 105039. [CrossRef]
90. Bhingardive, V.; Sharma, M.; Suwas, S.; Madras, G.; Bose, S. Polyvinylidene Fluoride Based Lightweight and Corrosion Resistant Electromagnetic Shielding Materials. *RSC Adv.* **2015**, *5*, 35909–35916. [CrossRef]
91. Jiao, Z.; Yao, Z.; Zhou, J.; Qian, K.; Lei, Y.; Wei, B.; Chen, W. Enhanced Microwave Absorption Properties of Nd-Doped NiZn Ferrite/Polyaniline Nanocomposites. *Ceram. Int.* **2020**, *46*, 25405–25414. [CrossRef]
92. Shu, R.; Li, W.; Zhou, X.; Tian, D.; Zhang, G.; Gan, Y.; Shi, J.; He, J. Facile Preparation and Microwave Absorption Properties of RGO/MWCNTs/$ZnFe_2O_4$ Hybrid Nanocomposites. *J. Alloy. Compd.* **2018**, *743*, 163–174. [CrossRef]
93. Wu, Y.; Shu, R.; Li, Z.; Guo, C.; Zhang, G.; Zhang, J.; Li, W. Design and Electromagnetic Wave Absorption Properties of Reduced Graphene Oxide/Multi-Walled Carbon Nanotubes/Nickel Ferrite Ternary Nanocomposites. *J. Alloy. Compd.* **2019**, *784*, 887–896. [CrossRef]
94. Liu, W.; Shao, Q.; Ji, G.; Liang, X.; Cheng, Y.; Quan, B.; Du, Y. Metal–Organic-Frameworks Derived Porous Carbon-Wrapped Ni Composites with Optimized Impedance Matching as Excellent Lightweight Electromagnetic Wave Absorber. *Chem. Eng. J.* **2017**, *313*, 734–744. [CrossRef]
95. Shen, W.; Ren, B.; Cai, K.; Song, Y.; Wang, W. Synthesis of Nonstoichiometric $Co_{0.8}Fe_{2.2}O_4$/Reduced Graphene Oxide (RGO) Nanocomposites and Their Excellent Electromagnetic Wave Absorption Property. *J. Alloy. Compd.* **2019**, *774*, 997–1008. [CrossRef]
96. Gao, S.; An, Q.; Xiao, Z.; Zhai, S.; Shi, Z. Significant promotion of porous architecture and magnetic Fe_3O_4 NPs inside honeycomb-like carbonaceous composites for enhanced microwave absorption. *RSC Adv.* **2018**, *8*, 19011. [CrossRef]
97. Liu, X.; Zhao, X.; Yan, J.; Huang, Y.; Li, T.; Liu, P. Enhanced electromagnetic wave absorption performance of core-shell Fe_3O_4@poly(3,4-ethylenedioxythiophene) microspheres/reduced graphene oxide composite. *Carbon* **2021**, *178*, 273–284. [CrossRef]
98. Ma, W.; He, P.; Wang, T.; Xu, J.; Liu, X.; Zhuang, Q.; Cui, Z.-K.; Lin, S. Microwave absorption of carbonization temperature-dependent uniform yolk-shell H-Fe_3O_4@C microspheres. *Chem. Eng. J.* **2021**, *420*, 129875. [CrossRef]
99. Xiong, L.; Yu, M.; Liu, J.; Li, S.; Xue, B. Preparation and evaluation of the microwave absorption properties of template-free graphene foam-supported Ni nanoparticles. *RSC Adv.* **2017**, *7*, 14733. [CrossRef]

100. Fang, C.; Zhang, Z.; Bing, X.; Lei, Y. Preparation, Characterization and Electrochemical Performance of Graphene from Microcrystalline Graphite. *J. Mater. Sci. Mater. Electron.* **2017**, *28*, 19174–19180. [CrossRef]
101. Naghdi, S.; Jaleh, B.; Eslamipanah, M.; Moradi, A.; Abdollahi, M.; Einali, N.; Rhee, K.Y. Graphene family, and their hybrid structures for electromagnetic interference shielding applications: Recent trends and prospects. *J. Alloy. Compd.* **2022**, *900*, 163176. [CrossRef]
102. Rasul, M.G.; Kiziltas, A.; Arfaei, B.; Shahbazian-Yassar, R. 2D boron nitride nanosheets for polymer composite materials. *NPJ 2D Mater. Appl.* **2021**, *5*, 56. [CrossRef]
103. Bustamante-Torres, M.; Romero-Fierro, D.; Arcentales-Vera, B.; Pardo, S.; Bucio, E. Interaction between Filler and Polymeric Matrix in Nanocomposites: Magnetic Approach and Applications. *Polymers* **2021**, *13*, 2998. [CrossRef] [PubMed]

International Journal of
Molecular Sciences

MDPI

Review

PLGA-Based Composites for Various Biomedical Applications

Cátia Vieira Rocha, Victor Gonçalves, Milene Costa da Silva, Manuel Bañobre-López and Juan Gallo *

Advanced (Magnetic) Theranostic Nanostructures Lab, Health Cluster, International Iberian Nanotechnology Laboratory, Av. Mestre José Veiga s/n, 4715-330 Braga, Portugal; catia.rocha@inl.int (C.V.R.); victor.goncalves@inl.int (V.G.); milene.silva@inl.int (M.C.d.S.); manuel.banobre@inl.int (M.B.-L.)

* Correspondence: juan.gallo@inl.int

Abstract: Polymeric materials have been extensively explored in the field of nanomedicine; within them, poly lactic-co-glycolic acid (PLGA) holds a prominent position in micro- and nanotechnology due to its biocompatibility and controllable biodegradability. In this review we focus on the combination of PLGA with different inorganic nanomaterials in the form of nanocomposites to overcome the polymer's limitations and extend its field of applications. We discuss their physicochemical properties and a variety of well-established synthesis methods for the preparation of different PLGA-based materials. Recent progress in the design and biomedical applications of PLGA-based materials are thoroughly discussed to provide a framework for future research.

Keywords: PLGA; composites; inorganic nanoparticles; scaffolds; biomedical applications

Citation: Rocha, C.V.; Gonçalves, V.; da Silva, M.C.; Bañobre-López, M.; Gallo, J. PLGA-Based Composites for Various Biomedical Applications. *Int. J. Mol. Sci.* **2022**, 23, 2034. https://doi.org/10.3390/ijms23042034

Academic Editor: Raghvendra Singh Yadav

Received: 14 January 2022
Accepted: 8 February 2022
Published: 12 February 2022

Publisher's Note: MDPI stays neutral with regard to jurisdictional claims in published maps and institutional affiliations.

Copyright: © 2022 by the authors. Licensee MDPI, Basel, Switzerland. This article is an open access article distributed under the terms and conditions of the Creative Commons Attribution (CC BY) license (https://creativecommons.org/licenses/by/4.0/).

1. Introduction

Polymeric materials have become extremely attractive materials for a wide range of applications. One of the areas where they are becoming increasingly important is biomedicine [1]. One extensively explored polymer is poly lactic-co-glycolic acid (PLGA), a biodegradable, biocompatible and FDA (U.S. Food and Drug Administration, USA)/EMA (European Medicines Agency)-approved copolymer [2,3]. It is commercially available with different molecular weights and copolymer ratios, which allow for the tuning of the final PLGA behavior to fit an application of interest [4,5]. Being the scope of this review of biomedical applications, PLGA features that can be useful in this area will be explored the most. PLGA can be processed into any shape and size, has great water solubility and allows for a tunable drug release. Its biodistribution and pharmacokinetics follow nonlinear and dose-dependent profiles [2,6]. Generally speaking, copolymers can be synthesized as random or block copolymers, presenting different intrinsic properties in this way. PLGA is generally synthesized through the ring-opening copolymerization of lactic acid (LA) and glycolic acid (GA), with the products of its degradation being nontoxic [7–9]. Normally, following the aforementioned synthesis technique, randomly distributed PLGA can be obtained in either: (i) an atactic configuration where the repeating units have no regular stereochemical configuration or (ii) a syndiotactic configuration where the repeating units have alternating stereochemical configurations [10]. Sequencing is important since it considerably influences the degradation rate of PLGA, with random PLGA degrading much quicker than sequenced PLGA [11].

Over the last decade PLGA copolymers have been extensively explored in combination with nanotechnology. Nanotechnology brings unique characteristics to the system, which, in combination with PLGA's unique features, allow for extensive biomedical applications. It is possible to encapsulate both organic and inorganic materials into PLGA, such as small-molecule drugs [2,12], vaccines [13], proteins [14,15] and metallic [16] as well as magnetic [17,18] nanoparticles (NPs). The methods of production of PLGA-based materials can be adapted to various types of drugs, making it possible to encapsulate hydrophobic and hydrophilic molecules and thus making this copolymer an ideal drug delivery system (DDS) [4].

PLGA-based technology has been used for many years for a myriad of applications in the biomedical arena, with many being already approved by the FDA and EMA (Table 1). Drug delivery is one of the main uses for PLGA, which can be accomplished in the form of macroscopic structures (scaffolds/gels), microparticles (MPs) or NPs. In general, in the biomedical field these PLGA structures have proved their potential for encapsulating several therapeutic agents towards different ends (e.g., antiseptics [19], antibiotics [20], anti-inflammatory [21] and antioxidant [22] drugs), and have shown promise for specific targeting when adequately functionalized [4,23]. On top of that, PLGA-based materials protect their cargo from degradation and provide a sustained drug release profile, ideal for long-term treatments [4,7].

Table 1. PLGA-based nano-, micro- and macromaterials approved through the years for different biomedical applications. MP = microparticle; NP = nanoparticle; and VIP = vasoactive intestinal peptide.

Approval Year	Brand	Form/Active Principle	Route of Administration	Synthesis	Application
1989	Zoladex®	Implant/goserelin acetate	Subcutaneous	NA	Prostate carcinoma
1995	Lupron®	MP/leuprolide acetate	Intramuscular	W/O emulsion	Central precocious puberty/endometriosis
1997	Sandostatin® LAR	MP/octreotide	Subcutaneous	Emulsification solvent evaporation method	Severe diarrhea associated with metastic tumors or VIP-secreting tumors
2002	Eligard®	Nanogel/leuprolide acetate	Subcutaneous	NA	Advanced prostate cancer
2002	Suprecur®	MP/buserelin acetate	Intramuscular	Spray-drying	Endometriosis
2003	Consta®	MP/risperidone	Intramuscular	Emulsification solvent evaporation method	Schizophrenia and bipolar disorder
2009	Ozurdex®	Implant/dexamethasone	Subcutaneous	NA	Macular edema
2014	Signifor® LAR	MP/pasireotide pamoate	Intramuscular	NA	Cushing's disease, acromegaly
2017	Zilretta®	MP/triamcinolone acetonide	Intra-articular	NA	Osteoarthritis
2017	Sublocade®	NP/buprenorphine	Subcutaneous	NA	Moderate to severe opioid addiction

The combination of PLGA with inorganic NPs (INPs) to form nanocomposite materials has the potential to extend the field of the application of this polymer even further and potentiate the effect of these materials while maintaining a reduced toxicity. There is a wide range of INPs available to be combined with PLGA, and by doing so some of the limitations of INPs, such as their toxicity and colloidal stability, can be circumvented. For these reasons the applications of PLGA-based nanocomposite materials in chemotherapy, cancer diagnosis and imaging, gene therapy and protein delivery, among others, have been thoroughly explored in recent years [4]. PLGA nanocomposites have also been explored for vaccination, brain targeting, cardiovascular and inflammatory diseases in addition to scaffolds for tissue engineering, as we will describe in this review. Nevertheless, research efforts are still required for the development and commercialization of PLGA nanocomposites for biomedical applications since they still present a number of limitations, such as poor drug loading, high burst release, high production costs and challenging scalability [4]. In this review we aim to explore recent advancements made in this area. It begins with a brief discussion on PLGA's properties, approved PLGA-based medicines and an overview of the main production methods of PLGA MPs and NPs. The main focus of this review is then the combination of PLGA with INPs to form hybrid PLGA-based composites (nano-, micro- and macro-), and it is organized into three principal subjects: types of hybrid PLGA/inorganic-based materials, synthesis methods of these composites and their applications in the biomedical area, mainly in the form of carriers and scaffolds. This review aims to provide a backdrop for future research.

2. PLGA Physicochemical Properties and Synthesis Methods

2.1. PLGA Properties

As already mentioned, PLGA is produced by the catalyzed ring-opening copolymerization of the LA and GA units [7,8]. During polymerization the monomeric units are consecutively linked together through ester linkages, resulting in the formation of the PLGA copolymer [3,24]. PLGA is widely used in nanomedicine due to its biocompatibility and effective biodegradability, which occurs through the hydrolysis of the ester bonds of lactate and glycolate [2]. These monomers can then be metabolized via the Krebs cycle, yielding nontoxic byproducts (H_2O and CO_2) [7,25].

Different forms of PLGA can be obtained by varying the Poly(lactic acid):Poly(glycolic acid) (PLA:PGA) ratio during polymerization; for example, PLGA 50:50, which is frequently used in nanotechnology, has a composition of 50% lactate and 50% glycolate [26]. PLGA copolymers inherit the intrinsic properties of their constituent monomers, where the PLA:PGA ratio, along with the polymer molecular weight, influence their hydrophobicity, crystallinity, mechanical properties, size and biodegradation rate [24,27–29]. PGA is a crystalline hydrophilic polymer, while PLA is a stiff and more hydrophobic polymer; therefore, PLGA copolymers with a higher PLA content are less hydrophilic, tend to absorb less water and consequently present longer degradation times [30]. The degradation time can vary from several months to several years depending on the molecular weight (Mw) and copolymer ratio [4]. PLGA is soluble in a variety of solvents, including organic solvents such as chloroform, acetone, ethyl acetate and tetrahydrofuran [2,31]. All the above-mentioned features of PLGA have been proven to be useful for several applications; controlled drug release in particular.

PLGA can be block-polymerized with other copolymers, which can alter its behavior and physicochemical properties [32]. Diblock or triblock copolymers have been developed to meet the need for better carrier functionality, both in terms of the variety of drugs incorporated and administration methods [2]. Block copolymers of PEG (poly(ethylene glycol)) and PLGA are the most reported both in diblock (PLGA-PEG) [33] and triblock conformations (PLGA-PEG-PLGA or PEG-PLGA-PEG) [34,35]. The creation of PEG layers can reduce interactions with foreign molecules, increasing shelf stability in this way. However, it can also decrease drug encapsulation efficiencies. When compared with PLGA alone diblock copolymers have shown improved release kinetics [2]. The random combination of other polymers with PLGA can also be beneficial; for example, combining biodegradable photoluminescent polyester (BPLP) with PLGA will make the system suitable for photoluminescence imaging [36]. Thus, it is important to consider the final purpose of a system when choosing a polymer conformation [37].

Regarding the physical characteristics of nano-PLGA structures, they can be controlled by parameters specific to the production method employed. For example, the size of PLGA NPs can be determined to a certain extent by the concentration of polymer used for their synthesis [7]. Surface functionalization is another important aspect that allows a certain control over particles' biocompatibility, biodegradation, blood half-life and, when applicable, targeting efficiency [7]. In fact, PEGylation has been shown to improve the pharmacokinetic properties of drugs encapsulated into PLGA composites [38]; coating PLGA NPs with biocompatible hydrophilic polymers (PEG or chitosan) can enhance stability and circulation time while diminishing toxicity [39,40].

In terms of biomedical applications, PLGA has been used in the clinic since 1989, being introduced mostly as microsphere formulations (Table 1); however, PLGA implants and nanocomposites are also a reality [41–43]. As mentioned above, PLGA is mostly utilized for drug delivery, having ~20 formulations that are FDA- and EMA-approved [43–47]. These are mainly administered by subcutaneous/intramuscular injections, yet they are versatile and present a plethora of applications. Table 1 presents a brief summary of these products as well as their respective production methods and applications [43,48–50].

Although many PLGA materials have already been commercialized, scientists are still trying to develop new methods of preparation that provide close control over their inherent characteristics. In the next section the most commonly used preparation methods for PLGA nanocomposites will be discussed.

2.2. PLGA MP/NP and Scaffold Preparation Methods

The physicochemical characteristics of PLGA particles can be controlled by manipulating specific parameters in their synthesis. It is of great importance to comprehend the different methods of PLGA particle preparation so that they can be used and manipulated to obtain optimized results. Although these are well-established methods each with different advantages, there are also limitations that should be considered in the preparation and development of PLGA materials. Common issues arise from mixing devices that usually do not allow for great batch-to-batch reproducibility, which leads to inconsistent particle behavior. The organic solvent used should be carefully chosen, since, similarly to the mixing device, it can create limitations in terms of emulsification, solvent extraction and particle homogeneity [32]. A wide range of techniques have been used for PLGA MP/NP synthesis, the most common one being the emulsification–solvent evaporation method, both single and double emulsion. Salting out, nanoprecipitation, emulsification–solvent diffusion and spray-drying are other common methods used in the synthesis of PLGA particles. For a better understanding these strategies will be discussed individually. Table 2 summarizes the main advantages and disadvantages of each method. A few techniques have also been reported for scaffold synthesis, with most of them resorting to molds to obtain a specific form. Electrospinning is one of the most used standard techniques for fibrous scaffold synthesis [51,52].

2.2.1. Emulsification–Solvent Evaporation (ESE) Method

Single emulsion

Oil-in-water (O/W) emulsification is the most popular method for the preparation of PLGA particles, mainly when their intended loading is hydrophobic [53]. Following this methodology, appropriate amounts of PLGA polymer are first dissolved in an organic solvent (e.g., dichloromethane (DCM), chloroform or ethyl acetate). Subsequently, this organic solution is emulsified in an aqueous solution in the presence of a surfactant (e.g., polyvinyl alcohol, PVA) under continuous stirring. Afterwards, the organic solvent is allowed to evaporate, either by stirring at an adequate temperature or by applying a reduced pressure. The resultant particles are washed multiple times to remove polymer/surfactant residues, and they can then be freeze-dried [2,8] for long-term storage.

To load other components (drugs, imaging probes and NPs) into the PLGA matrix, they are usually co-dissolved in the organic solvent [2,54] before the emulsion is formed. Although O/W emulsions are the most popular emulsions created with this method oil-in-oil (O/O) emulsions are also used to encapsulate water-insoluble drugs. Here, the process is the same as in O/W emulsions, but the organic solvent phase is emulsified in a continuous oil phase (e.g., liquid paraffin or vegetable oil) [32].

Double emulsion

The double emulsion method is a more complex variation of the single emulsion method. The water-in-oil-in-water (W/O/W) emulsification method is the most used double emulsion method. In the single emulsion method the encapsulation efficiency (EE%) of hydrophilic compounds is very limited. In double emulsion methodologies the EE% and particle size are affected by the solvent used and the stirring rate [2]. For the W/O/W method an aqueous solution containing the materials to be encapsulated is emulsified in a PLGA-containing organic phase under vigorous stirring. Next, this water-in-oil (W/O) emulsion is added to a surfactant-containing aqueous solution under continuous stirring.

Then, the organic solvent is allowed to evaporate via the same processes mentioned in the single emulsion method [24,32].

Even though W/O/W emulsions are by far the most used ones there are other double emulsion options available, such as water-in-oil-in-oil (W/O/O). Here, the second oil phase is made of an organic solvent that is miscible with the organic solvent of the first oil phase, but it is an antisolvent for PLGA [32].

2.2.2. Salting-Out Method

In this method a solution of a water-miscible organic solvent containing PLGA and the compounds to be loaded is added to an aqueous phase containing a salting-out agent (e.g., calcium chloride) and a stabilizer (e.g., PVA) [24]. Under continuous stirring an O/W emulsion is first formed. Then, a large volume of water is added to the emulsion until the organic solvent diffuses into the aqueous phase, which leads to particle formation [2]. Finally, salting-out agents are removed by filtration and the particles are washed several times to remove the excess stabilizer. This method is ideal for high concentrations of polymer and for the encapsulation of heat-sensitive compounds [24,55].

2.2.3. Emulsification–Solvent Diffusion (ESD) Method

This technique was first developed by Quintanar-Guerrero et al., and is a modification of the salting-out method [56,57]. Here, the organic solvent and water are mutually saturated at room temperature to attain a thermodynamic equilibrium. Afterwards, known amounts of polymer are dissolved into an organic solvent, and this solution is then emulsified in an aqueous solution containing a stabilizer (e.g., PVA) using a high-speed homogenizer [32,58,59]. Later, water is added to the O/W emulsion under regular stirring, which will allow the solvent to diffuse outwards from the internal phase. Then, the nanoprecipitation of the polymer occurs, leading to the formation of colloidal particles. Finally, the solvent can be removed either by evaporation or vacuum steam distillation [3,24,60]. This method presents high reproducibility and allows for a high EE% of hydrophobic drugs.

2.2.4. Nanoprecipitation Method

The nanoprecipitation method, also known as the solvent displacement method, is a simple one-step process with high reproducibility that is mainly used to entrap hydrophobic drugs [7] in PLGA NPs. Here, the physicochemical characteristics of the resulting NPs depend on the PLGA monomers ratio and Mw, the solvents and the mixing rate [7,61]. In this method the polymer and cargo are dissolved in a water-miscible organic solvent (e.g., acetone, ethanol or acetonitrile). Then, this solution is added drop-by-drop into an aqueous solution containing a surfactant or emulsifier. The rapid organic solvent diffusion into water leads to the immediate formation of PLGA NPs. Finally, the organic solvent is removed under reduced pressures [7,24,32]. Modifications to this method have been made to adapt it to the encapsulation of hydrophilic drugs, for example, by replacing water with cottonseed oil and Tween-80 [7,62,63]. A two-step method has also been reported to encapsulate enzymes, where a first protein nanoprecipitation step is introduced, followed by a second nanoprecipitation of PLGA, resulting in protein-loaded PLGA NPs with a high EE% [64].

2.2.5. Spray-Drying Method

The spray-drying technique is ideal for scaling-up the synthesis of PLGA particles. It is a rapid and convenient method with few processing parameters [2,8]. It sprays a W/O emulsion into a stream of hot air, which leads to the formation of particles. The solvent choice depends on the hydrophobicity of the cargo present in the W/O dispersion [8]. A major drawback for this method is the fact that the particles usually adhere to the inner walls of the spray-dryer [65].

Table 2. Advantages and disadvantages of PLGA particle synthesis methods.

Method	Advantages	Disadvantages
ESE [3,7,8,66]	-Encapsulates hydrophobic and hydrophilic agents. -Control of particle size. -Ease of scale-up.	-Time-consuming purification. -Needs heat or vacuum to remove solvent. -Biomacromolecule instability. -Instable W/O/W emulsions with poor drug EE%. -Batch-to-batch variability.
Salting out [7,24,29,32]	-Efficient encapsulation of heat-sensitive agents (proteins, DNA and RNA). -Low-energy mixing device.	-Time-consuming purification. -Not suitable for lipophilic drugs. -Use of large quantities of salting-out agents.
ESD [3,24,32,57]	-Simple and convenient. -Batch-to-batch reproducibility. -Ease of scale-up. -Monodisperse particle size. -High EE% (~70%).	-Leakage of water-soluble drugs into aqueous external phase, decreasing their EE%. -Large volumes of water to be removed.
Nanoprecipitation [7,8,24,32,67]	-Batch-to-batch reproducibility. -One-step process. -Smaller particle size. -Low-energy mixing device.	-Aggregation due to incomplete solvent removal. -Low EE% for hydrophilic drugs. -Negative effect of organic solvent on protein function.
Spray-drying [3,8,24,66,68,69]	-Encapsulates hydrophobic and hydrophilic agents. -Fast, convenient and few processing parameters. -Suitable for industrial scale-up.	-Adhesion of the particles to the spray-dryer wall. -Limited control of particle size.

2.2.6. Electrospinning Method

There are several techniques used for scaffold synthesis, such as gas foaming [70], porogen leaching [71], phase separation [72] and twin-screw extrusion [73]. However, electrospinning [51,52] is by far the most popular one. The electrospinning technique has gathered more interest in recent years due to its versatility, simplicity and low cost in production. This technology has been highly useful in fabricating scaffolds for tissue engineering [51]. Briefly, it consists of a simple setup where a PLGA aqueous solution is kept in a syringe and injected through a needle by applying a high electrical voltage. Consequently, the needle becomes unstable and nanofiber jet spinning is achieved. The nanofiber is later collected on a conducting substrate [74]. The diameter of the fibers can be controlled by adjusting parameters, such as the voltage used, injection rate and collector type. This method can usually produce nanofiber sheets ranging from a few nanometers to several micrometers to be used as scaffolds [75].

3. Types and Synthesis of Hybrid PLGA Composite Materials

The development of hybrid PLGA materials represents a growing area in nanomedicine that has been employed as an efficient strategy to improve the structural and functional properties of PLGA. Although PLGA presents many advantages, its poor mechanical properties, the release of acidic byproducts, its hydrophobicity and suboptimal bioactivity are major bottlenecks for its applications [76,77]. To improve the properties of PLGA, its combination with various inorganic materials has been studied [16,51,78–81]. On top of this, inorganic nanomaterials bring into play interesting capabilities beyond those of PLGA, whether in the form of imaging/sensing properties or as fundamental physical properties. Furthermore, inorganic materials such as metallic and magnetic NPs usually present low stability and a tendency to agglomerate, which can be circumvented via their incorporation into PLGA matrices [16].

PLGA is often used for drug delivery purposes, where the use of both drugs and inorganic nanomaterials can potentiate the drug effect and equip the particles with new abilities (e.g., imaging). Gold-, silver-, iron-, manganese- and titanium-based nanomaterials, among others, are often utilized in combination with PLGA for this purpose. Additionally, PLGA is consistently used to fabricate scaffolds for tissue engineering applications since it offers control over their degradation as well as excellent properties, such as tensile strength and elastic modulus [82]. The fabrication of these scaffolds can be made directly from PLGA polymers or PLGA MPs. Nonetheless, the use of PLGA in tissue engineering is still limited due to its poor osteoconductivity, the release of acidic byproducts, the hydrophobicity of PLGA scaffolds, which alienates cell infiltration, and its suboptimal mechanical properties [76].

Several different combinations of PLGA copolymers with INPs have been explored and will be discussed in the following sections. A wide range of emulsion-based synthesis techniques are used to prepare these PLGA particle composites [8,83]. Each of these techniques have advantages and limitations (refer to Table 2), and they should be chosen according to the intrinsic features of the polymer, the physicochemical properties of the loaded drugs and INPs as well as the nature of the final application [84]. Most of the methods share common features of dispersing the PLGA polymer in an organic phase and mixing the solution with an antisolvent or aqueous phase. The solvent is usually removed by evaporation and/or extraction [8,32,85]. Depending on the chemical nature of the materials to be encapsulated, they are usually added either to the organic or to the aqueous phase.

3.1. PLGA/Plasmonic Nanocomposites

Plasmonic materials have gathered great research attention in recent years due to their tunable optical properties, which are achieved through the exploitation of their surface plasmon resonance (SPR) effects [86]. Gold and silver are the most commonly used plasmonic materials. The wavelengths at which they interact with light can be tuned by changing their size and shape, and in this way bespoke materials can be prepared to match their intended application [87]. Figure 1a presents two different PLGA/plasmonic nanocomposites.

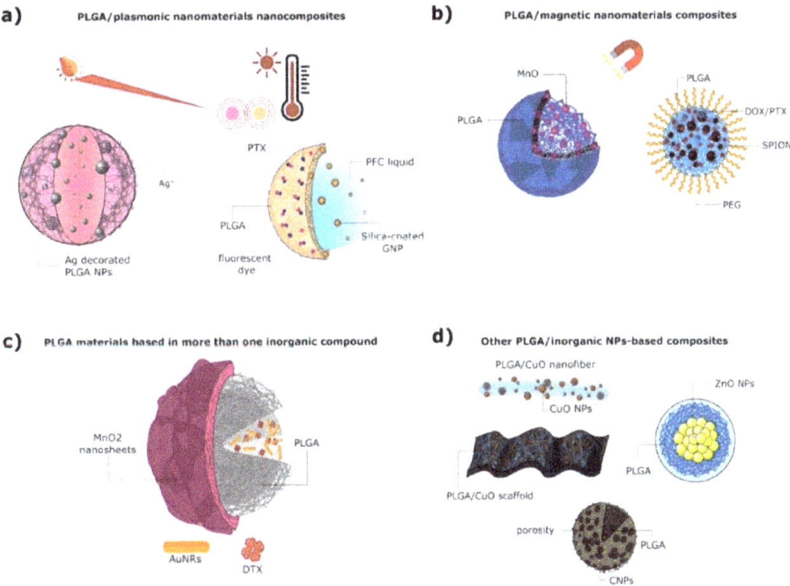

Figure 1. Different types of PLGA/INP nanocomposites: (**a**) plasmonic; (**b**) magnetic; (**c**) multifunctional (more than one inorganic compound); and (**d**) other nanocomposites.

Gold NPs (AuNPs) in the form of spherical nanoparticles, nanorods or nanocapsules are the most widely used plasmonic NPs in biomedical applications [88]. They present a localized SPR property, which allows the NPs to absorb light (in the visible to near infra-red range) and transform it into heat, ideal for thermal therapies [89]. AuNPs are frequently used due to their high chemical stability, easy functionalization, low toxicity and high optical absorption as well as photoacoustic (PA) signal [90,91]. Additionally, they have been proposed as a bone morphogenic substance, presenting not only inhibition properties to the formation of osteoclasts but also supporting osteoblast differentiation [92,93]. AuNPs have been widely studied in combination with PLGA, mainly for theranostic purposes, combining photothermal therapy with X-ray or PA imaging [16,78,91]. Several strategies to incorporate AuNPs into PLGA composites have been described, from encapsulation into the polymer core [91,94,95] to the decoration of and/or growth on their polymeric surface [78,96,97] to the coating of scaffolds [52].

Silver NPs (Ag NPs) are also known for their localized SPR, which allows control over their optical absorption and temperature profile [98]. However, in nanomedicine Ag NPs are mostly used due to their nontoxicity, biocompatibility, antimicrobial as well as antioxidant activity and ability to enhance the antibacterial activity of different drugs [99,100]. Several strategies have been presented to potentiate the antibacterial efficacy of the organic–silver NP combination, both as PLGA particles [19] and as functionalized scaffolds [51].

3.1.1. PLGA Materials with Plasmonic NPs in the Core

Fazio et al. [95] synthesized a PEG-PLGA copolymer nanocomposite loaded with AuNPs and silibinin (SLB), where the AuNPs were produced by laser ablation and immediately embedded into a previously prepared PEG-PLGA copolymer. After the bespoke preparation of the PEG-PLGA copolymer, the ESD method was followed for the preparation of the final nanocomposite: the PEG-PLGA copolymer and SLB were dissolved together in DCM, and this solution was immediately added into a AuNP solution and ultrasonicated, forming a W/O emulsion. Next, an aqueous solution was added to the W/O emulsion, which induced organic solvent diffusion outwards from the internal phase. The final emulsion was centrifuged, eliminating the low-molecular-weight polymer. The resulting SLB-loaded PEG-PLGA_Au nanocomposite presented a porous spherical form that was 200 nm in size. The encapsulated AuNPs ranged from 5–50 nm in size and enabled the use of this nanocomposite for photothermal therapy (PTT) and light-controlled drug release [95].

In another study by Deng et al. [94] small AuNPs (3–5 nm) were encapsulated together with the photosensitizer verteporfin (VP) into a PLGA matrix, at varying gold:VP ratios. These PLGA-AuNPs were synthesized by a single ESE method, obtaining spherical NPs approximately of 100 nm diameter. These NPs were suitable for cancer therapy through photodynamic therapy (PDT). Wang et al. [91] synthesized NPs containing silica-coated AuNPs and perfluorohexane (PFH) liquid in the core via the double emulsion method. The NPs were stabilized by a PLGA shell where the hydrophobic drug paclitaxel (PTX) and a fluorescent dye were incorporated (Figure 1a right). The AuNPs were produced via a single-phase aqueous reduction of tetrachloraunic acid by sodium citrate, further coated with silica and fluorinated [101–103]. PLGA particles were then prepared using a W/O/O double emulsion (Figure 2a) [91]. The resulting nanocomposites were spherical and had an approximate size of 550 nm. Due to their components, the particles are able to work as PA and fluorescent imaging agents as well as antitumoral agents.

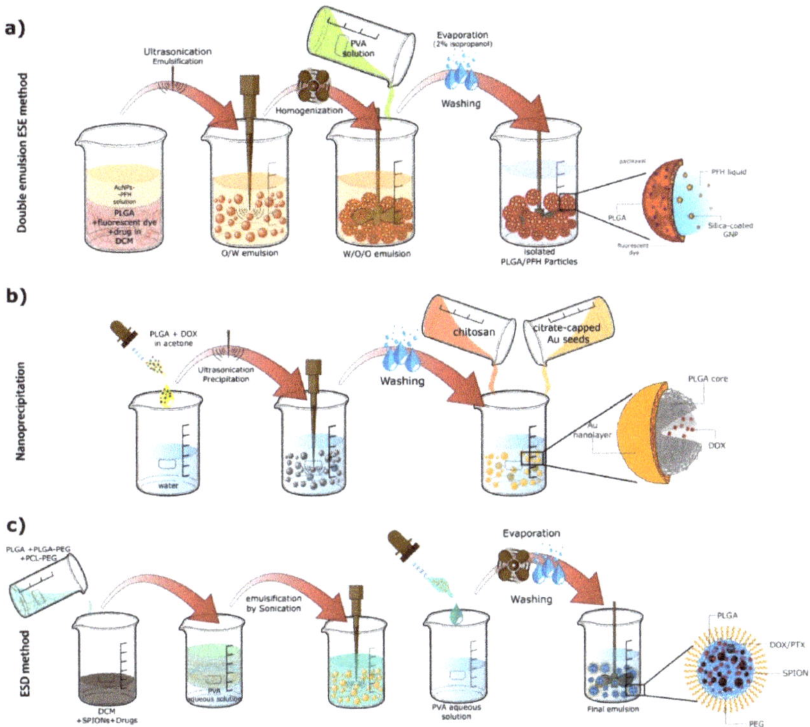

Figure 2. Illustration of (**a**) the double emulsion ESE method (**b**) the nanoprecipitation method and (**c**) the ESD method.

3.1.2. PLGA Materials with Plasmonic NPs at the Surface

Hao et al. used a modified emulsification solvent evaporation method [104] to synthesize PLGA-based NPs loaded with docetaxel (DTX), that were modified with polyethyleneimine (PEI) and a shell of AuNPs with the presence of the glioma-targeting peptide angiopep-2 (ANG/GS/PLGA/DTX NPs). First, PEI-modified PLGA/DTX NPs were prepared. Then, the incorporation of gold onto the surface of the NPs was performed. Negatively charged AuNPs were easily attracted to the PEI-PLGA NPs for subsequent growth of the gold overlays. Gold nanoseeds were formed on the surface of PLGA/DTX NPs in two steps: (i) the reduction of hydrogen tetrachloroaurate hydrate with sodium borohydride and then (ii) the reduction of hydrogen tetrachloroaurate hydrate with ascorbic acid, which allowed for a slow growth of the Au nanoseeds and the consequent formation of gold nanoshells. The final targeted particles were obtained via the incubation of the nanocomposites with angiopep-2 and HS-PEG2000 [78]. The obtained particles were spherical and had a size of approximately 200 nm. The nanocomposites were intended to be used as chemophotothermal agents (DTX plus PTT) and X-ray imaging agents in cancer therapy [78].

Likewise, Song et al. developed plasmonic composite PLGA vesicles [96] approximately 60 nm in size, where gold nanorods (AuNRs) were embedded in the vesicle shell formed by PLGA, with PEG molecules extending to the interior and exterior of the vesicle, providing stability. Inorganic cetyltrimethylammonium bromide (CTAB)-modified AuNRs were synthesized via a one-spot seedless method [105]. Then, for the synthesis of amphiphilic AuNRs attached with PEG and PLGA a "grafting to" reaction was performed [106–108]. This reaction happens by mixing thiolated PEG and PLGA (PEG-SH and PLGA-SH) with AuNR, where the polymers are "grafted" to the AuNRs' surface by forming Au-S covalent bonds [96]. Subsequently, the AuNR@PEG/PLGA vesicles were prepared using the emulsification solvent evaporation method. PLGA forms a vesicular shell embedded with AuNRs and PEG extended inside and outside the PLGA vesicle, stabilizing it in aqueous solution and preventing aggregation. The vesicle size can be controlled to some extend by controlling the concentration of AuNRs@PEG/PLGA in the organic phase. These AuNRs@PEG/PLGA nanocomposites were designed as theranostic agents using PA image-guided PTT [96].

In another study, Topete et al. [97] developed a multifunctional PLGA-based nanocomposite loaded with the anticancer drug doxorubicin (DOX), covered with a porous Au-branched shell (BGNSH) and functionalized with a human serum albumin/indocyanine green/folic acid complex (HSA-ISG-FA). The authors utilized the nanoprecipitation method to form these nanostructures. To form the Au shell around the PLGA core PLGA NPs were incubated with chitosan in order to invert the NPs' surface charge from negative to positive [109]. Next, the NPs were incubated with a Au seed solution followed by a Au growth solution to form the shell. Finally, this solution was reduced with ascorbic acid [97]. The resultant NPs were spherical and had an approximate size of 125 nm (Figure 2b). All the elements of this nanocomposite synergistically combined to obtain extended therapeutic effects and also serve as contrast agents for optical imaging due to the NPs' fluorescence [97].

Takahashi et al. prepared silver-decorated PLGA NPs (Ag PLGA) with an approximate size of 300 nm [19] (Figure 1a left) in a different manner: first, PLGA NPs were obtained by the ESD method, similarly to the above-mentioned studies; afterwards, a silver modification was performed. PLGA NPs were dispersed in water and added to an aqueous $AgNO_3$ solution. This solution was then mixed with $NaBH_4$. The resulting solution was freeze-dried and stored for further use [19]. Given the antimicrobial activity of silver, these particles were designed to act as a DDS against biofilm infections.

In a different approach Lee et al. developed PLGA nanofibrous scaffolds that were chemically coated with AuNPs by using electrospinning [52]. The PLGA was directly synthesized and functionalized with thiol ends to efficiently bind AuNPs (30 nm size). For the binding with AuNPs, the thiolated PLGA sheets prepared by electrospinning were immersed in AuNP solutions at different concentrations. AuNPs promote osteoblast differentiation and allow for the attachment of drugs to the system. Therefore, the synthesized composite scaffolds are promising tools as controlled release scaffolds for bone tissue regeneration.

In a different study by Almajhdi et al. PLGA nanofiber sheets were functionalized with Ag NPs at different percentages to evaluate their influence in the fibers' performance [51]. To form the Ag NPs $AgNO_3$ at different ratios was dissolved in tetrahydrofuran or dimethylformamide and reduced with PEG, which also served as a stabilizer. This solution was stirred for 30 min before PLGA was added to the solution. This mixture would later be electrospun to form PLGA@Ag NP nanofibers. The scaffolds synthesized by electrospinning were highly porous materials and had nanofiber diameters of around 100–200 nm, while the size of the Ag NPs went from 5–10 nm. Owing to the antibacterial and ROS (reactive oxygen species)-generating activity of the Ag NPs the scaffolds are ideal candidates for tissue engineering and suitable for anticancer as well as antibiotic drug delivery.

3.2. PLGA/Magnetic Nanocomposites

During recent decades nanotechnology has enabled the production of bespoke magnetic NPs (MNPs), characterized by a tailored response to an applied magnetic field. MNPs have been used in the clinic since the late 1990s as contrast agents (CAs) in magnetic resonance imaging (MRI) [110]. MNPs can also be used to generate heat under alternating magnetic fields (AMFs), useful for the direct ablation of tumors through magnetic hyperthermia and/or for the controlled delivery of drugs [111–116]. Additionally, some MNPs showcase a responsive behavior to changes in pH and redox states [117,118]. Figure 1b presents two different MNPs combined with PLGA to form functional magnetic composites.

Iron-oxide-based NPs (Fe_xO_y NPs) are one of the most researched structures in nanomedicine due to their relevant physicochemical properties and magnetic properties, facile synthesis and versatility of applications [119]. For most of the proposed applications in biomedicine the particles perform best below a critical diameter (<20 nm), being called superparamagnetic iron oxide nanoparticles (SPIONs) [119,120], where their magnetic properties are within the superparamagnetic regime.

Several strategies have been developed to encapsulate magnetic NPs onto PLGA MPs and NPs with the goal of obtaining multifunctional composites with diagnostic (mostly via PA or MRI [121–123]) and/or therapeutic purposes [36,121–129].

Sun et al. combined SPIONs with PLGA in such a way that the magnetic NPs were embedded in a polymer shell while the core was filled with liquid [124]. Here, a double ESE method (W/O/W) was used to synthesize microcapsules, which presented a nonuniform size of approximately 900 nm. The particles were designed as theranostic agents through the use of ultrasound (US) and MR imaging as well as to enhance the therapeutic efficiency of high-intensity focused ultrasound (HIFU) [124]. Based on a previous study [130], Schleich et al. [126] synthesized PEGylated PLGA-based NPs loaded with SPIONs and PTX or DOX (Figure 1b right) via the ESD method (Figure 2c). The SPIONs were first synthesized by a classical coprecipitation method of iron salts in an alkaline medium [131]. To fabricate the PLGA NPs, PLGA, PLGA-PEG, poly(ε-caprolactone-b-ethylene glycol) (PCL-PEG) and the drugs were dissolved in a SPION DCM solution. This organic phase was emulsified with a PVA aqueous solution. Then, this O/W emulsion was added dropwise to an aqueous solution containing PVA; DCM was subsequently evaporated [126]. The obtained PLGA NPs were spherical, and when containing PTX and DOX presented an approximate size of 250 nm and 290 nm, respectively. The repulsive forces between PEG molecules provided higher stability to the NPs in serum fluids, making them ideal for drug delivery applications [132,133]. The particles were also capable of producing T_2 contrast in MRI. The contrast observed in MR images is the result of local variations in longitudinal (T_1) and transverse (T_2) relaxation times of water molecules in regions adjacent to the injected particles. Likewise, Ruggiero et al. also synthesized PLGA NPs containing Fe_3O_4 NPs and PTX using the method described below [125]. A classical approach was used to fabricate Fe_3O_4 NPs via coprecipitation from a solution of iron(II) and iron (III) chloride in water using ammonium hydroxide as a base and later adding oleic acid [125,134,135]. In the reviewed studies the INPs and, when in cases where applicable, the drug were added into the organic phase of the single emulsion method [79,122,123,125]. For example, in a study by Ruggiero et al. a mixture of the produced Fe_3O_4 NPs, PLGA and PTX was constructed in chloroform [125]. This mixture was later added in a dropwise manner to an aqueous solution of PVA and sonicated. A rotatory evaporator was then used to evaporate the solvent from the obtained emulsion [125]. In this way PTX-loaded PLGA-Fe-NPs were obtained. The particle size was between 120–160 nm and their magnetic core size was around 15–20 nm. The PLGA-Fe nanocomposites were ideal for magnetic-hyperthermia-triggered drug release and could also be used as MRI contrast agents.

All reported studies where Fe_3O_4 NPs were encapsulated into PLGA propose very similar synthetic protocols both for the INPs and for the PLGA particles. Several studies used a thermal decomposition method [136,137] to synthesize iron NPs by heating up a mixture of an iron–oleate complex in the presence of oleic acid, with a high boiling point ether serving as a solvent [79,122,123]. This method usually produces high-quality and very monodisperse nanocrystals but is more cumbersome than, for example, co-precipitation. The above-mentioned studies used the single emulsion technique in a very similar way to Ruggiero et al., with relevant differences being the cargo, solvents used (methylene chloride [122,123] or DCM [79]) and evaporation method (stirring for 3 h [122,123] or agitating overnight [79]). Moreover, Nkansah et al. [123] used this same method to synthesize both PLGA MPs and NPs only by varying the %(w/v) of PVA in aqueous solutions and the homogenization method used.

ESE W/O/W emulsions are also very frequently used to prepare magnetic-PLGA composites and have different variations. The MNPs can be included in the first aqueous phase and emulsified with a PLGA-containing organic phase. Then, a PVA-containing second aqueous phase is added to the emulsion and emulsified, forming a W/O/W emulsion. Lastly, the emulsion is left stirring for several hours to evaporate the organic solvent [80,121,127]. Lee et al. [127] developed a more complex version of a multifunctional PLGA nanocomposite with a magnetite core using this approach. In this study, PLGA was linked with methoxy poly (ethylene glycol) (mPEG) and/or chlorin e6 (Ce6) through the Steglich esterification method [138]. Magnetite NPs were dispersed in water, making the first aqueous phase. PLGA-mPEG and/or PLGA-Ce6 were added to DCM, making the organic phase. The two phases were mixed together with vigorous vortexing, producing an emulsion. This emulsion was added to a second aqueous phase containing PVA and NaCl and emulsified. Finally, the solution was stirred to evaporate the organic solvent and collected by centrifugation [127]. The conjugation with mPEG was used to provide higher stability to the nanocomposites in biological fluids; with Ce6 the goal was to obtain fluorescent in vivo images and perform PDT tumor therapy, damaging tumor cells through ROS generation [139,140]. Due to the magnetic core it was also possible to perform MRI diagnoses.

Some researchers also proposed to include the iron-based NPs on the surface of PLGA NPs, or embedded into a polymer shell. Fang et al. [129] produced magnetic PLGA microspheres loaded with DOX (DOX-MMS), with an approximate size of 2.4 μm. In this example the drug was encapsulated in the core, and the iron NPs, in this case γ-Fe_2O_3, were electrostatically assembled on the microsphere's surface, which was precoated with PEI. The magnetic NPs were synthesized by coprecipitation and coated dimercaptosuccinic acid [141], while the polymeric particles were prepared via a modified W/O/W emulsion. The main significant difference between this and the above-mentioned double emulsion is that the inorganic material is not included when forming the W/O/W emulsion. Instead, the incorporation of the magnetic NPs to the PLGA microspheres happens as a postsynthetic modification through electrostatic deposition with the aid of a PEI coating [129]. The authors propose that in this construction the resulting microcomposites are more responsive and that under an external magnetic field the drug release will be enhanced, showing that the particles are ideal for magnetic-hyperthermia-induced drug release and chemothermal therapy.

In other studies where the double emulsion method was also used the MNPs were included together with PLGA in the organic phase of the W/O/W emulsions. Zhang et al. [36], for example, incorporated SPIONs into a polymer shell made of two different polymers: PLGA and BPLP. The photoluminescent (from the BPLP) nanocapsules where coated with PEG to give them an enhanced stealth effect, and BSA protein was encapsulated into their core. SPIONs coated with oleic acid were first synthesized by a microwave-assisted thermal decomposition method and then later incorporated into the polymer NPs' shell [142]. The PLGA-BPLP copolymer was synthesized by using BPLP as a macroinitiator that reacts with lactate and glycolate via ring-opening polymerization. PLGA-PEG was also used to obtain the PEGylated nanocapsules. For the synthesis of these nanocapsules an aqueous phase containing the encapsulant (bovine serum albumin, BSA) was emulsified in an organic phase containing SPIONs and different proportions of PLGA. Next, this W/O emulsion was emulsified with a second water phase containing PVA, and the second W/O/W was formed by sonication [36]. This DDS was proposed as a theranostic platform for MRI/photoluminescence dual-mode imaging and drug delivery. In a different example, Saengruengrit et al. [128] also developed core–shell magnetic PLGA nanocomposites, in which SPIONs were embedded in the polymer shell and the core was composed of an aqueous phase containing BSA (as a protein antigen model). Oleic-acid-coated SPIONs were synthesized by a classic thermal decomposition method [136]. SPIONs were also included in the PLGA NP shell and BSA was encapsulated in the particles' core, it being the case that the synthesis by double emulsion was performed in the same manner [128]. To encapsulate BSA the particles were separated into two size groups, one with 500 nm and another with 300 nm, to study the effect of NP size on immune modulation. The composites were envisioned as vehicles for antigen delivery to stimulate an adaptive immune response when combined with an alternating magnetic field.

Another widely studied type of MNPs is manganese-oxide-based NPs. These present a strong paramagnetic character, and therefore these particles do not retain any magnetization in the absence of an external magnetic field [118]. Mn NPs are mainly used as T_1-MRI contrast enhancers [143–145].

Bennewitz et al. [146] developed pH-sensitive PLGA NPs and MPs encapsulating MnO nanocrystals of 15–20 nm size (Figure 1b left). A single ESE with minor modifications was used to synthesize these PLGA NPs. The nanocrystals were synthesized via the controlled thermal decomposition of manganese(II) acetylacetonate in benzyl ether and oleic acid. Afterwards, the PLGA NPs were fabricated as follows: for the organic phase PLGA was dissolved in methylene chloride, and dried MnO NPs were added to this solution. The organic phase was added dropwise to an aqueous phase containing PVA as a stabilizer. This mixture was vortexed and sonicated to obtain a O/W emulsion. This emulsion was added to another aqueous PVA solution under rapid stirring. The resulting nanocomposites were left to stir to evaporate the residual organic solvent. The obtained NP and MP composites presented average diameters of 140 nm and 1.7 μm, respectively. This system is intended to be used for molecular and cellular MRI imaging [146].

3.3. Other PLGA/Inorganic-NP-Based Composites

Many other inorganic NPs have also been explored in combination with PLGA polymeric materials. Zinc-, cerium-, carbon-, copper-, hydroxyapatite (HAp)- and titanium-based nanoparticles are examples of different inorganic materials that have been investigated in combination with PLGA and will be addressed in this section. Zinc oxide (ZnO) presents great optical, catalytic and semiconducting properties; additionally, it can be used for a plethora of biomedical applications due to its antimicrobial activity, biocompatibility, chemical stability and nontoxicity [99,100,147]. ZnO NPs are also able to generate ROS, which help in their anticancer activity in addition to their antibacterial activity against both Gram-positive and Gram-negative bacteria [148,149]. Other interesting INPs are cerium oxide NPs (nanoceria, CeO_2), which display a recyclable antioxidative activity, being able to act as ROS-scavenging NPs due to their ability to regenerate their oxidation state under various environmental conditions [150]. Their antioxidative activity mimics the activity of superoxide dismutase (SOD) and catalase through the conversion of superoxide radicals and hydrogen peroxide into oxygen and water [151]. Carbon nanotubes (CNTs) are another example of a nanomaterial widely used in a variety of applications. They have high surface area, low density, high mechanical integrity and exhibit a high mechanical modulus (0.2–1 TPa) [152]. These properties make CNTs ideal nanomaterials to be incorporated into scaffolds, for example. Copper oxide (CuO) NPs have gathered great research attention since copper is one of the most used metals and presents useful optical, electrical and medicinal properties [153]. It is predominantly used in nanomedicine due to its antimicrobial activity against a range of bacterial strains, which mainly occurs through the release of Cu^{2+} ions [154,155]. Titanium dioxide (TiO_2) is a very important material used in the fabrication of bone implants and has received great attention due to its biocompatibility and appealing mechanical properties [156]. Scaffolds incorporating TiO_2 NPs showcase great physical properties, such as corrosion resistance, low weight and low toxicity, but also great bioactive properties, such as enhanced cell proliferation and cell adhesion [157–161]. Inorganic HAp NPs are another important class of NPs that are widely used in combination with PLGA [162,163]. HAp is a strong inorganic material that is naturally present in the bone, considered to be a vital component for scaffolds due to its excellent biocompatibility, bioactivity, lack of immunogenicity and osteoconductivity [77,164].

Stankovic et al. [165] prepared PLGA NPs loaded with inorganic ZnO NPs (Figure 1d) via a solvent/nonsolvent method. ZnO NPs were prepared via a modified microwave-assisted synthesis [166], while PLGA NPs were prepared by the aforementioned solvent/nonsolvent method. For the preparation of the final composite materials PLGA NPs were dissolved in acetone (or ethyl acetate), and subsequently a solution of ZnO in acetone was added dropwise under constant homogenization. Ethanol was then added to the mixture to form a precipitate, and this suspension was slowly poured into an aqueous solution containing polyvinylpyrrolidone (PVP) as a stabilizer. The resulting particles were spherical and uniform, presenting a size of approximately 200 nm. This system can be used for different biomedical applications, such as drug delivery and treatment of bacterial infections [165].

In a different study, Mehta et al. [167] fabricated PLGA MPs containing nanoceria and ROS-scavenging enzymes, either catalase or SOD, using a previously reported one-stage method [168]. These PLGA MPs are expected to showcase a synergistic effect with the enzymes and were synthesized via a standard double emulsion method [80], where the encapsulants (nanoceria and SOD) were dispersed together with PLGA in the organic phase [167]. The particles were uniform and presented a mean diameter of 800 nm. The synthesized PLGA microcomposites can be used to protect cells from oxidative stress. Singh et al. [80] also synthesized PLGA MPs encapsulating nanoceria (CNP-PLGA), with an average size of 60 µm and a porous surface, using the same standard double emulsion method (Figure 1d). They also incorporated nanoceria into PLGA scaffolds to be used in tissue engineering. The fabrication of PLGA scaffolds can be performed with a wide variety of techniques and depends on the desired scaffold structure and application. In this study the scaffolds were constructed by solution casting and rapid evaporation under reduced pressure. Since the nanoceria particles mimic the activity of SOD, their controlled and sustained release from PLGA MPs can make them useful for a variety of biomedical applications, such as mitigating damage from radiation and bacterial infection [80].

Other PLGA scaffolds combined with different INPs have been widely studied. Mikael et al. [169] combined multiwalled carbon nanotubes (MWCNTs) and PLGA to fabricate a mechanically enhanced and biodegradable composite scaffold from PLGA-MWCNT microspheres. PLGA-MWCNT microspheres were first synthesized using the single emulsion method (O/W). Afterwards, these MPs, in a size range of 425–600 µm, were used to fabricate circular scaffolds via thermal sintering. The particles were packed into a steel mold that was heated at 95 °C for 1 h, and then were cooled down and stored in a desiccator [169]. These scaffolds showed enhanced mechanical properties, indicated to be used in bone tissue engineering.

In another study Haider et al. [155] synthesized hybrid nanofiber PLGA scaffolds compositing CuO NPs by electrospinning (Figure 3a). For the synthesis of CuO NPs Cu powder was added to distilled water and sonicated. The reaction mixture was transferred into a glass bottle, sealed and autoclaved. CuO NPs were retrieved by centrifugation. To synthesize the scaffolds a PLGA/CuO solution was prepared and put into a syringe to be electrospun. The nanofibers were smooth and uniform, and their average diameter was around 550 nm. These scaffolds were proposed for antibacterial purposes due to the antimicrobial activity of CuO NPs.

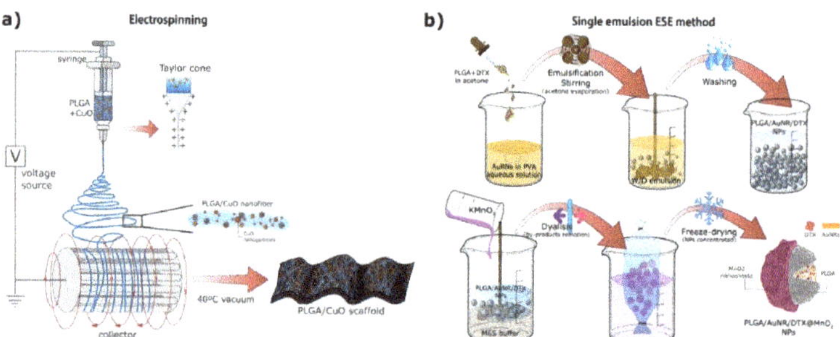

Figure 3. Illustration of (**a**) the electrospinning process and (**b**) the single emulsion ESE method as well as the further coating of the particles.

In a more recent study Pelaseyed et al. [170] fabricated PLGA nanocomposite scaffolds functionalized with commercially available titanium oxide (TiO_2) NPs at different ratios. These scaffolds were fabricated by an air–liquid foaming process that uses nucleation to create gas bubbles, which results in a porous microstructure. The functionalization with TiO_2 offers to improve the mechanical and bioactive properties of PLGA [157]. The scaffolds were highly porous and their pore size was reduced with higher contents of TiO_2, resulting in an improvement in the mechanical properties of the scaffolds, ideal for bone regenerative engineering.

Sheikh et al. [77] fabricated hybrid PLGA–silk fibroin composite scaffolds combined with HAp NPs (PLGA–silk–HAp scaffolds). According to the authors, both silk fibroin and HAp NPs cannot be used alone: silk fibroin has a slow degradation rate, making it difficult for the scaffold to be replaced by tissue; HAp NPs present a free needle-like particulate nature, which results in frail films that are inadequate to be used as bone grafts [77,171]. Hence, their combination with PLGA can improve the mechanical properties, biodegradability, hydrophilicity and osteoconductivity of the scaffolds, making them suitable for the intended application in bone tissue regeneration. Porous PLGA–silk–HAp scaffolds were fabricated by salt leaching and vacuum mixing. First, pristine PLGA scaffolds were prepared as follows: PLGA pellets were dissolved in methyl chloride, mixed with NaCl particles (180–250 mm) in a ratio of 1:9 and poured into silicon molds. The molds were pressed for 24 h at room temperature. Afterwards, the samples were kept in a vacuum oven at 25 °C for 1 week to remove residual solvents. A salt-leaching technique was carried out to porosify the PLGA by using distilled water. Finally, the samples were dried at room temperature to form porous PLGA scaffolds. Afterwards, to obtain the final PLGA–silk–HAp scaffolds, an aqueous solution of HAp NPs was sonicated and added to a 4% silk solution, blended and poured onto the pristine PLGA scaffolds. Vacuum mixing was performed by turning the vacuum pump ON and OFF; in this way the scaffold pores were filled with the solution. In a related study, Selvaraju et al. [172] synthesized collagen/PLGA-based composite scaffolds by also incorporating inorganic HAp NPs, using a different approach. The HAp nanopowder was synthesized through a modified chemical precipitation method [173]. Initially, an aqueous solution of H_3PO_4 was added dropwise to a $Ca(OH)_2$ solution. The pH of the final solution was adjusted via the addition of NH_4OH. Finally, the filtered cakes were oven-dried to remove free water. To fabricate the scaffolds a PLGA/HAp composite was first synthesized, where an in situ preparation of NPs in a polymeric matrix was performed. For this synthesis the PLGA copolymer was prepared from scratch using the individual monomers in its constitution. While this method is not a common technique for the synthesis of PLGA composites, the HAp NPs can grow inside the polymer matrix, reducing aggregation and maintaining a good spatial distribution. The in situ polymerization was conducted by mixing HAp nanopowder with monomers, D,L lactide and glycolide under a N_2 atmosphere. An initiator of the polymerization, stannous octoate, was added, and the reaction was performed for 12 h. Collagen was then dissolved in an acetate buffer and mixed with the PLGA/HAp composite. For the scaffold fabrication a vacuum drying technique was used [172]. The incorporation of HAp allows for an improvement in the composite physicochemical properties and stability, making it possible to form a scaffold of a defined pore structure. In this way it is possible to use this composite as a scaffold for regenerative engineering.

3.4. PLGA Composites Incorporating More Than One Inorganic Nanomaterial

Composite PLGA-based materials combining more than one inorganic nanomaterial have also been investigated [79,174,175]. Numerous types of combinations and designs can be made, e.g., including both INPs in the PLGA shell, only in the core or both in the core and shell. When combining together different types of INPs it is important to take into account their physicochemical properties to obtain a viable particle design. The concentrations of each material need to be thoughtfully optimized in a way that their abilities are not compromised and can perform synergistically [73,176]. With the use of these combined inorganic materials, it is possible to incorporate additional advantages into the system that will be able to perform multiple functions.

Ye et al. [79] combined inorganic imaging agents of manganese-doped zinc sulfide (Mn:ZnS) quantum dots (QDs, optical imaging) and SPIONs (MRI) by encapsulating them into PLGA vesicles together with the anticancer drug busulfan (PLGA-SPION-Mn:ZnS). The SPIONs were synthesized via thermal decomposition [136] and the QDs were synthesized via a nucleation-doping strategy [177]. These composites were fabricated via the single ESE method, where the INPs were included in the organic phase and worked as DDS theranostic agents.

In another study by Wang et al. [175] PLGA NPs were loaded with AuNRs coated with CTAB and the anticancer drug DTX, and then further coated with manganese dioxide (MnO_2) ultrathin nanofilms (PLGA/AuNR/DTX@MnO_2 NPs). These particles are illustrated in Figure 1c. For their synthesis AuNRs were synthetized through the CTAB-induced seed-mediated growth method [178,179]. AuNRs and DTX-loaded PLGA NPs were prepared using the simple ESE method. Lastly, to obtain the final MnO_2-nanofilm-coated PLGA nanocomposites a 2-(*N*-morpholino) ethanesulfonic acid (MES) oxidation–reduction method was employed [180]. The final nanocomposites presented a spherical form with a rough surface due to the MnO_2 shell, and their average size was around 280 nm (Figure 3b). These PLGA/AuNR/DTX@MnO_2 structures were promising nanocomposites for theranostic applications, being able to perform dual-mode diagnosis (MRI from the Mn and X-ray CT from the Au) and radiofrequency-induced hyperthermia combined with chemotherapy [175].

Li et al. [174] developed core–shell nanocomposites based on PLGA copolymers using a stabilizer-free method. Here, the PLGA was loaded with PTX, resulting in spherical, ~200 nm size NPs that were further modified with the surface growth of a silver/gold nanoshell around the PLGA core (PLGA@Ag-AuNPs). Although two different materials are used this case differs from the other examples since both the materials have the same function. The particles' nanoshell allows for a greatly improved surface-enhanced Raman spectroscopy (SERS) signal, making this platform very attractive for biodetection and controlled drug release, serving as a promising theranostic tool.

4. Applications

PLGA copolymers tailored in combination with INPs help to broaden the range of applications in biomedicine for which PLGA can be used. Some of the most explored uses for these hybrid polymer composites are therapy [97,129,181,182] (e.g., PTT, chemotherapy), diagnostics [121–123,146] (e.g., MRI, X-ray and PA), theranostics [78,79,96,175] and tissue engineering [77,155,169,172]. In this section we will briefly discuss several representative studies of each of these biomedical applications, each presenting different combinations of PLGA with INPs, in the form of MPs, NPs and macroscopic structures. Different targeting ligands and drugs are also considered.

4.1. Therapy

Several PLGA-only nanoparticles are already used in the clinic for the treatment of a variety of conditions (see Table 1). The combination of biocompatible, nontoxic and biodegradable PLGA nanoparticles with INPs can improve and expand the functionalities of each individual nanoparticulated counterpart in this area.

A good example is the combination of PLGA nanoparticles with gold nanoparticles for cancer treatment. In the work by Fazio et al. [95], the authors used a PEG-PLGA copolymer nanoplatform incorporating AuNPs and SLB for cancer therapy (Figure 4a). In this example, AuNPs absorb energy and transform it into heat, resulting in hyperthermal cancer therapy and photothermal drug delivery. Laser-irradiated NPs showed higher drug release when compared to nonirradiated samples, with a 75% relative release increment in the first 5 h of drug release (Figure 4b), proving that these NPs have great potential for biomedical applications. Nevertheless, in vitro and in vivo studies are still required to assess their biocompatibility and real anticancer potential. In a similar example, Deng et al. [94]

incorporated AuNPs and the photosensitizer verteprofin (VP) in PLGA nanocomposites (PLGA@AuNPs-VP). In this case, AuNPs in combination with a photosensitizer can be used for PDT. In fact, the nanocomposite decreased the viability of pancreatic tumor cells after laser exposure (31% of live cells treated with PLGA-VP-AuNPs, compared to 44% of live cells treated only with PLGA-VP nanoparticles, Figure 4c), with an increase in the production of singlet oxygen, helping in the killing of tumor cells.

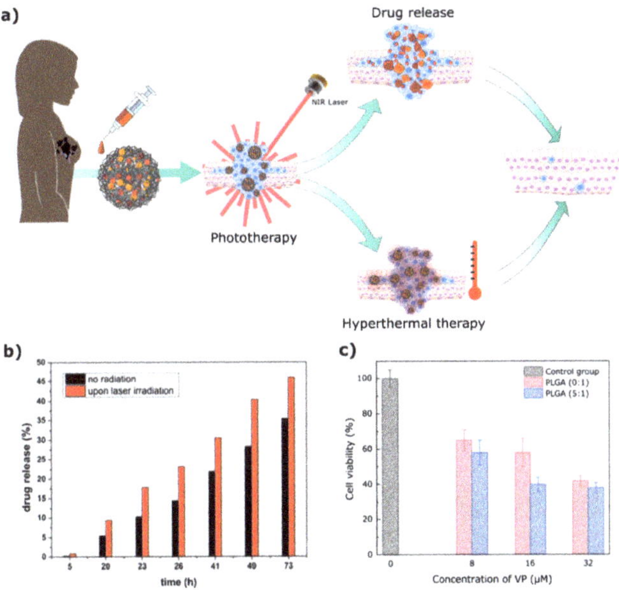

Figure 4. (**a**) Schematic representation of the applications of plasmonic PLGA NPs by Fazio et al. [95]. (**b**) Cumulative release from plasmonic PLGA NPs with and without laser irradiation. Reproduced with permission from ref. [95]. Copyright (2015), Royal Society of Chemistry. (**c**) Cytotoxicity of PLGA@AuNPs-VP by Deng et al. in PANC-1 cells after 5 min of 405 nm laser illumination. The molar ratios of Au and VP molecules in PLGA samples were 0:1 and 5:1, respectively. Adapted from ref. [94]. Copyright (2016), Royal Society of Chemistry.

Besides the use of AuNPs, iron oxide NPs have become a suitable partner for PLGA-based particles. Due to their superparamagnetic properties iron oxide nanoparticles can respond to an AMF, resulting in the production of heat in a process called magnetic hyperthermia. Fang et al. [129] synthesized DOX-containing PLGA microspheres coated with Fe_3O_4 NPs (DOX-MMS). In vitro and in vivo studies demonstrated that DOX release from the microsphere increased after exposure to an AMF, resulting in the killing of tumor cells (81.6% of viable 4T1 cells when treated with DOX-MMS versus 10% of viable cells when treated with DOX-MMS and exposed to an AMF). In vivo studies using 4T1 tumor models have shown that mice treated with DOX-MMS and exposed to an AMF had reduced tumor growth (3.4 $\pm$ 0.6-fold, compared to 6.2 $\pm$ 0.8-fold in the control group and 4.0 $\pm$ 0.8-fold in the group treated with DOX-MMS without an AMF). Interestingly, PLGA coated with Fe_3O_4 NPs without a drug (MMS) also resulted in impaired tumor growth after exposure to an AMF, with tumors being smaller than the DOX-MMS without an AMF group, showing that the effect of the final formulation is due to a combination of magnetic hyperthermia (thermal therapy) and enhanced drug release. Fe_3O_4/PLGA nanoparticles can also be helpful for immunotherapy purposes, as shown by Saengruengrit et al. [128]. Under an AMF, Fe_3O_4/PLGA nanoparticles encapsulating BSA were internalized by macrophages and bone-marrow-derived dendritic cells. Furthermore, the polymeric particles stimulated the differentiation of immature dendritic cells in vitro, promoting the expression of proinflammatory surface markers and cytokines.

In principle, these hybrid PLGA nanocomposites can be used for vaccination purposes since BSA could be replaced by any real antigen of interest.

Besides applications in the oncology arena, PLGA composites were also investigated for the treatment of infections. Stankovic et al. [165] and others [19,183] have shown that PLGA NPs with nano-ZnO and organic-Ag hybrid materials display a bacteriostatic effect against a wide range of Gram-positive and Gram-negative bacteria, as well as against the yeast Candida albicans. Therefore, the combination of Ag NPs with PLGA polymers can be very beneficial in this arena. Takahashi et al. [19] tested the efficacy of PLGA-Ag NPs against Staphylococcus epidermidis biofilms covered with a thick film of extracellular polymeric substances (EPS). The antibacterial activity of Ag salts in general and Ag NPs in particular has been well-characterized [155,184]. The NPs were able to adhere to the biofilm, and as PLGA degrades the Ag NPs promote the dissolution of the EPS film. The bacteria were damaged by the interaction with Ag NPs and voids were induced in the biofilms. Additionally, Ag NPs can cause the separation of the bacterial cell wall, leading to their death. Furthermore, the interaction of silver ions from the degradation of the Ag NPs with enzymes and proteins induces the production of ROS, which inhibits bacterial growth and colony formation, mainly through their reaction with DNA. Furthermore, PLGA was found to reduce the metal NPs' aggregation and the Ag toxicity, which potentiates their antibacterial action. Ag-PLGA nanocomposites were shown to be suitable as a drug delivery system with high antibacterial activity, treating biofilm infections [19].

In a different scenario, PLGA/INP nanocomposites were shown to reduce ROS and counteract oxidative stress. In the work of Mehta et al. [167] PLGA MPs were used to encapsulate cerium oxide NPs (nanoceria—CNPs) together with SOD and catalase enzymes. Both the organic and the inorganic part of the system act synergistically to help prevent oxidative stress by scavenging ROS. In vitro studies have shown that the PLGA–nanoceria–SOD composites were efficiently delivered to macrophages and clearly displayed protection against environmental oxidative stress. These composites can potentially be applied to several diseases, such as neurodegenerative disorders and cardiovascular diseases, where oxidative stress plays a key role in the progression of the disease.

4.2. Diagnostic

Although theranosis is the most sought-after application involving imaging for PLGA-based materials, they have also been explored for purely diagnostic purposes, with one of the most common applications being MRI. This is made possible through the combination of PLGA with INPs with a magnetic character, which can either result in enhanced T_1 or T_2 MRI signals.

A good example was presented by Bennewitz et al. [146], which used PLGA-encapsulated MnO nanocrystals for cellular MR imaging. The longitudinal relaxivity, r_1 (a measure of the goodness of a contrast agent for T_1-weighted MR imaging), of these composites was poor when the particles were intact. However, after incubation in acidic media the r_1 increased (35-fold) due to the reduction and dissolution of MnO to free Mn^{2+}. In the extracellular space the NPs do not release Mn^{2+} and consequently do not display a significative T_1 contrast. On the other hand, when the composites were internalized by cells MR images showed enhanced T_1 contrast due to the evolution of free Mn^{2+}, allowing for the tracking of endocytosis [146]. In a different study by Nkansah et al. [123] PLGA micro- and nanoparticles were used for MRI-based cell tracking, this time through the encapsulation of magnetite. The magnetic PLGA composites' degradation is faster in the first two weeks, displaying a quick decline in the r_2* relaxivity and then a gradual decline during the following 50 days. The composites were nontoxic, and mesenchymal stem cells labeled with magnetic PLGA composites presented a significant loss of T_2* signal as opposed to Feridex® (a commercial MRI contrast agent) and blank PLGA-labeled cells. This study showed the utility of the presented composites as contrast agents for MRI-based cell tracking, being great candidates for clinical translation [123]. A similar study by Tang et al. [122] also proved the labeling efficiency of Fe_3O_4-encapsulating PLGA NPs.

In a different study Lu et al. [121] utilized PLGA MPs containing iron oxide for dual-modal PA/MRI tracking of tendon stem cells (TSCs) (Figure 5a). To enhance the cell internalization of the particles they were coated with poly-L-lysine, leading to a positive surface charge. It was demonstrated that the higher the Fe concentration the higher the PA and MRI signal achieved (Figure 5b). However, high concentrations of iron (>200 μg/mL) significantly reduced cell viability. Thus, an optimal concentration to maintain good PA and MRI signals for cell tracking without significant cytotoxicity was determined to be around 100 μg Fe/mL. It was observed that TSCs' position could be tracked by monitoring the signal intensity at the anatomical sites where they accumulate. The main limitation of this study was that the labeling of TSCs was not detected in vivo. Nevertheless, the produced magnetic PLGA MPs demonstrated great potential to be used as a dual-mode imaging contrast agents.

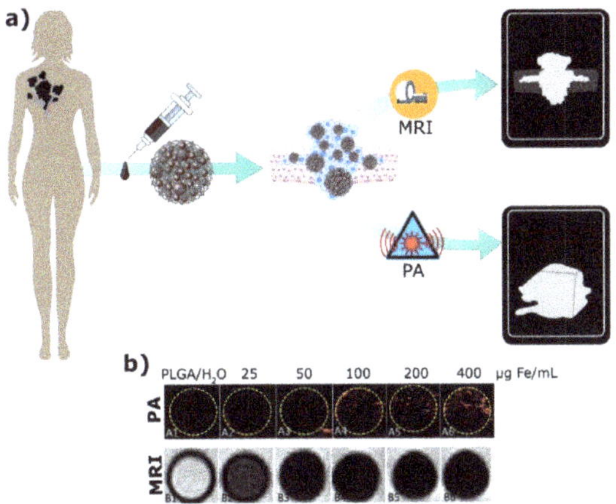

Figure 5. (**a**) Schematic representation of the applications of magnetic PLGA MPs by Lu et al. [121]. (**b**) Dual-modal PA/MRI imaging ability of magnetic PLGA MPs. Reproduced with permission from ref. [121]. Copyright (2018), Public Library of Science.

The PLGA/INP combination for diagnosis is an ongrowing field that can be helpful for the study of various diseases where cell degeneration, mutation and growth is a determinant factor in disease evolution (e.g., cancer, neurodegenerative diseases). Extensive preclinical studies are already being developed in this field [185]; hence, it is of great importance to continue to explore it and eventually achieve the clinical translation of viable PLGA/INP formulations.

4.3. Theranostics

The effective prevention and treatment of diseases require the development of new strategies, with the aim of diagnosing and treating diseases at an early stage. PLGA/INPs composites that are able to act as multifunctional agents to achieve these ends have been extensively studied.

Lee et al. [127] performed an interesting conjugation of PLGA with complex molecules (mPEG and Ce6) and magnetite (core) to achieve a theranostic PLGA/INP composite. Ce6 and magnetite were the main functional parts of the system, with Ce6 working as a fluorescent and PDT agent able to generate ROS and magnetite working as an MRI contrast agent. It was observed that PLGA/INP composites with Ce6 molecules produced more ROS than free Ce6 when illuminated with a laser, resulting in a significant inhibition of the growth of tumors, it being the case that the tumor volume of mice treated with PLGA/INP composites was between ~1.5–3 times smaller than the ones treated with free Ce6. Even at low-dose administrations, PLGA composites with Ce6 were more efficient to cause tumor regression than free Ce6 administered at high doses. Consistent with this finding, PLGA/INPs showed a strong fluorescence signal at the tumor site, while free Ce6 had a very weak signal, proving the ability of multifunctional PLGA/INPs to accumulate in a tumor site. The encapsulation of Fe_3O_4 made the polymeric particles suitable for MRI imaging. In vivo studies showed an enhanced T_2* contrast in the tumor when compared with Feridex®.

Once again, magnetite was the INP of choice in a study by Ye et al. [79], where PLGA vesicles containing SPIONs, Mn:ZnS QDs and a chemotherapeutic drug were tailored to be used as cancer theranostic agents. PLGA-SPION-Mn:ZnS vesicles exhibited a high r_2* relaxivity and prominently enhanced the T_2* signal in MR imaging. The signal decreased with the increase in the concentration of PLGA/INPs vesicles (and consequently of iron), showing that these hybrid composites can be used as negative contrast agents in MRI. Fluorescence imaging studies with macrophages facilitated by the presence of QDs showed that the composites had high uptake efficiency. The PLGA/INP vesicles could also be used as a DDS for the delivery of the lipophilic drug busulfan. The EE% of these drug was $89 \pm 2\%$, and about 70–80% of the drug was released after the first 5 h of dialysis. Studies on the degradation of PLGA vesicles have suggested that 32% of PLGA degrades into LA and GA in 5 weeks. This relatively slow degradation behavior of PLGA vesicles combined with the high entrapment efficiency of busulfan makes this system ideal for sustained drug release, although further studies are needed to optimize the system release properties to match the degradation rate of PLGA.

In a different approach, Hao et al. [78] designed PEI-functionalized PLGA NPs combined with gold nanoshells, a targeting ligand (angiopep-2) and DTX, also to be used for cancer theranostics. The remote activation stimulus of an 808 nm laser on the Au nanoshell results in a heat-induced release of the cargo. The percentage of released drug increased by 17% compared with the release from nonirradiated composites. The PLGA/INP composites promoted efficient tumor inhibition with the help of angiogep-2 in addition to the enhanced permeability and retention effect. The nanocomposites also exhibited the potential for X-ray imaging applications. CT signal intensity increased with an increase in the concentration of composites. Overall, the core–shell DTX-loaded PLGA@Au nanocomposites represent a promising drug delivery system for tumor-targeted chemophotothermal therapy and X-ray imaging.

Song et al. [96] also proposed the conjugation of PLGA with gold in the form of AuNRs, where the gold (functionalized with PEG) was embedded in a PLGA shell. The AuNRs@PEG/PLGA nanocomposites showed good performance in PA image-guided photothermal therapy due to their optical properties, increased photothermal conversion efficiency and enhanced light scattering as well as PA signal (Figure 6a). The 60 nm composites showed an absorbance peak around 800 nm, which made them suitable for irradiation with an 808 nm laser. The photothermal conversion efficiency of the composites was up to twofold higher than that of AuNRs alone, due to the strong plasmonic coupling of the nanorods with the PLGA shell. When irradiated with the 808 nm laser the composites generated a PA signal and a temperature increase higher than that of free AuNRs. This temperature increase is essential for the photothermal destruction of the tumor, and the strong PA signal allows for the monitoring of this process. PA and PET (after labeling with Cu) imaging confirmed that after an IV injection the nanocomposites accumulated in the tumor region. Additionally, laser irradiation leads to the disassembly of the nanocomposites, where the PLGA shell degrades over time through the hydrolysis of PLGA ester bonds, resulting in AuNRs coated only with PEG. At day 10 postinjection the composites were already cleared from the body, which is essential for future clinical translations. These AuNR/PLGA nanocomposites were able to completely ablate tumors without reoccurrence in vivo (Figure 6b), presenting themselves as promising agents to be translated to the clinic in the near future for image-guided cancer therapy.

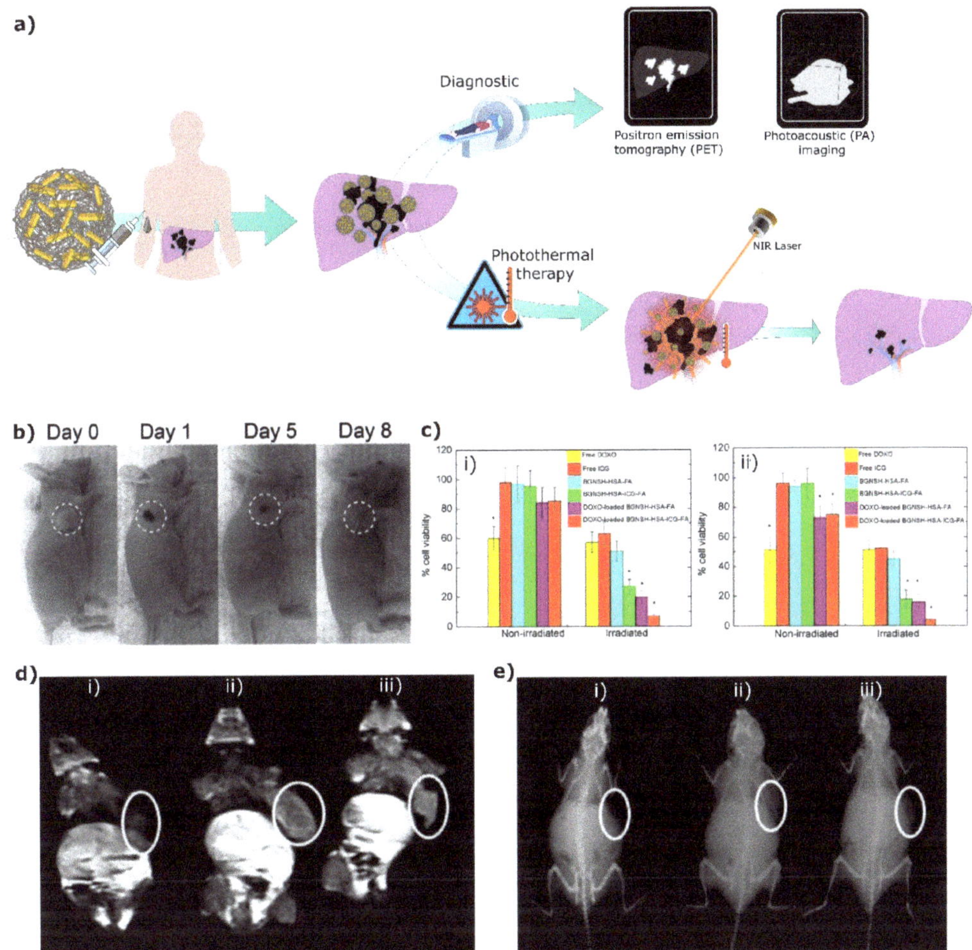

Figure 6. (**a**) Schematic representation of the applications of the AuNR vesicles from Song et al. [96]. (**b**) Photographs of the tumor-bearing mice at days 0, 1, 5 and 8 after being treated with the AuNR vesicles. Reproduced with permission from [95]. Copyright (2015), Wiley VCH. C. (**c**) Cell viability of (i) HeLa and (ii) MDA-MB-231cells after 24 h of incubation in the absence and presence of NIR light irradiation. Reproduced with permission from [97]. Copyright (2015), Wiley VCH; dual-mode imaging in vivo. (**d**) T_1-weighted MR images and (**e**) X-ray CT images of (i) control, (ii) 4 h and (iii) 8 h after treatment. Reproduced with permission from [175]. Copyright (2017), Dovepress. * $p < 0.05$.

In the work by Topete et al. [97] a core–shell PLGA–gold nanocomposite was loaded with DOX and functionalized with HSA, noncovalently conjugated with indocyanine green (ICG) and covalently conjugated with folic acid (DOXO-loaded BGNSH-HSA-ICG-FA). The produced composites function as a drug release system of a chemotherapeutic drug through PTT. The gold nanoshell allows for the absorption of NIR (near-infrared) light, resulting in an increase in the localized heat that promotes efficient drug release, with irradiated samples releasing approximately 80% of the drug while nonirradiated samples only release approximately 20%. HSA grants stealth, while folic acid provides specific targeting. ICG is a NIR probe that can be used for PDT, where the absorption of NIR light results in the production of singlet oxygen that kills cancer cells. In addition, NIR probes can be used for in vivo imaging. Altogether, these nanoconjugates function as imaging agents and deliver a chemotherapeutic drug, heat and singlet oxygen directly to tumors. The in vitro treatment of cancer cells with the nanocomposites resulted in increased tumor cell death in the presence of NIR irradiation when compared to the respective controls (Figure 6c). This enhanced killing of tumor cells was the result of the synergistic effect of PDT/PTT. Furthermore, in vivo and in vitro studies showed the potential application of the nanoplatform as an imaging agent.

In another study, Wang et al. [175] also used AuNRs, but this time they were co-loaded with DTX in PLGA NPs coated with MnO_2 nanofilms. The presence of Au and MnO_2 in the PLGA/AuNR/DTX@MnO_2 composites made them suitable for dual-mode imaging, with the Mn being able to work as a T_1 contrast agent in MRI and the gold working as a positive X-ray CT imaging contrast agent (Figure 6d,e, respectively). Signal generation in both CT and MRI was indeed confirmed to be PLGA/INP-concentration-dependent. In addition, the MR signal derived from the MnO_2 was enhanced in the presence of higher levels of glutathione (GSH), indicative of a responsive behavior that will specifically increase the contrast in the tumor (due to altered redox homeostasis). In terms of therapy, this study showed that the release rates from the PLGA/INPs were higher in acidic environments and with higher levels of GSH (from 39.7% to 78.1%). As with the responsive MR signal, this is due to the surface degradation of MnO_2 and is a useful feature to add specificity to chemotherapeutic treatments. The presence of AuNRs in the particles also made them suitable for radiofrequency (RF) hyperthermia, which helps in the effectiveness of drug release. The combination of chemotherapy with the heat generated by the Au during RF hyperthermia allows for a reduction in the hypoxic area in the center of the tumor, since hypoxic cells are more sensitive to hyperthermia. The role of AuNRs in RF hyperthermia was proven in vitro, where it was demonstrated that cell viability significantly decreases when RF hyperthermia was applied in combination with the PLGA/AuNR/DTX@MnO_2 (from approximately 60% to 20%). The same effect was observed in vivo. This indicates that AuNRs play a key role in this treatment and can in fact produce heat when irradiated with an RF pulse. Furthermore, AuNR-induced hyperthermia in combination with the effect of DTX responsive release effectively induced cellular apoptosis.

In summary, the combination of PLGA with INPs is a promising area, particularly in oncology theranostics, but it is of great importance to continue to optimize and develop new smart theranostic concepts and agents that have the potential to be translated to the clinic.

4.4. Tissue Engineering

The main application of PLGA in tissue engineering is through the development of PLGA scaffolds that are capable of inducing bone tissue regeneration to correct orthopedic defects, treat bone neoplasia and pseudoarthrosis or to stimulate recovery after surgery [11,186,187]. PLGA implants are growing rapidly as an alternative to traditional implants in many orthopedic applications. Since PLGA boasts a tunable degradation rate, it is possible to avoid further surgery to remove the implants, as in current clinical practice.

Mikael et al. [169] fabricated PLGA-MP-based scaffolds functionalized with inorganic MWCNT that render mechanically stronger scaffolds while maintaining good cellular compatibility for bone tissue engineering and regenerative engineering. The mechanical strength and compressive modulus of the scaffolds increased significantly with the addition of 1 and 3 wt% MWCNTs, since the nanotubes reinforced the junctions between microparticles. Furthermore, scaffolds made with a higher percentage of PLA (85:15 ratio) were mechanically stronger and degraded at a slower rate. PLGA/MWCNT composite scaffolds demonstrated their ability to act as substrates for HAp crystal growth, showing a higher deposition rate than that of pure PLGA scaffolds. In vivo studies did not show systemic or neurological toxicity. Furthermore, all implants maintained their structural integrity until week four, when they started to degrade. By weeks eight–twelve biological tissue started invading the implants. However, at this time point an inflammatory response was observed, which should be carefully studied before the translation of these composites scaffolds can move on.

Sheikh et al. [77] fabricated hybrid PLGA–silk fibroin scaffolds embedded with HAp NPs, which bring a beneficial synergistic effect from each of their components. PLGA–silk–HAp scaffolds present an improved swelling and water uptake capacity, which indicates that they can retain biological fluids, a beneficial feature, upon implantation. The inclusion of silk in the scaffolds presents several advantages: (i) it increases the hydrophilic character of PLGA, hence increasing the affinity towards biological fluids, allowing the cells to permeate deeper into the scaffold upon implantation; (ii) it improves the mechanical stability of the scaffold by penetrating PLGA micropores. Moreover, the incorporation of HAp NPs gives a higher stress-bearing capacity to the scaffold, improving its mechanical properties. In vitro studies have shown that the scaffolds had great cell infiltration, favorable for the intended application. In vivo studies have demonstrated that these scaffolds were able to complete intramembranous ossification at the site of a cavity due to the bone-inducing agent HAp. Following a very similar strategy, Selvaraju et al. [172] also prepared a hybrid composite scaffold with collagen (instead of silk) and PLGA embedded with HAp NPs. This study showed that HAp gives rigidity and stiffness to the nanocomposite. Besides enhancing the mechanical properties of the PLGA polymer, HAp also increases the thermal and conformational stability of collagen. The composite was nontoxic to healthy cells, being ideal to act as a scaffold. The degradation rate of the scaffold was studied for a period of 45 days and showed a minimum weight loss of 2%, which remained constant over time. Overall, both of these PLGA/INP scaffolds are great candidates for use in bone tissue engineering and clinical translation.

In a recent study, Pelaseyed et al. [170] created PLGA scaffolds (prepared as a foam) embedded with a TiO_2 nanopowder. For applications in bone tissue regeneration the scaffolds should be highly porous (exhibiting micro- and macroporosity) to support cell seeding, adhesion and ingrowth proliferation. The incorporation of TiO_2 into a PLGA matrix can improve these features, since PLGA itself is restricted by poor osteoconductivity. The prepared composite foam presented highly interconnected porous structures, which can be tuned by controlling PLGA concentration, freezing rate and cooling temperature. TiO_2 NPs led to a decrease in pore size in the scaffolds. Additionally, the uniformly dispersed TiO_2 particles in PLGA and the interaction between these two components improved the mechanical properties of the scaffolds. TiO_2 in low concentrations (5 and 10 wt%) contributed to the decrease in the degradation rate of the polymer and the presence of this ceramic nanomaterial enhanced the bioactivity of PLGA. The prepared scaffolds were suitable for cell attachment, proliferation and nutrient transfer, and presented good mechanical properties, ideal again for applications in bone regeneration.

With a different goal in mind, Haider et al. [155] synthesized an antibacterial PLGA nanofiber scaffold composite incorporating CuO NPs. The fabricated hybrid PLGA/CuO nanocomposite fibbers were first studied as antibacterial materials through disc diffusion and optical density methods. In both methods the scaffolds produced zones of inhibition against *E. coli* and *S. aureus*. The hybrid scaffolds showed efficient inhibition of bacterial growth through the penetration of CuO or Cu^{2+} ions across the cell membrane as well as the generation of ROS; the adhesion of CuO NPs to proteins present in the bacterial cell wall caused bacterial death. The studies also proved that Cu^{2+} ions were released from the scaffolds in a sustained manner, enabling the antibacterial activity for longer periods of time. In vitro studies have shown that fibroblast cells had a good adherence, spreading and proliferating in the hybrid composite scaffolds, proving their good cytocompatibility and nontoxic nature. The results demonstrated that the PLGA/CuO nanofiber scaffolds have great potential to be used as internal and/or external wound dressing material.

Singh et al. [80] used PLGA MPs to produce hybrid PLGA scaffolds containing nanoceria, an INP that mimics the activity of SOD and catalase enzymes. In vitro studies have shown that the release of nanoceria from PLGA is slower than that of most drugs, since it relies on polymer degradation. The released INPs displayed a higher SOD activity in acidic media due to a higher presence of Ce ions. The produced MPs and scaffolds were biocompatible and the nanoceria particles were released in a slow and controlled fashion, being ideal as antioxidant treatments.

5. Conclusions

Currently, PLGA-based materials designed for biomedical applications are a subject of great interest, both at the research and industrial levels. The present review provides a compilation of several hybrid inorganic nanoparticle/PLGA-based composites in the nano-, micro- and macrorange and showcases the versatility of this family of materials. These hybrid materials have proved their superiority in terms of physicochemical features, biocompatibility, morphology and multifunctionality. Most of the discussed hybrid materials present a simple and straightforward synthesis, offering a great advantage over other kinds of materials since the synthesis can be easily adapted to fit a plethora of applications. The inorganic materials to be combined with PLGA often determine their final purposes, as do the organic therapeutic agents (proteins, DNA and drugs) that can be loaded onto the PLGA platform. Multifunctional PLGA-based composite materials allow for the combination of different applications that result in enhanced diagnostic and therapeutic outcomes. PLGA materials and nanocomposites represent a step forward in the biomedicine field and show great promise for further improvements in the theranostic, therapeutic, diagnostic and tissue engineering areas.

Author Contributions: C.V.R., V.G. and M.C.d.S. wrote the manuscript, J.G. supervised and edited and M.B.-L. and J.G. commented on the paper. All authors have read and agreed to the published version of the manuscript.

Funding: The authors would like to acknowledge the financial support from the 2014–2020 INTERREG Cooperation Programme Spain–Portugal (POCTEP) through the project 0624_2IQBIONEURO_6_E. C.R. and V.G. also thank the Foundation for Science and Technology (FCT-Fundação para a Ciência e a Tecnologia) for their PhD fellowships (2020.08240.BD and 2020.10155.BD).

Institutional Review Board Statement: Not applicable.

Informed Consent Statement: Not applicable.

Data Availability Statement: Not applicable.

Conflicts of Interest: The authors declare no conflict of interest.

Abbreviations

Ag NPs	Silver NPs;
AMF	Alternating magnetic field;
AuNPs	Gold NPs;
AuNRs	Gold nanorods;
BPLP	Biodegradable photoluminescent polyester;
CAs	Contrast agents;
Ce6	Chlorin e6;
CNPs	Cerium oxide NPs;
CNTs	Carbon nanotubes;
CTAB	Cetyltrimethylammonium bromide;
DCM	Dichloromethane;
DDS	Drug delivery system;
DOX	Doxorubicin;
DTX	Docetaxel;
EE%	Encapsulation efficiency;
EMA	European Medicines Agency;
ESD	Emulsification–solvent diffusion;
ESE	Emulsification–solvent evaporation;

FDA	U.S. Food and Drug Administration;
GA	Glycolic acid;
GSH	Glutathione;
HAp	Hydroxyapatite;
HIFU	High-intensity focused ultrasound;
INPs	Inorganic NPs;
LA	Lactic acid;
MES	2-(*N*-morpholino) ethanesulfonic acid;
MNPs	Magnetic NPs;
mPEG	Methoxy poly(ethylene glycol);
MPs	Microparticles;
MRI	Magnetic resonance imaging;
Mw	Molecular weight;
MWCNTs	Multiwalled carbon nanotubes;
NIR	Near-infrared;
NPs	Nanoparticles;
O/O	Oil-in-oil;
O/W	Oil-in-water;
PA	Photoacoustic;
PCL-PEG	Poly(ε-caprolactone-b-ethylene glycol);
PDT	Photodynamic therapy;
PEG	Poly(ethylene glycol);
PEI	Polyethyleneimine;
PFH	Perfluorohexane;
PGA	Poly(glycolic acid);
PLA	Poly(lactic acid);
PLGA	Poly lactic-co-glycolic acid;
PTT	Photothermal therapy;
PTX	Paclitaxel;
PVA	Polyvinyl alcohol;
PVP	Polyvinylpyrrolidone;
QDs	Quantum dots;
RF	Radiofrequency;
ROS	Reactive oxygen species;
SLB	Silibinin;
SOD	Superoxide dismutase;
SPIONs	Superparamagnetic iron oxide nanoparticles;
SPR	Surface plasmon resonance;
TSC	Tendon stem cells;
US	Ultrasound;
VIP	Vasoactive intestinal peptide;
VP	Verteporfin;
W/O	Water-in-oil;
W/O/O	Water-in-oil-in-oil;
W/O/W	Water-in-oil-in-water.

References

1. Roy, I.; Gupta, M.N. Smart Polymeric Materials: Emerging Biochemical Applications. *Chem. Biol.* **2003**, *10*, 1161–1171. [CrossRef]
2. Makadia, H.K.; Siegel, S.J. Poly lactic-*co*-glycolic Acid (PLGA) as Biodegradable Controlled Drug Delivery Carrier. *Polymers* **2011**, *3*, 1377–1397. [CrossRef] [PubMed]
3. Mirakabad, F.S.T.; Nejati-Koshki, K.; Akbarzadeh, A.; Yamchi, M.R.; Milani, M.; Zarghami, N.; Zeighamian, V.; Rahimzadeh, A.; Alimohammadi, S.; Hanifehpour, Y.; et al. PLGA-Based Nanoparticles as Cancer Drug Delivery Systems. *Asian Pac. J. Cancer Prev.* **2014**, *15*, 517–535. [CrossRef] [PubMed]
4. Danhier, F.; Ansorena, E.; Silva, J.M.; Coco, R.; Le Breton, A.; Préat, V. PLGA-Based Nanoparticles: An Overview of Biomedical Applications. *J. Control. Release* **2012**, *161*, 505–522. [CrossRef] [PubMed]
5. Prokop, A.; Davidson, J.M. Nanovehicular Intracellular Delivery Systems. *J. Pharm. Sci.* **2008**, *97*, 3518–3590. [CrossRef] [PubMed]
6. Yang, Y.Y.; Chung, T.S.; Ping Ng, N. Morphology, Drug Distribution, and in Vitro Release Profiles of Biodegradable Polymeric Microspheres Containing Protein Fabricated by Double-Emulsion Solvent Extraction/Evaporation Method. *Biomaterials* **2001**, *22*, 231–241. [CrossRef]

7. Rezvantalab, S.; Drude, N.I.; Moraveji, M.K.; Güvener, N.; Koons, E.K.; Shi, Y.; Lammers, T.; Kiessling, F. PLGA-Based Nanoparticles in Cancer Treatment. *Front. Pharmacol.* **2018**, *9*, 1260. [CrossRef]
8. Ding, D.; Zhu, Q. Recent Advances of PLGA Micro/Nanoparticles for the Delivery of Biomacromolecular Therapeutics. *Mater. Sci. Eng. C* **2018**, *92*, 1041–1060. [CrossRef]
9. Graf, N.; Bielenberg, D.R.; Kolishetti, N.; Muus, C.; Banyard, J.; Farokhzad, O.C.; Lippard, S.J. Avβ 3 Integrin-Targeted PLGA-PEG Nanoparticles for Enhanced Anti-Tumor Efficacy of a Pt(IV) Prodrug. *ACS Nano* **2012**, *6*, 4530–4539. [CrossRef]
10. Dechy-Cabaret, O.; Martin-Vaca, B.; Bourissou, D. Controlled Ring-Opening Polymerization of Lactide and Glycolide. *Chem. Rev.* **2004**, *104*, 6147–6176. [CrossRef]
11. Gentile, P.; Chiono, V.; Carmagnola, I.; Hatton, P.V. An Overview of Poly(Lactic-Co-Glycolic) Acid (PLGA)-Based Biomaterials for Bone Tissue Engineering. *Int. J. Mol. Sci.* **2014**, *15*, 3640–3659. [CrossRef] [PubMed]
12. Choi, Y.; Joo, J.R.; Hong, A.; Park, J.S. Development of Drug-Loaded PLGA Microparticles with Different Release Patterns for Prolonged Drug Delivery. *Bull. Korean Chem. Soc.* **2011**, *32*, 867–872. [CrossRef]
13. Cappellano, G.; Comi, C.; Chiocchetti, A.; Dianzani, U. Exploiting PLGA-Based Biocompatible Nanoparticles for next-Generation Tolerogenic Vaccines against Autoimmune Disease. *Int. J. Mol. Sci.* **2019**, *20*, 204. [CrossRef]
14. Ospina-Villa, J.D.; Gómez-Hoyos, C.; Zuluaga-Gallego, R.; Triana-Chávez, O. Encapsulation of Proteins from Leishmania Panamensis into PLGA Particles by a Single Emulsion-Solvent Evaporation Method. *J. Microbiol. Methods* **2019**, *162*, 1–7. [CrossRef] [PubMed]
15. Roces, C.B.; Christensen, D.; Perrie, Y. Translating the Fabrication of Protein-Loaded Poly(Lactic-Co-Glycolic Acid) Nanoparticles from Bench to Scale-Independent Production Using Microfluidics. *Drug Deliv. Transl. Res.* **2020**, *10*, 582–593. [CrossRef] [PubMed]
16. Zare, E.N.; Jamaledin, R.; Naserzadeh, P.; Afjeh-Dana, E.; Ashtari, B.; Hosseinzadeh, M.; Vecchione, R.; Wu, A.; Tay, F.R.; Borzacchiello, A.; et al. Metal-Based Nanostructures/PLGA Nanocomposites: Antimicrobial Activity, Cytotoxicity, and Their Biomedical Applications. *ACS Appl. Mater. Interfaces* **2019**, *12*, 3279–3300. [CrossRef] [PubMed]
17. Ghitman, J.; Biru, E.I.; Stan, R.; Iovu, H. Review of Hybrid PLGA Nanoparticles: Future of Smart Drug Delivery and Theranostics Medicine. *Mater. Des.* **2020**, *193*, 108805. [CrossRef]
18. Liang, C.; Li, N.; Cai, Z.; Liang, R.; Zheng, X.; Deng, L.; Feng, L.; Guo, R.; Wei, B. Co-Encapsulation of Magnetic Fe3O4 Nanoparticles and Doxorubicin into Biocompatible PLGA-PEG Nanocarriers for Early Detection and Treatment of Tumours. *Artif. Cells Nanomed. Biotechnol.* **2019**, *47*, 4211–4221. [CrossRef]
19. Takahashi, C.; Matsubara, N.; Akachi, Y.; Ogawa, N.; Kalita, G.; Asaka, T.; Tanemura, M.; Kawashima, Y.; Yamamoto, H. Visualization of Silver-Decorated Poly (D,L-Lactide-Co-Glycolide) Nanoparticles and Their Efficacy against Staphylococcus Epidermidis. *Mater. Sci. Eng. C* **2017**, *72*, 143–149. [CrossRef]
20. Toti, U.S.; Guru, B.R.; Hali, M.; McPharlin, C.M.; Wykes, S.M.; Panyam, J.; Whittum-Hudson, J.A. Targeted Delivery of Antibiotics to Intracellular Chlamydial Infections Using PLGA Nanoparticles. *Biomaterials* **2011**, *32*, 6606–6613. [CrossRef]
21. Arafa, M.G.; Girgis, G.N.; El-Dahan, M.S. Chitosan-Coated PLGA Nanoparticles for Enhanced Ocular Anti-Inflammatory Efficacy of Atorvastatin Calcium. *Int. J. Nanomed.* **2020**, *15*, 1335–1347. [CrossRef]
22. Gao, M.; Long, X.; Du, J.; Teng, M.; Zhang, W.; Wang, Y.; Wang, X.; Wang, Z.; Zhang, P.; Li, J. Enhanced Curcumin Solubility and Antibacterial Activity by Encapsulation in PLGA Oily Core Nanocapsules. *Food Funct.* **2020**, *11*, 448–455. [CrossRef] [PubMed]
23. Berthet, M.; Gauthier, Y.; Lacroix, C.; Verrier, B.; Monge, C. Nanoparticle-Based Dressing: The Future of Wound Treatment? *Trends Biotechnol.* **2017**, *35*, 770–784. [CrossRef] [PubMed]
24. Sharma, S.; Parmar, A.; Kori, S.; Sandhir, R. PLGA-Based Nanoparticles: A New Paradigm in Biomedical Applications. *TrAC Trends Anal. Chem.* **2016**, *80*, 30–40. [CrossRef]
25. Acharya, S.; Sahoo, S.K. PLGA Nanoparticles Containing Various Anticancer Agents and Tumour Delivery by EPR Effect. *Adv. Drug Deliv. Rev.* **2011**, *63*, 170–183. [CrossRef] [PubMed]
26. Lü, J.M.; Wang, X.; Marin-Muller, C.; Wang, H.; Lin, P.H.; Yao, Q.; Chen, C. Current Advances in Research and Clinical Applications of PLGA-Based Nanotechnology. *Expert Rev. Mol. Diagn.* **2009**, *9*, 325–341. [CrossRef]
27. Park, T.G. Degradation of Poly(D,L-Lactic Acid) Microspheres: Effect of Molecular Weight. *J. Control. Release* **1994**, *30*, 161–173. [CrossRef]
28. Schliecker, G.; Schmidt, C.; Fuchs, S.; Wombacher, R.; Kissel, T. Hydrolytic Degradation of Poly(Lactide-Co-Glycolide) Films: Effect of Oligomers on Degradation Rate and Crystallinity. *Int. J. Pharm.* **2003**, *266*, 39–49. [CrossRef]
29. Dinarvand, R.; Sepehri, N.; Manoochehri, S.; Rouhani, H.; Atyabi, F. Polylactide-Co-Glycolide Nanoparticles for Controlled Delivery of Anticancer Agents. *Int. J. Nanomed.* **2011**, *6*, 877–895. [CrossRef]
30. Schliecker, G.; Schmidt, C.; Fuchs, S.; Kissel, T. Characterization of a Homologous Series of D,L-Lactic Acid Oligomers; a Mechanistic Study on the Degradation Kinetics in Vitro. *Biomaterials* **2003**, *24*, 3835–3844. [CrossRef]
31. Wu, X.S.; Wang, N. Synthesis, Characterization, Biodegradation, and Drug Delivery Application of Biodegradable Lactic/Glycolic Acid Polymers. Part II: Biodegradation. *J. Biomater. Sci. Polym. Ed.* **2001**, *12*, 21–34. [CrossRef] [PubMed]
32. Sah, E.; Sah, H. Recent Trends in Preparation of Poly(lactide-*co*-glycolide) Nanoparticles by Mixing Polymeric Organic Solution with Antisolvent. *J. Nanomater.* **2015**, *2015*, 794601. [CrossRef]
33. Cheng, J.; Teply, B.A.; Sherifi, I.; Sung, J.; Luther, G.; Gu, F.X.; Levy-Nissenbaum, E.; Radovic-Moreno, A.F.; Langer, R.; Farokhzad, O.C. Formulation of Functionalized PLGA-PEG Nanoparticles for in Vivo Targeted Drug Delivery. *Biomaterials* **2007**, *28*, 869–876. [CrossRef] [PubMed]

34. Ghahremankhani, A.A.; Dorkoosh, F.; Dinarvand, R. PLGA-PEG-PLGA Tri-Block Copolymers as in Situ Gel-Forming Peptide Delivery System: Effect of Formulation Properties on Peptide Release. *Pharm. Dev. Technol.* **2008**, *13*, 49–55. [CrossRef]
35. Jeong, B.; Bae, Y.H.; Kim, S.W. In Situ Gelation of PEG-PLGA-PEG Triblock Copolymer Aqueous Solutions and Degradation Thereof. *J. Biomed. Mater. Res.* **2000**, *50*, 171–177. [CrossRef]
36. Zhang, Y.; García-Gabilondo, M.; Rosell, A.; Roig, A. MRI/Photoluminescence Dual-Modal Imaging Magnetic PLGA Nanocapsules for Theranostics. *Pharmaceutics* **2020**, *12*, 16. [CrossRef] [PubMed]
37. Tobío, M.; Gref, R.; Sánchez, A.; Langer, R.; Alonso, M.J. Stealth PLA-PEG Nanoparticles as Protein Carriers for Nasal Administration. *Pharm. Res.* **1998**, *15*, 270–275. [CrossRef]
38. Khalil, N.M.; do Nascimento, T.C.F.; Casa, D.M.; Dalmolin, L.F.; de Mattos, A.C.; Hoss, I.; Romano, M.A.; Mainardes, R.M. Pharmacokinetics of Curcumin-Loaded PLGA and PLGA-PEG Blend Nanoparticles after Oral Administration in Rats. *Colloids Surf. B Biointerfaces* **2013**, *101*, 353–360. [CrossRef]
39. Koopaei, M.N.; Khoshayand, M.R.; Mostafavi, S.H.; Amini, M.; Khorramizadeh, M.R.; Tehrani, M.J.; Atyabi, F.; Dinarvand, R. Docetaxel Loaded PEG-PLGA Nanoparticles: Optimized Drug Loading, in-Vitro Cytotoxicity and in-Vivo Antitumor Effect. *Iran. J. Pharm. Res.* **2014**, *13*, 819–833. [CrossRef]
40. Tahara, K.; Sakai, T.; Yamamoto, H.; Takeuchi, H.; Hirashima, N.; Kawashima, Y. Improved Cellular Uptake of Chitosan-Modified PLGA Nanospheres by A549 Cells. *Int. J. Pharm.* **2009**, *382*, 198–204. [CrossRef]
41. Garner, J.; Skidmore, S.; Park, H.; Park, K.; Choi, S.; Wang, Y. A Protocol for Assay of Poly(Lactide-Co-Glycolide) in Clinical Products. *Int. J. Pharm.* **2015**, *495*, 87–92. [CrossRef] [PubMed]
42. Park, K.; Skidmore, S.; Hadar, J.; Garner, J.; Park, H.; Otte, A.; Soh, B.K.; Yoon, G.; Yu, D.; Yun, Y.; et al. Injectable, Long-Acting PLGA Formulations: Analyzing PLGA and Understanding Microparticle Formation. *J. Control. Release* **2019**, *304*, 125–134. [CrossRef] [PubMed]
43. Wang, Y.; Qu, W.; Choi, S.H. FDA's Regulatory Science Program for Generic PLA/PLGA-Based Drug Products. Available online: https://www.americanpharmaceuticalreview.com/Featured-Articles/188841-FDA-s-Regulatory-Science-Program-for-Generic-PLA-PLGA-Based-Drug-Products/ (accessed on 1 March 2021).
44. Silverman, B.L.; Blethen, S.L.; Reiter, E.O.; Attie, K.M.; Neuwirth, R.B.; Ford, K.M. A Long-Acting Human Growth Hormone (Nutropin Depot®): Efficacy and Safety Following Two Years of Treatment in Children with Growth Hormone Deficiency. *J. Pediatric Endocrinol. Metab.* **2002**, *15*, 715–722. [CrossRef] [PubMed]
45. Guo, S.; Fu, D.; Utupova, A.; Sun, D.; Zhou, M.; Jin, Z.; Zhao, K. Applications of polymer-based nanoparticles in vaccine field. *Nanotechnol. Rev.* **2019**, *8*, 143–155. [CrossRef]
46. Qi, F.; Wu, J.; Li, H.; Ma, G. Recent Research and Development of PLGA/PLA Microspheres/Nanoparticles: A Review in Scientific and Industrial Aspects. *Front. Chem. Sci. Eng.* **2019**, *13*, 14–27. [CrossRef]
47. WCG. CenterWatch FDA Approved Drugs. Available online: https://www.centerwatch.com/directories/1067 (accessed on 1 March 2021).
48. Schoubben, A.; Ricci, M.; Giovagnoli, S. Meeting the Unmet: From Traditional to Cutting-Edge Techniques for Poly Lactide and Poly Lactide-Co-Glycolide Microparticle Manufacturing. *J. Pharm. Investig.* **2019**, *49*, 381–404. [CrossRef]
49. Bobo, D.; Robinson, K.J.; Islam, J.; Thurecht, K.J.; Corrie, S.R. Nanoparticle-Based Medicines: A Review of FDA-Approved Materials and Clinical Trials to Date. *Pharm. Res.* **2016**, *33*, 2373–2387. [CrossRef]
50. Zhong, H.; Chan, G.; Hu, Y.; Hu, H.; Ouyang, D. A Comprehensive Map of FDA-Approved Pharmaceutical Products. *Pharmaceutics* **2018**, *10*, 263. [CrossRef]
51. Almajhdi, F.N.; Fouad, H.; Khalil, K.A.; Awad, H.M.; Mohamed, S.H.S.; Elsarnagawy, T.; Albarrag, A.M.; Al-Jassir, F.F.; Abdo, H.S. In-Vitro Anticancer and Antimicrobial Activities of PLGA/Silver Nanofiber Composites Prepared by Electrospinning. *J. Mater. Sci. Mater. Med.* **2014**, *25*, 1045–1053. [CrossRef]
52. Lee, D.; Heo, D.N.; Lee, S.J.; Heo, M.; Kim, J.; Choi, S.; Park, H.K.; Park, Y.G.; Lim, H.N.; Kwon, I.K. Poly(Lactide-Co-Glycolide) Nanofibrous Scaffolds Chemically Coated with Gold-Nanoparticles as Osteoinductive Agents for Osteogenesis. *Appl. Surf. Sci.* **2017**, *432*, 300–307. [CrossRef]
53. Jain, R.A. The Manufacturing Techniques of Various Drug Loaded Biodegradable Poly(Lactide-Co-Glycolide) (PLGA) Devices. *Biomaterials* **2000**, *21*, 2475–2490. [CrossRef]
54. Tansik, G.; Yakar, A.; Gündüz, U. Tailoring Magnetic PLGA Nanoparticles Suitable for Doxorubicin Delivery. *J. Nanoparticle Res.* **2014**, *16*, 2171. [CrossRef]
55. Mir, M.; Ahmed, N.; Rehman, A.U. Recent Applications of PLGA Based Nanostructures in Drug Delivery. *Colloids Surf. B Biointerfaces* **2017**, *159*, 217–231. [CrossRef] [PubMed]
56. Yezhelyev, M.V.; Gao, X.; Xing, Y.; Al-Hajj, A.; Nie, S.; O'Regan, R.M. Emerging Use of Nanoparticles in Diagnosis and Treatment of Breast Cancer. *Lancet Oncol.* **2006**, *7*, 657–667. [CrossRef]
57. Quintanar-Guerrero, D.; Allémann, E.; Fessi, H.; Doelker, E. Preparation Techniques and Mechanisms of Formation of Biodegradable Nanoparticles from Preformed Polymers. *Drug Dev. Ind. Pharm.* **1998**, *24*, 1113–1128. [CrossRef]
58. Esmaeili, F.; Atyabi, F.; Dinarvand, R. Preparation of PLGA Nanoparticles Using TPGS in the Spontaneous Emulsification Solvent Diffusion Method. *J. Exp. Nanosci.* **2007**, *2*, 183–192. [CrossRef]
59. Murakami, H.; Kobayashi, M.; Takeuchi, H.; Kawashima, Y. Preparation of Poly(D,L-Lactide-Co-Glycolide) Nanoparticles by Modified Spontaneous Emulsification Solvent Diffusion Method. *Int. J. Pharm.* **1999**, *187*, 143–152. [CrossRef]

60. D'Mello, S.R.; Das, S.K.; Das, N.G. Polymeric Nanoparticles for Small-Molecule Drugs: Biodegradation of Polymers and Fabrication of Nanoparticles. In *Drug Delivery Nanoparticles Formulation and Characterization*; Pathak, Y., Thassu, D., Eds.; Informa Healthcare: Indianapolis, IN, USA, 2006; pp. 16–30.
61. Miladi, K.; Sfar, S.; Fessi, H.; Elaissari, A. Nanoprecipitation Process: From Particle Preparation to In Vivo Applications. In *Polymer Nanoparticles for Nanomedicines*; Vauthier, C., Ponchel, G., Eds.; Springer International Publishing: Berlin/Heidelberg, Germany, 2016; pp. 17–53.
62. Dalpiaz, A.; Sacchetti, F.; Baldisserotto, A.; Pavan, B.; Maretti, E.; Iannuccelli, V.; Leo, E. Application of the "in-Oil Nanoprecipitation" Method in the Encapsulation of Hydrophilic Drugs in PLGA Nanoparticles. *J. Drug Deliv. Sci. Technol.* **2016**, *32*, 283–290. [CrossRef]
63. Dalpiaz, A.; Vighi, E.; Pavan, B.; Leo, E. Fabrication via a Nonaqueous Nanoprecipitation Method, Characterization and in Vitro Biological Behavior of N6-Cyclopentyladenosine-Loaded Nanoparticles. *J. Pharm. Sci.* **2009**, *98*, 4272–4284. [CrossRef]
64. Morales-Cruz, M.; Flores-Fernández, G.M.; Morales-Cruz, M.; Orellano, E.A.; Rodriguez-Martinez, J.A.; Ruiz, M.; Griebenow, K. Two-Step Nanoprecipitation for the Production of Protein-Loaded PLGA Nanospheres. *Results Pharma Sci.* **2012**, *2*, 79–85. [CrossRef]
65. Nie, H.; Lee, L.Y.; Tong, H.; Wang, C.H. PLGA/Chitosan Composites from a Combination of Spray Drying and Supercritical Fluid Foaming Techniques: New Carriers for DNA Delivery. *J. Control. Release* **2008**, *129*, 207–214. [CrossRef] [PubMed]
66. Patel, K.; Atkinson, C.; Tran, D.; Nadig, S.N. Nanotechnological Approaches to Immunosuppression and Tolerance Induction. *Curr. Transplant. Rep.* **2017**, *4*, 159–168. [CrossRef] [PubMed]
67. Bilati, U.; Allémann, E.; Doelker, E. Development of a Nanoprecipitation Method Intended for the Entrapment of Hydrophilic Drugs into Nanoparticles. *Eur. J. Pharm. Sci.* **2005**, *24*, 67–75. [CrossRef] [PubMed]
68. Arpagaus, C. PLA/PLGA nanoparticles prepared by nano spray drying. *J. Pharm. Investig.* **2019**, *49*, 405–426. [CrossRef]
69. Freitas, S.; Merkle, H.P.; Gander, B. Microencapsulation by Solvent Extraction/Evaporation: Reviewing the State of the Art of Microsphere Preparation Process Technology. *J. Control. Release* **2005**, *102*, 313–332. [CrossRef] [PubMed]
70. Harris, L.D.; Kim, B.S.; Mooney, D.J. Open Pore Biodegradable Matrices Formed with Gas Foaming. *J. Biomed. Mater. Res.* **1998**, *42*, 396–402. [CrossRef]
71. Mikos, A.G.; Thorsen, A.J.; Czerwonka, L.A.; Bao, Y.; Langer, R.; Winslow, D.N.; Vacanti, J.P. Preparation and Characterization of poly(L-lactic acid) Foams. *Polymer* **1994**, *35*, 1068–1077. [CrossRef]
72. Zhang, R.; Ma, P.X. Poly(α-hydroxyl acids)/Hydroxyapatite Porous Composites for Bone- Tissue Engineering. I. Preparation and Morphology. *J. Biomed. Mater. Res.* **1999**, *44*, 446–455. [CrossRef]
73. Go, E.J.; Kang, E.Y.; Lee, S.K.; Park, S.; Kim, J.H.; Park, W.; Kim, I.H.; Choi, B.; Han, D.K. An Osteoconductive PLGA Scaffold with Bioactive β-TCP and Anti-Inflammatory Mg(OH)2 to Improve: In Vivo Bone Regeneration. *Biomater. Sci.* **2019**, *8*, 937–948. [CrossRef]
74. Shin, S.H.; Purevdorj, O.; Castano, O.; Planell, J.A.; Kim, H.W. A Short Review: Recent Advances in Electrospinning for Bone Tissue Regeneration. *J. Tissue Eng.* **2012**, *3*, 1–11. [CrossRef]
75. Yu, D.-G.; Zhu, L.-M.; White, K.; Branford-White, C. Electrospun Nanofiber-Based Drug Delivery Systems. *Health* **2009**, *1*, 67–75. [CrossRef]
76. Mura, S.; Hillaireau, H.; Nicolas, J.; Le Droumaguet, B.; Gueutin, C.; Zanna, S.; Tsapis, N.; Fattal, E. Influence of Surface Charge on the Potential Toxicity of PLGA Nanoparticles towards Calu-3 Cells. *Int. J. Nanomed.* **2011**, *6*, 2591–2605. [CrossRef]
77. Sheikh, F.A.; Ju, H.W.; Moon, B.M.; Lee, O.J.; Kim, J.-H.; Park, H.J.; Kim, D.W.; Kim, D.-K.; Jang, J.E.; Khang, G.; et al. Hybrid Scaffolds Based on PLGA and Silk for Bone Tissue Engineering. *J. Tissue Eng. Regen. Med.* **2015**, *10*, 209–221. [CrossRef] [PubMed]
78. Hao, Y.; Zhang, B.; Zheng, C.; Ji, R.; Ren, X.; Guo, F.; Sun, S.; Shi, J.; Zhang, H.; Zhang, Z.; et al. The Tumor-Targeting Core-Shell Structured DTX-Loaded PLGA@Au Nanoparticles for Chemo-Photothermal Therapy and X-Ray Imaging. *J. Control. Release* **2015**, *220*, 545–555. [CrossRef]
79. Ye, F.; Barrefelt, Å.; Asem, H.; Abedi-Valugerdi, M.; El-Serafi, I.; Saghafian, M.; Abu-Salah, K.; Alrokayan, S.; Muhammed, M.; Hassan, M. Biodegradable Polymeric Vesicles Containing Magnetic Nanoparticles, Quantum Dots and Anticancer Drugs for Drug Delivery and Imaging. *Biomaterials* **2014**, *35*, 3885–3894. [CrossRef]
80. Singh, V.; Singh, S.; Das, S.; Kumar, A.; Self, W.T.; Seal, S. A Facile Synthesis of PLGA Encapsulated Cerium Oxide Nanoparticles: Release Kinetics and Biological Activity. *Nanoscale* **2012**, *4*, 2597–2605. [CrossRef]
81. Lee, S.K.; Han, C.M.; Park, W.; Kim, I.H.; Joung, Y.K.; Han, D.K. Synergistically Enhanced Osteoconductivity and Anti-Inflammation of PLGA/β-TCP/$Mg(OH)_2$ Composite for Orthopedic Applications. *Mater. Sci. Eng. C* **2019**, *94*, 65–75. [CrossRef]
82. Dhandayuthapani, B.; Yoshida, Y.; Maekawa, T.; Kumar, D.S. Polymeric Scaffolds in Tissue Engineering Application: A Review. *Int. J. Polym. Sci.* **2011**, *2011*, 290602. [CrossRef]
83. Anton, N.; Benoit, J.P.; Saulnier, P. Design and Production of Nanoparticles Formulated from Nano-Emulsion Templates-A Review. *J. Control. Release* **2008**, *128*, 185–199. [CrossRef]
84. Allahyari, M.; Mohit, E. Peptide/Protein Vaccine Delivery System Based on PLGA Particles. *Hum. Vaccines Immunother.* **2016**, *12*, 806–828. [CrossRef]
85. Mundargi, R.C.; Babu, V.R.; Rangaswamy, V.; Patel, P.; Aminabhavi, T.M. Nano/Micro Technologies for Delivering Macromolecular Therapeutics Using Poly(D,L-lactide-*co*-glycolide) and Its Derivatives. *J. Control. Release* **2008**, *125*, 193–209. [CrossRef] [PubMed]

86. Wang, G.; Fang, N. Detecting and Tracking Nonfluorescent Nanoparticle Probes in Live Cells. In *Methods in Enzymology*; Hochstrasser, M., Ed.; Academic Press Inc.: Cambridge, MA, USA, 2012; Volume 504, pp. 83–108.
87. Nolan, J.P.; Sebba, D.S. Surface-Enhanced Raman Scattering (SERS) Cytometry. In *Recent Advances in Cytometry, Part A*; Darzynkiewicz, Z., Holden, E., Orfao, A., Telford, W., Wlodkowic, D., Eds.; Academic Press: Cambridge, MA, USA, 2011; Volume 102, pp. 515–532.
88. Pissuwan, D. Monitoring and Tracking Metallic Nanobiomaterials In Vivo. In *Monitoring and Evaluation of Biomaterials and Their Performance In Vivo*; Narayan, R.J., Ed.; Elsevier Inc.: Amsterdam, The Netherlands, 2017; pp. 135–149. ISBN 9780081006047.
89. Dreaden, E.C.; Alkilany, A.M.; Huang, X.; Murphy, C.J.; El-Sayed, M.A. The Golden Age: Gold Nanoparticles for Biomedicine. *Chem. Soc. Rev.* **2012**, *41*, 2740–2779. [CrossRef] [PubMed]
90. Swider, E.; Koshkina, O.; Tel, J.; Cruz, L.J.; de Vries, I.J.M.; Srinivas, M. Customizing Poly(Lactic-Co-Glycolic Acid) Particles for Biomedical Applications. *Acta Biomater.* **2018**, *73*, 38–51. [CrossRef] [PubMed]
91. Wang, Y.; Strohm, E.M.; Sun, Y.; Wang, Z.; Zheng, Y.; Wang, Z.; Kolios, M.C. Biodegradable Polymeric Nanoparticles Containing Gold Nanoparticles and Paclitaxel for Cancer Imaging and Drug Delivery Using Photoacoustic Methods. *Biomed. Opt. Express* **2016**, *7*, 4125. [CrossRef] [PubMed]
92. Lee, D.; Heo, D.N.; Kim, H.J.; Ko, W.K.; Lee, S.J.; Heo, M.; Bang, J.B.; Lee, J.B.; Hwang, D.S.; Do, S.H.; et al. Inhibition of Osteoclast Differentiation and Bone Resorption by Bisphosphonate-Conjugated Gold Nanoparticles. *Sci. Rep.* **2016**, *6*, 27336. [CrossRef]
93. Yi, C.; Liu, D.; Fong, C.C.; Zhang, J.; Yang, M. Gold Nanoparticles Promote Osteogenic Differentiation of Mesenchymal Stem Cells through P38 MAPK Pathway. *ACS Nano* **2010**, *4*, 6439–6448. [CrossRef] [PubMed]
94. Deng, W.; Kautzka, Z.; Chen, W.; Goldys, E.M. PLGA Nanocomposites Loaded with Verteporfin and Gold Nanoparticles for Enhanced Photodynamic Therapy of Cancer Cells. *RSC Adv.* **2016**, *6*, 112393–112402. [CrossRef]
95. Fazio, E.; Scala, A.; Grimato, S.; Ridolfo, A.; Grassi, G.; Neri, F. Laser Light Triggered Smart Release of Silibinin from a PEGylated-PLGA Gold Nanocomposite. *J. Mater. Chem. B* **2015**, *3*, 9023–9032. [CrossRef]
96. Song, J.; Yang, X.; Jacobson, O.; Huang, P.; Sun, X.; Lin, L.; Yan, X.; Niu, G.; Ma, Q.; Chen, X. Ultrasmall Gold Nanorod Vesicles with Enhanced Tumor Accumulation and Fast Excretion from the Body for Cancer Therapy. *Adv. Mater.* **2015**, *27*, 4910–4917. [CrossRef]
97. Topete, A.; Alatorre-Meda, M.; Iglesias, P.; Villar-Alvarez, E.M.; Barbosa, S.; Costoya, J.A.; Taboada, P.; Mosquera, V. Fluorescent Drug-Loaded, Polymeric-Based, Branched Gold Nanoshells for Localized Multimodal Therapy and Imaging of Tumoral Cells. *ACS Nano* **2014**, *8*, 2725–2738. [CrossRef]
98. West, J.; Sears, J.W.; Smith, S.; Carter, M. Photonic Sintering—An Example: Photonic Curing of Silver Nanoparticles. In *Sintering of Advanced Materials*; Zang, Z.Z.F., Ed.; Woodhead Publishing: Cambridge, UK, 2010; pp. 275–288. ISBN 978-1-84569-562-0.
99. Hong, H.; Shi, J.; Yang, Y.; Zhang, Y.; Engle, J.W.; Nickles, R.J.; Wang, X.; Cai, W. Cancer-Targeted Optical Imaging with Fluorescent Zinc Oxide Nanowires. *Nano Lett.* **2011**, *11*, 3744–3750. [CrossRef] [PubMed]
100. Zhang, H.J.; Xiong, H.M.; Ren, Q.G.; Xia, Y.Y.; Kong, J.L. ZnO@silica Core-Shell Nanoparticles with Remarkable Luminescence and Stability in Cell Imaging. *J. Mater. Chem.* **2012**, *22*, 13159–13165. [CrossRef]
101. Bastús, N.G.; Comenge, J.; Puntes, V. Kinetically Controlled Seeded Growth Synthesis of Citrate-Stabilized Gold Nanoparticles of up to 200 Nm: Size Focusing versus Ostwald Ripening. *Langmuir* **2011**, *27*, 11098–11105. [CrossRef] [PubMed]
102. Shuhua, L.; Mingyong, H. Synthesis, Functionalization, and Bioconjugation of Monodisperse, Silica-Coated Gold Nanoparticles: Robust Bioprobes. *Adv. Funct. Mater.* **2005**, *15*, 961–967.
103. Gorelikov, I.; Martin, A.L.; Seo, M.; Matsuura, N. Silica-Coated Quantum Dots for Optical Evaluation of Perfluorocarbon Droplet Interactions with Cells. *Langmuir* **2011**, *27*, 15024–15033. [CrossRef]
104. Esmaeili, F.; Ghahremani, M.H.; Ostad, S.N.; Atyabi, F.; Seyedabadi, M.; Malekshahi, M.R.; Amini, M.; Dinarvand, R. Folate-Receptor Targeted Delivery of Docetaxel Nanoparticles Prepared by PLGA PEG Folate Conjugate. *J. Drug Target.* **2008**, *16*, 415–423. [CrossRef]
105. Ali, M.R.K.; Snyder, B.; El-Sayed, M.A. Synthesis and Optical Properties of Small Au Nanorods Using a Seedless Growth Technique. *Langmuir* **2012**, *28*, 9807–9815. [CrossRef]
106. Song, J.; Cheng, L.; Liu, A.; Yin, J.; Kuang, M.; Duan, H. Plasmonic Vesicles of Amphiphilic Gold Nanocrystals: Self-Assembly and External-Stimuli-Triggered Destruction. *J. Am. Chem. Soc.* **2011**, *133*, 10760–10763. [CrossRef]
107. Song, J.; Zhou, J.; Duan, H. Self-Assembled Plasmonic Vesicles of SERS-Encoded Amphiphilic Gold Nanoparticles for Cancer Cell Targeting and Traceable Intracellular Drug Delivery. *J. Am. Chem. Soc.* **2012**, *134*, 13458–13469. [CrossRef]
108. Song, J.; Pu, L.; Zhou, J.; Duan, B.; Duan, H. Biodegradable Theranostic Plasmonic Vesicles of Amphiphilic Gold Nanorods. *ACS Nano* **2013**, *7*, 9947–9960. [CrossRef]
109. Jana, N.R.; Gearheart, L.; Murphy, C.J. Wet Chemical Synthesis of High Aspect Ratio Cylindrical Gold Nanorods. *J. Phys. Chem. B* **2001**, *105*, 4065–4067. [CrossRef]
110. Gallo, J.; Long, N.J.; Aboagye, E.O. Magnetic Nanoparticles as Contrast Agents in the Diagnosis and Treatment of Cancer. *Chem. Soc. Rev.* **2013**, *42*, 7816–7833. [CrossRef] [PubMed]
111. Bañobre-López, M.; Teijeiro, A.; Rivas, J. Magnetic Nanoparticle-Based Hyperthermia for Cancer Treatment. *Rep. Pract. Oncol. Radiother.* **2013**, *18*, 397–400. [CrossRef] [PubMed]

112. Meikle, S.T.; Piñeiro, Y.; Bañobre López, M.; Rivas, J.; Santin, M. Surface Functionalization Superparamagnetic Nanoparticles Conjugated with Thermoresponsive Poly(Epsilon-Lysine) Dendrons Tethered with Carboxybetaine for the Mild Hyperthermia-Controlled Delivery of VEGF. *Acta Biomater.* **2016**, *40*, 235–242. [CrossRef] [PubMed]
113. De Moura, C.L.; Gallo, J.; García-Hevia, L.; Pessoa, O.D.L.; Ricardo, N.M.P.S.; Bañobre-López, M. Magnetic Hybrid Wax Nanocomposites as Externally Controlled Theranostic Vehicles: High MRI Enhancement and Synergistic Magnetically Assisted Thermo/Chemo Therapy. *Chem. A Eur. J.* **2020**, *26*, 4531–4538. [CrossRef]
114. Ong, Y.S.; Bañobre-López, M.; Costa Lima, S.A.; Reis, S. A Multifunctional Nanomedicine Platform for Co-Delivery of Methotrexate and Mild Hyperthermia towards Breast Cancer Therapy. *Mater. Sci. Eng. C* **2020**, *116*, 111255. [CrossRef]
115. García-Hevia, L.; Casafont, Í.; Oliveira, J.; Terán, N.; Fanarraga, M.L.; Gallo, J.; Bañobre-López, M. Magnetic Lipid Nanovehicles Synergize the Controlled Thermal Release of Chemotherapeutics with Magnetic Ablation While Enabling Non-Invasive Monitoring by MRI for Melanoma Theranostics. *Bioact. Mater.* **2022**, *8*, 153–164. [CrossRef]
116. Ribeiro, M.; Boudoukhani, M.; Belmonte-Reche, E.; Genicio, N.; Sillankorva, S.; Gallo, J.; Rodríguez-Abreu, C.; Moulai-Mostefa, N.; Bañobre-López, M. Xanthan-Fe_3O_4Nanoparticle Composite Hydrogels for Non-Invasive Magnetic Resonance Imaging and Magnetically Assisted Drug Delivery. *ACS Appl. Nano Mater.* **2021**, *4*, 7712–7729. [CrossRef]
117. Bañobre-Lopez, M.; Garcia-Hevia, L.; Cerqueira, F.; Rivadulla, F.; Gallo, J. Tunable Performance of Manganese Oxide Nanostructures as MRI Contrast Agents. *Chem. A Eur. J.* **2018**, *24*, 1295. [CrossRef]
118. Gallo, J.; Vasimalai, N.; Fernandez-Arguelles, M.T.; Bañobre-López, M. Green Synthesis of Multimodal "OFF-ON" Activatable MRI/Optical Probes. *Dalton Trans.* **2016**, *45*, 17672–17680. [CrossRef]
119. Gupta, A.K.; Gupta, M. Synthesis and Surface Engineering of Iron Oxide Nanoparticles for Biomedical Applications. *Biomaterials* **2005**, *26*, 3995–4021. [CrossRef] [PubMed]
120. Lu, A.H.; Salabas, E.L.; Schüth, F. Magnetic Nanoparticles: Synthesis, Protection, Functionalization, and Application. *Angew. Chem. Int. Ed.* **2007**, *46*, 1222–1244. [CrossRef] [PubMed]
121. Lu, M.; Cheng, X.; Jiang, J.; Li, T.; Zhang, Z.; Tsauo, C.; Liu, Y.; Wang, Z. Dual-Modal Photoacoustic and Magnetic Resonance Tracking of Tendon Stem Cells with PLGA/Iron Oxide Microparticles in Vitro. *PLoS ONE* **2018**, *13*, e0193362. [CrossRef] [PubMed]
122. Tang, K.S.; Hashmi, S.M.; Shapiro, E.M. The Effect of Cryoprotection on the Use of PLGA Encapsulated Iron Oxide Nanoparticles for Magnetic Cell Labeling. *Nanotechnology* **2013**, *24*, 125101. [CrossRef] [PubMed]
123. Nkansah, M.K.; Thakral, D.; Shapiro, E.M. Magnetic Poly(Lactide-Co-Glycolide) (PLGA) and Cellulose Particles for MRI-Based Cell Tracking. *Magn. Reson. Med.* **2011**, *65*, 1776–1785. [CrossRef]
124. Sun, Y.; Zheng, Y.; Ran, H.; Zhou, Y.; Shen, H.; Chen, Y.; Chen, H.; Krupka, T.M.; Li, A.; Li, P.; et al. Superparamagnetic PLGA-Iron Oxide Microcapsules for Dual-Modality US/MR Imaging and High Intensity Focused US Breast Cancer Ablation. *Biomaterials* **2012**, *33*, 5854–5864. [CrossRef]
125. Ruggiero, M.R.; Crich, S.G.; Sieni, E.; Sgarbossa, P.; Forzan, M.; Cavallari, E.; Stefania, R.; Dughiero, F.; Aime, S. Magnetic Hyperthermia Efficiency and 1H-NMR Relaxation Properties of Iron Oxide/Paclitaxel-Loaded PLGA Nanoparticles. *Nanotechnology* **2016**, *27*, 285104. [CrossRef]
126. Schleich, N.; Sibret, P.; Danhier, P.; Ucakar, B.; Laurent, S.; Muller, R.N.; Jérôme, C.; Gallez, B.; Préat, V.; Danhier, F. Dual Anticancer Drug/Superparamagnetic Iron Oxide-Loaded PLGA-Based Nanoparticles for Cancer Therapy and Magnetic Resonance Imaging. *Int. J. Pharm.* **2013**, *447*, 94–101. [CrossRef]
127. Lee, D.J.; Park, G.Y.; Oh, K.T.; Oh, N.M.; Kwag, D.S.; Youn, Y.S.; Oh, Y.T.; Park, J.W.; Lee, E.S. Multifunctional poly(lactide-*co*-glycolide) Nanoparticles for Luminescence/Magnetic Resonance Imaging and Photodynamic Therapy. *Int. J. Pharm.* **2012**, *434*, 257–263. [CrossRef]
128. Saengruengrit, C.; Ritprajak, P.; Wanichwecharungruang, S.; Sharma, A.; Salvan, G.; Zahn, D.R.T.; Insin, N. The Combined Magnetic Field and Iron Oxide-PLGA Composite Particles: Effective Protein Antigen Delivery and Immune Stimulation in Dendritic Cells. *J. Colloid Interface Sci.* **2018**, *520*, 101–111. [CrossRef]
129. Fang, K.; Song, L.; Gu, Z.; Yang, F.; Zhang, Y.; Gu, N. Magnetic Field Activated Drug Release System Based on Magnetic PLGA Microspheres for Chemo-Thermal Therapy. *Colloids Surf. B Biointerfaces* **2015**, *136*, 712–720. [CrossRef]
130. Danhier, F.; Vroman, B.; Lecouturier, N.; Crokart, N.; Pourcelle, V.; Freichels, H.; Jérôme, C.; Marchand-Brynaert, J.; Feron, O.; Préat, V. Targeting of Tumor Endothelium by RGD-Grafted PLGA-Nanoparticles Loaded with Paclitaxel. *J. Control. Release* **2009**, *140*, 166–173. [CrossRef]
131. Massart, R. Preparation of Aqueous Magnetic Liquids in Alkaline and Acidic Media. *IEEE Trans. Magn.* **1981**, *17*, 1247–1248. [CrossRef]
132. Pourcelle, V.; Devouge, S.; Garinot, M.; Préat, V.; Marchand-Brynaert, J. PCL-PEG-Based Nanoparticles Grafted with GRGDS Peptide: Preparation and Surface Analysis by XPS. *Biomacromolecules* **2007**, *8*, 3977–3983. [CrossRef] [PubMed]
133. Garinot, M.; Fiévez, V.; Pourcelle, V.; Stoffelbach, F.; des Rieux, A.; Plapied, L.; Theate, I.; Freichels, H.; Jérôme, C.; Marchand-Brynaert, J.; et al. PEGylated PLGA-Based Nanoparticles Targeting M Cells for Oral Vaccination. *J. Control. Release* **2007**, *120*, 195–204. [CrossRef] [PubMed]
134. Ghasemi, E.; Mirhabibi, A.; Edrissi, M. Synthesis and Rheological Properties of an Iron Oxide Ferrofluid. *J. Magn. Magn. Mater.* **2008**, *320*, 2635–2639. [CrossRef]

135. Laurent, S.; Forge, D.; Port, M.; Roch, A.; Robic, C.; Vander Elst, L.; Muller, R.N. Magnetic Iron Oxide Nanoparticles: Synthesis, Stabilization, Vectorization, Physicochemical Characterizations, and Biological Applications. *Chem. Rev.* **2008**, *108*, 2064–2110. [CrossRef]
136. Park, J.; An, K.; Hwang, Y.; Park, J.E.G.; Noh, H.J.; Kim, J.Y.; Park, J.H.; Hwang, N.M.; Hyeon, T. Ultra-Large-Scale Syntheses of Monodisperse Nanocrystals. *Nat. Mater.* **2004**, *3*, 891–895. [CrossRef]
137. Park, J.; Joo, J.; Soon, G.K.; Jang, Y.; Hyeon, T. Synthesis of Monodisperse Spherical Nanocrystals. *Angew. Chem. Int. Ed.* **2007**, *46*, 4630–4660. [CrossRef]
138. Neises, B.; Steglich, W. Simple Method for the Esterification of Carboxylic Acids. *Angew. Chem. Int. Ed. Engl.* **1978**, *17*, 522–524. [CrossRef]
139. Park, S.Y.; Baik, H.J.; Oh, Y.T.; Oh, K.T.; Youn, Y.S.; Lee, E.S. A Smart Polysaccharide/Drug Conjugate for Photodynamic Therapy. *Angew. Chem. Int. Ed.* **2011**, *50*, 1644–1647. [CrossRef] [PubMed]
140. Oh, N.M.; Kwag, D.S.; Oh, K.T.; Youn, Y.S.; Lee, E.S. Electrostatic Charge Conversion Processes in Engineered Tumor-Identifying Polypeptides for Targeted Chemotherapy. *Biomaterials* **2012**, *33*, 1884–1893. [CrossRef] [PubMed]
141. Zhang, S.; Bian, Z.; Gu, C.; Zhang, Y.; He, S.; Gu, N.; Zhang, J. Preparation of Anti-Human Cardiac Troponin I Immunomagnetic Nanoparticles and Biological Activity Assays. *Colloids Surf. B Biointerfaces* **2007**, *55*, 143–148. [CrossRef] [PubMed]
142. Gonzalez-Moragas, L.; Yu, S.M.; Murillo-Cremaes, N.; Laromaine, A.; Roig, A. Scale-up Synthesis of Iron Oxide Nanoparticles by Microwave-Assisted Thermal Decomposition. *Chem. Eng. J.* **2015**, *281*, 87–95. [CrossRef]
143. Huang, J.; Xie, J.; Chen, K.; Bu, L.; Lee, S.; Cheng, Z.; Li, X.; Chen, X. HSA Coated MnO Nanoparticles with Prominent MRI Contrast for Tumor Imaging. *Chem. Commun.* **2010**, *46*, 6684–6686. [CrossRef]
144. Na, H.B.; Lee, J.H.; An, K.; Park, Y.I.; Park, M.; Lee, I.S.; Nam, D.H.; Kim, S.T.; Kim, S.H.; Kim, S.W.; et al. Development of a T_1 Contrast Agent for Magnetic Resonance Imaging Using MnO Nanoparticles. *Angew. Chem. Int. Ed.* **2007**, 5397–5401. [CrossRef]
145. Gallo, J.; Alam, I.S.; Lavdas, I.; Wylezinska-Arridge, M.; Aboagye, E.O.; Long, N.J. RGD-Targeted MnO Nanoparticles as T_1 Contrast Agents for Cancer Imaging—The Effect of PEG Length in Vivo. *J. Mater. Chem. B* **2014**, *2*, 868–876. [CrossRef]
146. Bennewitz, M.F.; Lobo, T.L.; Nkansah, M.K.; Ulas, G.; Brudvig, G.W.; Shapiro, E.M. Biocompatible and PH-Sensitive PLGA Encapsulated MnO Nanocrystals for Molecular and Cellular MRI. *ACS Nano* **2011**, *5*, 3438–3446. [CrossRef]
147. Zhang, Y.; Ram, M.K.; Stefanakos, E.K.; Goswami, D.Y. Synthesis, characterization, and applications of ZnO nanowires. *J. Nanomater.* **2012**, *2012*, 624520. [CrossRef]
148. Zhang, H.; Chen, B.; Jiang, H.; Wang, C.; Wang, H.; Wang, X. A Strategy for ZnO Nanorod Mediated Multi-Mode Cancer Treatment. *Biomaterials* **2011**, *32*, 1906–1914. [CrossRef]
149. Tang, E.; Dong, S. Preparation of Styrene Polymer/ZnO Nanocomposite Latex via Miniemulsion Polymerization and Its Antibacterial Property. *Colloid Polym. Sci.* **2009**, *287*, 1025–1032. [CrossRef]
150. Karakoti, A.S.; Monteiro-Riviere, N.A.; Aggarwal, R.; Davis, J.P.; Narayan, R.J.; Seif, W.T.; McGinnis, J.; Seal, S. Nanoceria as Antioxidant: Synthesis and Biomedical Applications. *JOM* **2008**, *60*, 33–37. [CrossRef] [PubMed]
151. Shcherbakov, A.B.; Zholobak, N.M.; Ivanov, V.K. Biological, Biomedical and Pharmaceutical Applications of Cerium Oxide. In *Cerium Oxide (CeO_2): Synthesis, Properties and Applications*; Scirè, S., Palmisano, L., Eds.; Elsevier: Amsterdam, The Netherlands, 2020; pp. 279–358.
152. Kis, A.; Zettl, A. Nanomechanics of Carbon Nanotubes. *Philos. Trans. R. Soc. A Math. Phys. Eng. Sci.* **2008**, *366*, 1591–1611. [CrossRef] [PubMed]
153. Ahamed, M.; Alhadlaq, H.A.; Khan, M.A.M.; Karuppiah, P.; Al-Dhabi, N.A. Synthesis, Characterization, and Antimicrobial Activity of Copper Oxide Nanoparticles. *J. Nanomater.* **2014**, *2014*, 637858. [CrossRef]
154. Ren, G.; Hu, D.; Cheng, E.W.C.; Vargas-Reus, M.A.; Reip, P.; Allaker, R.P. Characterisation of Copper Oxide Nanoparticles for Antimicrobial Applications. *Int. J. Antimicrob. Agents* **2009**, *33*, 587–590. [CrossRef]
155. Haider, A.; Kwak, S.; Gupta, K.C.; Kang, I.K. Antibacterial Activity and Cytocompatibility of PLGA/CuO Hybrid Nanofiber Scaffolds Prepared by Electrospinning. *J. Nanomater.* **2015**, *2015*, 832762. [CrossRef]
156. Sengottuvelan, A.; Balasubramanian, P.; Will, J.; Boccaccini, A.R. Bioactivation of Titanium Dioxide Scaffolds by ALP-Functionalization. *Bioact. Mater.* **2017**, *2*, 108–115. [CrossRef]
157. Savaiano, J.K.; Webster, T.J. Altered Responses of Chondrocytes to Nanophase PLGA/Nanophase Titania Composites. *Biomaterials* **2004**, *25*, 1205–1213. [CrossRef]
158. Dhand, C.; Dwivedi, N.; Loh, X.J.; Ying, A.N.J.; Verma, N.K.; Beuerman, R.W.; Lakshminarayanan, R.; Ramakrishna, S. Methods and Strategies for the Synthesis of Diverse Nanoparticles and Their Applications: A Comprehensive Overview. *RSC Adv.* **2015**, *5*, 105003–105037. [CrossRef]
159. Jokinen, M.; Pätsi, M.; Rahiala, H.; Peltola, T.; Ritala, M.; Rosenholm, J.B. Influence of Sol and Surface Properties on in Vitro Bioactivity of Sol-gel-derived TiO_2 and TiO_2-SiO_2 Films Deposited by Dip-coating Method. *J. Biomed. Mater. Res.* **1998**, *42*, 295–302. [CrossRef]
160. Long, M.; Rack, H.J. Titanium Alloys in Total Joint Replacement—A Materials Science Perspective. *Biomaterials* **1998**, *19*, 1621–1639. [CrossRef]
161. López-Huerta, F.; Cervantes, B.; González, O.; Hernández-Torres, J.; García-González, L.; Vega, R.; Herrera-May, A.L.; Soto, E. Biocompatibility and Surface Properties of TiO_2 Thin Films Deposited by DC Magnetron Sputtering. *Materials* **2014**, *7*, 4105–4117. [CrossRef] [PubMed]

162. Luo, H.; Zhang, Y.; Gan, D.; Yang, Z.; Ao, H.; Zhang, Q.; Yao, F.; Wan, Y. Incorporation of Hydroxyapatite into Nanofibrous PLGA Scaffold towards Improved Breast Cancer Cell Behavior. *Mater. Chem. Phys.* **2019**, *226*, 177–183. [CrossRef]
163. Sokolova, V.; Kostka, K.; Shalumon, K.T.; Prymak, O.; Chen, J.P.; Epple, M. Synthesis and Characterization of PLGA/HAP Scaffolds with DNA-Functionalised Calcium Phosphate Nanoparticles for Bone Tissue Engineering. *J. Mater. Sci. Mater. Med.* **2020**, *31*, 1–12. [CrossRef]
164. Costa, D.O.; Prowse, P.D.H.; Chrones, T.; Sims, S.M.; Hamilton, D.W.; Rizkalla, A.S.; Dixon, S.J. The Differential Regulation of Osteoblast and Osteoclast Activity Bysurface Topography of Hydroxyapatite Coatings. *Biomaterials* **2013**, *34*, 7215–7226. [CrossRef]
165. Stanković, A.; Sezen, M.; Milenković, M.; Kaišarević, S.; Andrić, N.; Stevanović, M. PLGA/Nano-ZnO Composite Particles for Use in Biomedical Applications: Preparation, Characterization, and Antimicrobial Activity. *J. Nanomater.* **2016**, *2016*, 9425289. [CrossRef]
166. Marković, S.; Rajić, V.; Stanković, A.; Veselinović, L.; Belošević-Čavor, J.; Batalović, K.; Abazović, N.; Škapin, S.D.; Uskoković, D. Effect of PEO Molecular Weight on Sunlight Induced Photocatalytic Activity of ZnO/PEO Composites. *Sol. Energy* **2016**, *127*, 124–135. [CrossRef]
167. Mehta, A.; Scammon, B.; Shrake, K.; Bredikhin, M.; Gil, D.; Shekunova, T.; Baranchikov, A.; Ivanov, V.; Reukov, V. Nanoceria: Metabolic Interactions and Delivery through PLGA-Encapsulation. *Mater. Sci. Eng. C* **2020**, *114*, 111003. [CrossRef]
168. Ivanova, O.S.; Shekunova, T.O.; Ivanov, V.K.; Shcherbakov, A.B.; Popov, A.L.; Davydova, G.A.; Selezneva, I.I.; Kopitsa, G.P.; Tret'yakov, Y.D. One-Stage Synthesis of Ceria Colloid Solutions for Biomedical Use. *Dokl. Chem.* **2011**, *437*, 103–106. [CrossRef]
169. Mikael, P.E.; Amini, A.R.; Basu, J.; Arellano-Jimenez, M.J.; Laurencin, C.T.; Sanders, M.M.; Carter, C.B.; Nukavarapu, S.P. Functionalized Carbon Nanotube Reinforced Scaffolds for Bone Regenerative Engineering: Fabrication, in Vitro and in Vivo Evaluation. *Biomed. Mater.* **2014**, *9*, 1–13. [CrossRef]
170. Pelaseyed, S.S.; Madaah Hosseini, H.R.; Samadikuchaksaraei, A. A Novel Pathway to Produce Biodegradable and Bioactive PLGA/TiO2 Nanocomposite Scaffolds for Tissue Engineering: Air–Liquid Foaming. *J. Biomed. Mater. Res.* **2020**, *108*, 1390–1407. [CrossRef] [PubMed]
171. Roohani-Esfahani, S.I.; Nouri-Khorasani, S.; Lu, Z.; Appleyard, R.; Zreiqat, H. The Influence Hydroxyapatite Nanoparticle Shape and Size on the Properties of Biphasic Calcium Phosphate Scaffolds Coated with Hydroxyapatite-PCL Composites. *Biomaterials* **2010**, *31*, 5498–5509. [CrossRef] [PubMed]
172. Selvaraju, S.; Ramalingam, S.; Rao, J.R. Inorganic Apatite Nanomaterial: Modified Surface Phenomena and Its Role in Developing Collagen Based Polymeric Bio-Composite (Coll-PLGA/HAp) for Biological Applications. *Colloids Surf. B Biointerfaces* **2018**, *172*, 734–742. [CrossRef]
173. Binnaz, A.; Yoruç, H.; Koca, Y. Double Step Stirring: A Novel Method for Precipitation of Nano-Sized Hydroxyapatite Powder. *Dig. J. Nanomater. Biostructures* **2009**, *4*, 73–81.
174. Li, S.Y.; Wang, M. Novel Core-Shell Structured Paclitaxel-Loaded PLGA@Ag-Au Nanoparticles. *Mater. Lett.* **2013**, *92*, 350–353. [CrossRef]
175. Wang, L.; Li, D.; Hao, Y.; Niu, M.; Hu, Y.; Zhao, H.; Chang, J.; Zhang, Z.; Zhang, Y. Gold Nanorod-Based Poly(Lactic-Co-Glycolic Acid) with Manganese Dioxide Core-Shell Structured Multifunctional Nanoplatform for Cancer Theranostic Applications. *Int. J. Nanomed.* **2017**, *12*, 3059–3075. [CrossRef] [PubMed]
176. Vieira Rocha, C.; Costa da Silva, M.; Bañobre-López, M.; Gallo, J. (Para)Magnetic Hybrid Nanocomposites for Dual MRI Detection and Treatment of Solid Tumours. *Chem. Commun.* **2020**, *56*, 8695–8698. [CrossRef]
177. Zhang, W.; Li, Y.; Zhang, H.; Zhou, X.; Zhong, X. Facile Synthesis of Highly Luminescent Mn-Doped ZnS Nanocrystals. *Inorg. Chem.* **2011**, *50*, 10432–10438. [CrossRef]
178. Nguyen, S.C.; Zhang, Q.; Manthiram, K.; Ye, X.; Lomont, J.P.; Harris, C.B.; Weller, H.; Alivisatos, A.P. Study of Heat Transfer Dynamics from Gold Nanorods to the Environment via Time-Resolved Infrared Spectroscopy. *ACS Nano* **2016**, *10*, 2144–2151. [CrossRef]
179. Ren, F.; Bhana, S.; Norman, D.D.; Johnson, J.; Xu, L.; Baker, D.L.; Parrill, A.L.; Huang, X. Gold Nanorods Carrying Paclitaxel for Photothermal-Chemotherapy of Cancer. *Bioconjugate Chem.* **2013**, *24*, 376–386. [CrossRef]
180. Yang, X.; He, D.; He, X.; Wang, K.; Zou, Z.; Li, X.; Shi, H.; Luo, J.; Yang, X. Glutathione-Mediated Degradation of Surface-Capped MnO_2 for Drug Release from Mesoporous Silica Nanoparticles to Cancer Cells. *Part. Part. Syst. Charact.* **2015**, *32*, 205–212. [CrossRef]
181. Takeuchi, T.; Jyonotsuka, T.; Kamemori, N.; Kawano, G.; Shimizu, H.; Ando, K.; Harada, E. Enteric-Formulated Lactoferrin Was More Effectively Transported into Blood Circulation from Gastrointestinal Tract in Adult Rats. *Exp. Physiol.* **2006**, *91*, 1033–1040. [CrossRef] [PubMed]
182. Mehta, R. The Potential for the Use of Cell Proliferation and Oncogene Expression as Intermediate Markers during Liver Carcinogenesis. *Cancer Lett.* **1995**, *93*, 85–102. [CrossRef]
183. Arvizo, R.R.; Bhattacharyya, S.; Kudgus, R.A.; Giri, K.; Bhattacharya, R.; Mukherjee, P. Intrinsic Therapeutic Applications of Noble Metal Nanoparticles: Past, Present and Future. *Chem. Soc. Rev.* **2012**, *41*, 2943–2970. [CrossRef]
184. Durán, N.; Durán, M.; de Jesus, M.B.; Seabra, A.B.; Fávaro, W.J.; Nakazato, G. Silver Nanoparticles: A New View on Mechanistic Aspects on Antimicrobial Activity. *Nanomed. Nanotechnol. Biol. Med.* **2016**, *12*, 789–799. [CrossRef]

185. Granot, D.; Nkansah, M.K.; Bennewitz, M.F.; Tang, K.S.; Markakis, E.A.; Shapiro, E.M. Clinically Viable Magnetic Poly(Lactide-Co-Glycolide) Particles for MRI-Based Cell Tracking. *Magn. Reson. Med.* **2014**, *71*, 1238–1250. [CrossRef]
186. Ferrone, M.L.; Raut, C.P. Modern Surgical Therapy: Limb Salvage and the Role of Amputation for Extremity Soft-Tissue Sarcomas. *Surg. Oncol. Clin. N. Am.* **2012**, *21*, 201–213. [CrossRef]
187. Martou, G.; Antonyshyn, O.M. Advances in Surgical Approaches to the Upper Facial Skeleton. *Curr. Opin. Otolaryngol. Head Neck Surg.* **2011**, *19*, 242–247. [CrossRef]

International Journal of
Molecular Sciences

MDPI

Article

Mucoadhesive Electrospun Nanofiber-Based Hybrid System with Controlled and Unidirectional Release of Desmopressin

Mai Bay Stie [1,2], Johan Ring Gätke [1,2], Ioannis S. Chronakis [3], Jette Jacobsen [1] and Hanne Mørck Nielsen [1,2,*]

1 Department of Pharmacy, University of Copenhagen, Universitetsparken 2, 2100 Copenhagen, Denmark; mai.bay.stie@sund.ku.dk (M.B.S.); kjn376@alumni.ku.dk (J.R.G.); jette.jacobsen@sund.ku.dk (J.J.)
2 Center for Biopharmaceuticals and Biobarriers in Drug Delivery, Department of Pharmacy, University of Copenhagen, Universitetsparken 2, 2100 Copenhagen, Denmark
3 DTU-Food, Technical University of Denmark, Kemitorvet, B202, 2800 Kongens Lyngby, Denmark; ioach@food.dtu.dk
* Correspondence: hanne.morck@sund.ku.dk

Abstract: The sublingual mucosa is an attractive route for drug delivery, although challenged by a continuous flow of saliva that leads to a loss of drug by swallowing. It is of great benefit that drugs absorbed across the sublingual mucosa avoid exposure to the harsh environment of the gastro-intestinal lumen; this is especially beneficial for drugs of low physicochemical stability such as therapeutic peptides. In this study, a two-layered hybrid drug delivery system was developed for the sublingual delivery of the therapeutic peptide desmopressin. It consisted of peptide-loaded mucoadhesive electrospun chitosan/polyethylene oxide-based nanofibers (mean diameter of 183 ± 20 nm) and a saliva-repelling backing film to promote unidirectional release towards the mucosa. Desmopressin was released from the nanofiber-based hybrid system (approximately 80% of the loaded peptide was released within 45 min) in a unidirectional manner in vitro. Importantly, the nanofiber–film hybrid system protected the peptide from wash-out, as demonstrated in an ex vivo flow retention model with porcine sublingual mucosal tissue. Approximately 90% of the loaded desmopressin was retained at the surface of the ex vivo porcine sublingual mucosa after 15 min of exposure to flow rates representing salivary flow.

Keywords: sublingual delivery; biopharmaceuticals; peptide drug delivery; electrospinning; mucoadhesion; ex vivo flow retention model

Citation: Stie, M.B.; Gätke, J.R.; Chronakis, I.S.; Jacobsen, J.; Nielsen, H.M. Mucoadhesive Electrospun Nanofiber-Based Hybrid System with Controlled and Unidirectional Release of Desmopressin. *Int. J. Mol. Sci.* **2022**, 23, 1458. https://doi.org/10.3390/ijms23031458

Academic Editor: Raghvendra Singh Yadav

Received: 20 December 2021
Accepted: 23 January 2022
Published: 27 January 2022

Publisher's Note: MDPI stays neutral with regard to jurisdictional claims in published maps and institutional affiliations.

Copyright: © 2022 by the authors. Licensee MDPI, Basel, Switzerland. This article is an open access article distributed under the terms and conditions of the Creative Commons Attribution (CC BY) license (https://creativecommons.org/licenses/by/4.0/).

1. Introduction

Biopharmaceuticals—e.g., peptides and proteins—are the fastest growing group of drugs [1]. Peptides and proteins are highly potent and specific, but suffer from a low physicochemical stability; poor absorption across biological membranes; and, as a consequence, low bioavailability [2]. The primary route of administration of peptides and proteins is therefore via injections, which is inconvenient and often associated with poor patient compliance; new approaches for non-injectable formulations of peptides and proteins are therefore needed. The sublingual route is an attractive, non-invasive alternative site of administration as the mucosa is non-keratinized, highly vascularized, and consists of only 8–12 cell layers and thus has a reasonably low thickness (100–200 μm) compared to the much thicker buccal mucosa [3]. Furthermore, drugs delivered via the oral cavity avoid the harsh environment of the gastro-intestinal tract and bypass the hepatic first-pass metabolism. These factors are beneficial for drugs, which are prone to enzymatic degradation and sensitive to the low pH of the stomach [4]. Formulations for sublingual administration can easily be handled by the patient as the ventral side of the tongue is readily accessible and the quick removal of the formulation is possible if needed. Further, such formulations can be administered without water, which is beneficial for patient groups with swallowing difficulties—e.g., the elderly or small children [5].

Sublingual administration of peptides and proteins is challenged by a continuous flow of saliva (total volume of saliva of 0.5–2 L/day) that may lead to the swallowing of the drug or drug delivery system and the subsequent degradation of the drug in the gastro-intestinal tract [6]. Even if not swallowed, the dilution of drug in the saliva impairs absorption by decreasing the amount of drug available at the site of absorption and thus also the drug concentration gradient across the mucosal epithelium, the most significant biological barrier to the absorption of the drug. Through a rational design, mucoadhesive drug delivery systems can protect the drug by encapsulation, retain the drug at the site of application for a prolonged period by controlled drug release, and thus not only increase, but also maintain the drug concentration gradient across the tissue, leading to the improved absorption of drug and therefore higher systemic bioavailability.

Electrospun nanofibers are especially suitable for topical administration on, e.g., the sublingual mucosa because of their high surface area to volume ratio and tunable mechanical properties, which can make them flexible enough to bend to the curved mucosal surfaces in the mouth. We have previously demonstrated a robust protocol for the electrospinning of biocompatible chitosan/polyethylene oxide (PEO) nanofibers [7] and demonstrated their suitability for sublingual administration because of their mucoadhesive properties [8]. The use of electrospun nanofibers has been explored for the oromucosal delivery of peptides and proteins, including insulin [9–11], FITC-labelled albumin [12], and lysozyme [13]. Reports aiming to achieve peptide delivery by the sublingual route are rare [9].

The therapeutic peptide desmopressin is used for the treatment of, amongst others, nocturnal enuresis, diabetes insipidus and hemophilia A. This peptide is currently administered either as a solution by intravenous or subcutaneous injection, as a nasal spray, or as a freeze-dried sublingual tablet. The bioavailability of desmopressin after sublingual administration of a freeze-dried fast-dissolving tablet is, however, reported to be only ~0.25% [14], leaving room for improvement. Desmopressin is a relatively small peptide; it is modified from the structure of vasopressin (an endogenous hormone with a half-life in plasma of 10–35 min) to display a higher stability (half-life in plasma ~160 min) and 10 times higher antidiuretic potency and 1500 times lower vasoconstricting potency compared to its natural analogue [15].

In this work, a combination of mucoadhesive electrospun chitosan/PEO nanofibers that facilitate the controlled release of desmopressin and a local high concentration at the site of application, and a water-repelling backing film that ensures unidirectional release and prevents the wash-out of peptide by saliva, were assessed for the sublingual delivery of the therapeutic peptide desmopressin. The protection of desmopressin from wash-out by saliva subsequent to the sublingual administration of the nanofiber–film hybrid system was also evaluated ex vivo.

2. Results and Discussion

2.1. Clinically Relevant Doses of Desmopressin Loaded in Nanofiber–Film Hybrid System

A two-layered hybrid drug delivery system, which consisted of a mucoadhesive electrospun chitosan-based nanofiber mat and a saliva-repelling backing film (Figure 1a), was developed. It is important to note that the peptide delivery system is considered biocompatible, as all selected excipients are biocompatible, and the electrospun nanofibers were prepared using a minimum amount of acetic acid (0.7% (w/w)). The therapeutic peptide desmopressin was encapsulated within the nanofibers by co-electrospinning. Then, a saliva-repelling backing layer was sprayed onto the fiber mat as a thin film based on ethyl-cellulose. The food coloring iron oxide, a red pigment insoluble in water, was also included in the film to provide a visual discrepancy in the orientation of the nanofiber–film hybrid system (Figure 1b). As visualized by scanning electron microscopy (SEM), close connection between the fibers and the thin coherent film was achieved (Figure 1c). The peptide-loaded nanofibers were smooth, uniform, and without artifacts such as beading (Figure 1d), with an average diameter of 183 $\pm$ 20 nm (mean $\pm$ standard deviation (SD)) and with a narrow size distribution. The average diameter of the nanofibers is similar

(168 ± 38 nm) to that of electrospun chitosan/PEO nanofibers prepared under the same conditions without the peptide [8].

The loading of desmopressin in the nanofibers was 8% (*w/w*). The weight of a 10 mm nanofiber disc (Figure 1b) was in the range of 1.5–3 mg; hence, the fiber discs provide a theoretical loading of desmopressin of 120–240 µg per disc, which is equivalent to the dose of desmopressin in the freeze-dried, sublingual tablets already on the market. Marketed freeze-dried tablets with desmopressin for sublingual administration (Minirin® by Ferring Pharmaceuticals) contain 60, 120, or 240 µg of desmopressin per dose. Thus, clinically relevant doses of desmopressin per nanofiber patch were achieved.

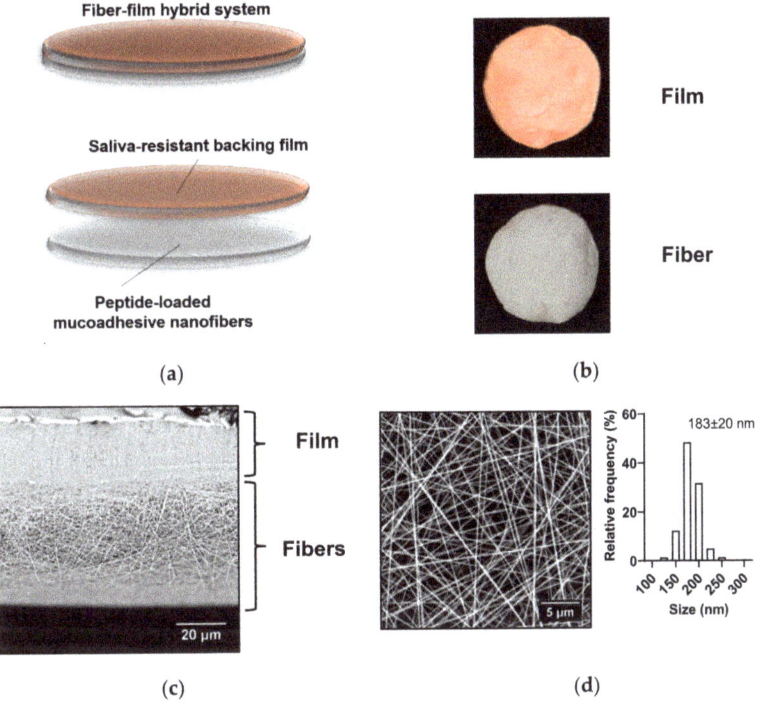

Figure 1. Properties of the desmopressin-loaded nanofiber–film hybrid system. (**a**) Schematic representation of the nanofiber–film hybrid system. (**b**) Image of a 10 mm disc of the two-sided nanofiber–film hybrid system. (**c**) Cross-section of peptide-loaded nanofiber–film hybrid system visualized by SEM. (**d**) Desmopressin-loaded chitosan/PEO nanofibers visualized by SEM. The size distribution of the nanofibers is given. The diameter is presented as mean ± SD. $N = 3$, $n = 100$, where N is the number of individual batches of nanofibers produced, and n is the number of nanofibers measured per batch.

2.2. Nanofiber–Film Hybrid System Ensures Controlled and Unidirectional Release of Desmopressin

A thin water-repelling backing film was applied to the mucoadhesive nanofiber layer to facilitate the unidirectional release of desmopressin, and, furthermore, to protect the drug delivery system and released therapeutic peptide from wash-out by saliva. The unidirectional and controlled release of desmopressin from the nanofibers and the barrier function of the water-repelling backing film were evaluated by determining the release of desmopressin into both sides of a diffusion cell (Ussing chamber) separated by the nanofiber–film hybrid system (Figure 2a). In the absence of the backing-membrane used as a control, approximately 50% of the loaded peptide was released from the nanofibers in the left and right chambers—i.e., it did not show a unidirectional release. We have previously demonstrated that electrospun chitosan/PEO nanofibers display extraordinary swelling properties and can swell >1000% (*w/w*) within 15 min of exposure to water [1]. Thus, a fast release of the highly water-soluble peptide desmopressin was therefore expected. Accordingly, approximately 80% of the encapsulated desmopressin was released fast (within 20 min) from the nanofibers, and the complete release of desmopressin, represented by the amount detected in both chambers, was observed after approximately 60 min (Figure 2b). In the presence of a water-repelling backing film, the unidirectional release of desmopressin from the nanofiber–film hybrid system was achieved, as <1% of the encapsulated desmopressin was detected in the left chamber fronting the backing membrane and approximately 80% of the loaded desmopressin was released within 45 min to the right receiver chamber fronting the nanofiber layer (Figure 2b). In total, 100% of desmopressin was released from the fiber–film hybrid system within 1 h (Figure 2b). This confirms that the nanofiber–film hybrid system indeed ensures the unidirectional and complete release of the encapsulated therapeutic peptide. The average cumulative release of desmopressin exceeded 100% after ≥90 min, but the cumulative release was not statistically significant different from 100% at any time point. A slight evaporation of the release medium over time at 37 °C can cause the average cumulative release of desmopressin to exceed >100% for some samples. The release of peptide was determined from 2–3 areas of the same electrospun nanofiber mat, and three individual nanofiber mats were evaluated. In general, no significant difference in peptide release was found between the individual areas nor mats, which is indicative of a homogenous distribution of peptide in the electrospun nanofibers.

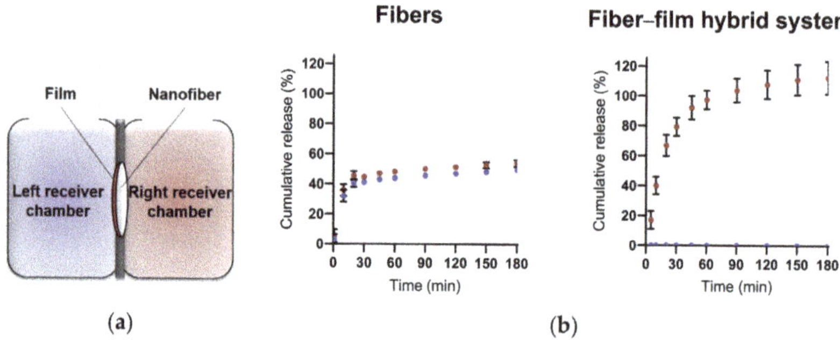

Figure 2. Nanofiber–film hybrid system facilitates unidirectional release of desmopressin. (**a**) Experimental setup used for studying the release of desmopressin from the nanofibers. The release of desmopressin into both chambers (in the Ussing chamber setup) separated either by the nanofibers or by the nanofiber–film hybrid system was determined. (**b**) Release of desmopressin from nanofibers or the nanofiber–film hybrid system in the left (blue) and right (red) Ussing chamber, respectively, as depicted in (**a**). The cumulative release was based on a loading of 8% (*w/w*) desmopressin in the chitosan/PEO nanofibers. The data are presented as mean ± SD. $N = 3$, $n = 2–3$, where N represent the number of individually prepared nanofibers/hybrid systems and n is the number of measurements per system.

2.3. Nanofiber–Film Hybrid System Protects Peptide from Wash-Out

Sublingual drug delivery is challenged by a continuous flow of saliva that leads to the wash-out of the drug delivery system and/or drug, which are subsequently swallowed. We have previously demonstrated the good mucoadhesive properties of electrospun chitosan/PEO nanofibers for the sublingual mucosa [8]. We hypothesize that a combination of the mucoadhesive properties of the electrospun chitosan-based nanofibers and the protective function of the backing film will facilitate the close adhesion of the nanofiber–film hybrid system to the sublingual region, and furthermore, protect desmopressin from wash-out by saliva. The ability of the nanofiber–film hybrid system to prohibit the wash-out of desmopressin was investigated by a flow retention model using ex vivo porcine sublingual mucosa (Figure 3a). The flow rate of whole saliva in humans is reported to be around 0.3 mL/min without stimulation and up to 7 mL/min during stimulation by mastication, etc. [16]. The flow rate used in this experimental setup was 0.5 mL/min, and thus was within the average salivary flow rate in vivo. Isotonic buffer with 0.05% (*w*/*v*) bovine serum albumin (BSA) with a pH of 6.8, which is the average pH of saliva secreted without stimulation [17], was used as the medium. The mucosa from the ventral side of the porcine tongue was chosen as the surface of adhesion as it, like the human sublingual mucosa, is non-keratinized, it has rete ridges, and its epithelium thickness is similar to that found in man [18,19].

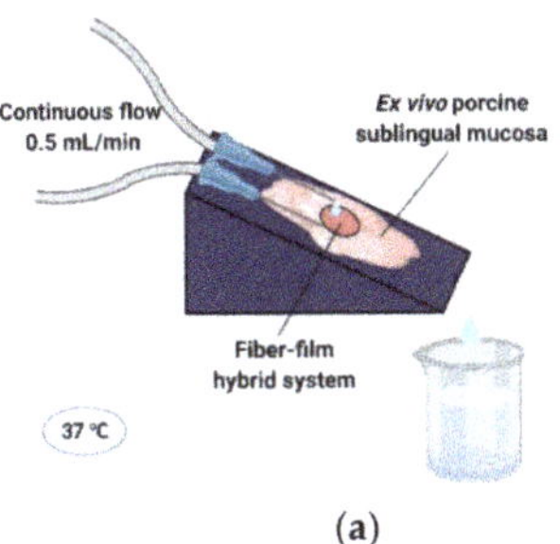

(a)

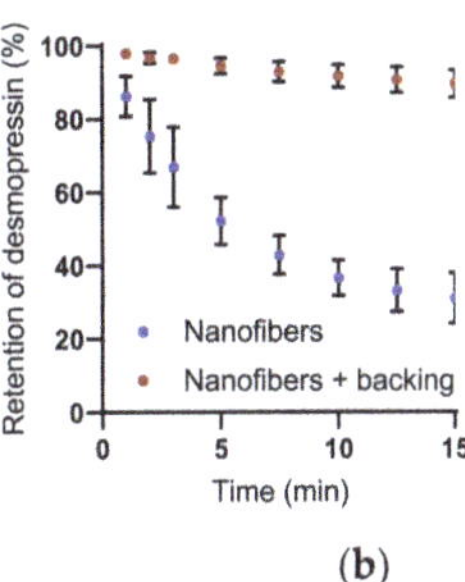

(b)

Figure 3. Nanofiber–film hybrid system protects desmopressin from wash-out by saliva. (**a**) Ex vivo porcine sublingual mucosa was mounted on a support with an angle of 16° and the retention of desmopressin on the tissue was evaluated over time. Created by Biorender.com. (**b**) Retention of desmopressin dosed in nanofibers with and without a saliva-repelling backing film on ex vivo porcine sublingual mucosa. *N* = 4, where N specifies the number of biological replicates.

MiniRin® desmopressin freeze-dried tablets dissolved immediately when exposed to moisture, and thus the tablet disintegrated within seconds and was quickly washed off the ex vivo porcine sublingual mucosa after initiating the flow. The content of desmopressin released from the commercial tablets was not quantified due to the presence of gelatin in the tablets, which interfered with the analysis method. In contrast, the nanofibers, irrespective of the presence of a protective backing film, adhered strongly to the ex vivo porcine sublingual mucosa throughout the duration of the experiment. Desmopressin was quickly released and washed out of the nanofibers without a protective backing-film, al-though it was lost in significantly less time than desmopressin released from MiniRin® freeze-dried tablets. Hence, the mucoadhesive nanofibers alone retained desmopressin significantly longer on the tissue in comparison to the marketed tablet. In accordance with the hypothesis, the nanofiber–film hybrid system further protected desmopressin from wash-out and retained approximately 90% of the loaded desmopressin on the tissue after 15 min of exposure to flow.

Mucoadhesive drug delivery systems, ensuring unidirectional release, thus not only limit loss of drug by wash-out, but also provide an advantageous increase in the residence time of the drug on the mucosal tissue, the site of absorption. Furthermore, a local environment with a high drug concentration can be achieved between the adhesive drug delivery system and the sublingual mucosa. This increases the concentration gradient of the drug across the mucosal barrier and can lead to improved absorption. Furthermore, the local treatment of, e.g., lesions in the oral mucosa is possible if wash-out is prevented. Site-specific drug delivery can limit exposure to other sites in the oral cavity in general, which is beneficial for drugs with an unpleasant taste and mouthfeel or teeth staining. Some drugs may indeed induce side-effects upon swallowing, which can be avoided if they are efficiently absorbed from the oral mucosa. Thus, electrospinning is an interesting approach to produce solid dosage forms for mucosal application, as this method benefits from short processing times, mild drying conditions and prospects for industrial scalabi-lity, continuous manufacturing, as well as reduced cost [20,21]. Additional benefits are that biopharmaceuticals, formulated as solid dosage forms, display improved drug stabi-lity and offer easier handling for patients or healthcare personnel, as well as a reduced cost of transportation [20].

3. Materials and Methods

3.1. Materials

Chitoceuticals chitosan 95/100 (degree of deacetylation 96%, Mw 100–250 kDa, chitosan-96) was purchased from Heppe Medical Chitosan (Halle, Germany). Polyethylene oxide (Mw 900 kDa, PEO), bovine serum albumin (BSA), acetic acid anhydride, Hank's balanced salt solution (HBSS), Dulbecco's phosphate buffered saline (PBS), glycerol (≥99%), tributyl citrate and ethyl cellulose, trifluoroacetic acid (TFA), and acetonitrile were obtained from Sigma Aldrich (St. Louis, MO, USA). N-2-hydroxyethylpiperazine-N′-2-ethanesulfonic acid (HEPES) was obtained from PanReac AppliChem (Darmstadt, Germany). Iron(III) oxide (Secovit® E172) was from BASF (Copenhagen, Denmark). Desmopressin as TFA salt (purity > 98%) was obtained from SynPeptide (Shanghai, China). MiniRin® freeze-dryed tablets-60 µg desmopressin (Ferring Pharmaceuticals, Copenhagen, Denmark) were purchased through the Association of Danish Pharmacies. Ultrapure water (18.2 MΩ × cm) purified by a PURELAB flex 4 (ELGA LabWater, High Wycombe, UK) was used.

3.2. Electrospinning of Chitosan/PEO Nanofibers with Desmopressin

The chitosan/PEO nanofibers were produced according to Stie et al., 2019 [7] (Figure 4a). Briefly, a 2% (*w*/*w*) clear solution of chitosan in 0.7% (*w*/*w*) acetic acid was prepared in ultrapure water and 4% (*w*/*w*) PEO was dissolved in ultrapure water. Both solutions were stirred for two days at room temperature (RT) to ensure the complete hydration of the polymers. To obtain a 1:1 (*w*/*w*) ratio of chitosan:PEO and a 8% (*w*/*w*) loading efficiency of the peptide, 1.8 g of 2% (*w*/*w*) chitosan and 0.9 g of 4% (*w*/*w*) PEO were mixed and stirred for at least 15 min, and hereafter, mixed with 8 mg desmopressin-TFA (corresponding to 6.6 mg desmopressin). The polymer-peptide solution was stirred for at least 30 min before electrospinning from a 1 mL syringe with a 20 G blunt needle (Photo-Advantage, Ancaster, ON, Canada) by an ES50P-10W high voltage source set to 20 kV. The temperature was controlled in the range 23–25 °C and a low humidity (<25% relative humidity) was maintained by a continuous flow of dry air. The nanofibers were collected on aluminum foil on a stainless steel collector plate placed 15 cm from the tip of the needle. The fibers were stored in a desiccator at 4 °C to avoid absorption of water and to preserve the stability of desmopressin. Further experiments were conducted within two weeks of the preparation of the electrospun nanofibers.

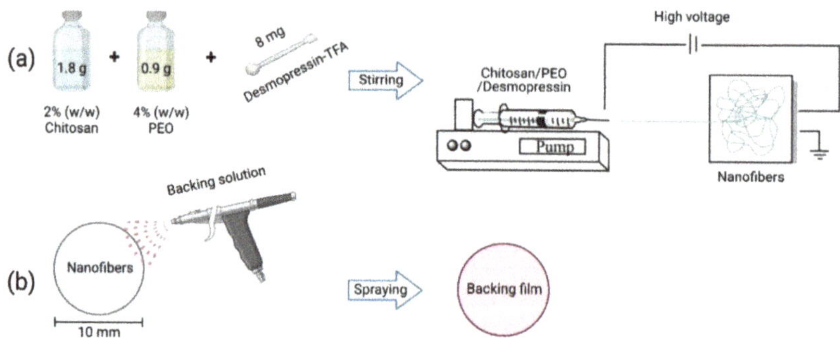

Figure 4. Preparation of the nanofiber–film hybrid system. (**a**) Electrospinning of chitosan/PEO na-nofibers with desmopressin as described in Section 3.2. (**b**) Spraying of film on nanofibers as described in Section 3.3. Created by Biorender.com.

3.3. Preparation of Nanofiber–Film Hybrid System

Ten-millimeter discs of peptide-loaded nanofibers were punched out from the main mat and weighed to determine the exact loading of desmopressin per disc. Amounts of 141 mg of acetyl tributyl citrate, 47 mg glycerin, and 750 mg ethyl cellulose were dissolved in 15 mL ethanol (absolute) and stirred overnight at room temperature. Iron oxide was added and the solution was stirred for at least 30 min. The film was sprayed directly on the 10 mm discs of nanofibers employing an airbrush (Model BD-134, Custom Colors, Jyderup, Denmark) (Figure 4b). The fibers were fixed on an aluminum plate with holes under vacuum to avoid movement of the fibers as a result of the air flow from the airbrush during the application of the film-forming material by spraying.

3.4. Visualization of Nanofiber Morphology by Scanning Electron Microscopy

The morphology of the nanofibers was visualized by scanning electron microscopy (SEM), as previously described by Stie et al., 2019 [7]. The fiber–film hybrid system was cut with a scalpel and mounted vertically on the carbon tape to achieve an image of the cross-section of the fiber–film hybrid system. The average diameter of the nanofibers was determined using the Image J version 1.51j8 software (National Institute of Health, Bethesda, MD, USA) by measuring 25 individual fibers from four areas of interest from a total of three individual batches of electrospun nanofibers loaded with desmopressin.

3.5. Release of Desmopressin from Fiber–Film Hybrid System in Ussing Chambers

Discs of nanofibers (diameter 10 mm, weight 2–3 mg) or the nanofiber-film hybrid system were fixed in Ussing sliders with a diffusion area of 0.4 cm^2 and placed in a side-by-side diffusion cells apparatus EM-CSY-8 Ussing chambers (Physiologic Instruments, San Diego, CA, USA), with 2 mL of 10 mM HEPES in HBSS pH 6.8 with 0.05% (*w*/*v*) BSA in each chamber, and incubated at 37 °C for 3 h. Samples of 100 µL were collected from each chamber at various time intervals and replenished with 100 µL warm 10 mM HEPES in HBSS pH 6.8 with 0.05% (*w*/*v*) BSA. The samples were centrifuged (10,000 rpm/9279× *g*, 10 min, 5 °C) and the concentration of desmopressin was determined by high-performance liquid chromatography with UV detection (HPLC-UV, Experimental Section 3.7). The cumulative release of desmopressin (M) was determined according to Equation (1).

$$M = V_S \cdot \left(\sum_{n=1}^{n} C_{n-1} \right) + C_n \cdot V_T \quad (1)$$

where V_S is the sample volume (100 µL), C_n is the concentration of desmopressin at time point *n*, and V_T is the total volume of the receiver Ussing chambers (2 mL). The cumulative peptide release in percent (%) was calculated based on the theoretical loading of desmopressin of 8% (*w*/*w*) and a diffusion area of 0.4 cm^2.

3.6. Ex Vivo Flow Retention Model

The ex vivo flow retention setup was inspired by Madsen et al. [17]. Sublingual porcine tissue from healthy pigs (approximately 30–60 kg, Danish Landrace/Yorkshire/Duroc) was collected immediately after euthanization and kept in PBS on ice until use on the same day as harvesting of the tissue. Thin sections (0.5–0.7 mm) of the ventral side of the tongue were obtained by means of an electric dermatome (Zimmer Biomet, Alberts-lund, Denmark). The sublingual mucosa was mounted on a rubber pad with pins at an angle of 16° and placed on a heating plate to achieve a temperature of approximately 37 °C of the tissue. The ex vivo sublingual mucosa was equilibrated for 10–15 min in warm (37 °C) 10 mM HEPES in HBSS pH 6.8 with 0.05% (*w*/*v*) BSA with a flow of 0.5 mL/min from two 13 G needles placed above the tissue and the flow was controlled by a syringe pump (11 Elite, Harvard Apparatus, Holliston, MA, USA). 10 mm discs of electrospun nanofibers, with or without backing, or a MiniRin® freeze-dried tablet (60 µg desmopressin) were placed on the mucosal tissue and flushed with warm (37 °C) 10 mM HEPES in HBSS with 0.05% (*w*/*v*)

BSA, pH 6.8 with a flow of 0.5 mL/min for 15 min controlled by a syringe pump (11 Elite, Harvard Apparatus, Holliston, MA, USA). Samples of 100 μL of the eluate were collected after 1, 2, 3, 5, 7.5, 10, 12.5 and 15 min and centrifuged (10,000 rpm/9279× *g*, 10 min, 5 °C). The concentration of desmopressin was determined by HPLC-UV (Experimental Section 3.7). As gelatin from the MiniRin® tablets interfered with the HPLC-UV method, the concentration of desmopressin released from the freeze-dried tablets was not determined. The retention of desmopressin was determined according to Equation (2).

$$\text{Peptide wash out (mg)} = V_S \cdot \sum_{n=1}^{n} C_{n-1} + C_n \cdot \left(V_T - \sum_{n=1}^{n} V_{n-1} \right) \quad (2)$$

where V_S is the eluate sample volume (100 μL), C_n is the concentration of desmopressin at time point n, V_T is the total volume of eluate at time point n based on a flow of 0.5 mL/min, and $\sum_{n=1}^{n} V_{n-1}$ is the sum of the eluate volumes sampled at time point $n-1$. The retention of desmopressin in percent (%) was determined based on a theoretical loading of 8% (100% peptide retention) (*w/w*).

3.7. Quantification of Desmopressin by HPLC-UV

Desmopressin was quantified by HPLC-UV (λ218 nm) on a Shimadzu Prominence system (Kyoto, Japan) with an Aeris peptide XB-C18 column (100 × 2.1 mm, 3.6 μm, Phenomenex, Torrance, CA, USA). Desmopressin was eluted from 10 μL samples with a gradient 0→40% eluent B in eluent A over 8 min at 0.8 mL/min at 40 °C, where eluent A consisted of 95:5:0.1% (*v*/*v*) acetonitrile:water:TFA and eluent B of 5:95:0.1% (*v*/*v*) acetonitrile:water:TFA. The samples were stored at 4 °C during analysis. The limit of detection and limit of quantification were 0.09 μg/mL and 0.26 μg/mL, respectively.

4. Conclusions

The therapeutic peptide desmopressin was encapsulated within mucoadhesive electrospun chitosan/PEO nanofibers intended for sublingual delivery. A two-layered hybrid drug delivery system was developed by combining a saliva-repelling backing film with the nanofibers to ensure unidirectional drug release. The nanofibers displayed a unidirectional, controlled, and fast release of desmopressin with an approximately 80% release of the loaded peptide from the nanofiber-based hybrid system within 45 min. Importantly, the nanofiber–film hybrid system showed resilience to saliva flow and retained approximately 90% of the desmopressin loaded on the tissue after 15 min of exposure to flow. This can potentially improve the absorption of peptide, but also potentially improve the absorption of small molecular drugs across the mucosae. Mucoadhesive electrospun na-nofibers are considered promising carriers for peptide delivery via mucosal routes upon, e.g., sublingual administration.

Author Contributions: Conceptualization, M.B.S., I.S.C., J.J. and H.M.N.; methodology, M.B.S., J.R.G., I.S.C., J.J. and H.M.N.; investigation, M.B.S. and J.R.G.; resources, I.S.C., J.J. and H.M.N.; writing—original draft preparation, M.B.S.; writing—review and editing, J.R.G., I.S.C., J.J. and H.M.N.; funding acquisition, I.S.C., J.J. and H.M.N. All authors have read and agreed to the published version of the manuscript.

Funding: This research was funded by The Danish Council for Independent Research, Technology and Production, grant number DFF-6111-00333; the Novo Nordisk Foundation Grand Challenge Program, NNF16OC0021948; and Innovation Fund Denmark PROBIO, project-7076-00053B.

Institutional Review Board Statement: Not applicable.

Informed Consent Statement: Not applicable.

Data Availability Statement: The data presented in this study are available on request from the corresponding author.

Acknowledgments: Department of Experimental Medicine and the Large Animal Teaching Hospital at University of Copenhagen are acknowledged for assisting with the porcine tissue used for the ex vivo studies. The authors acknowledge LEO Pharma (Ballerup, Denmark) for granting the use of the dermatome equipment.

Conflicts of Interest: The authors declare no conflict of interest.

References

1. Walsh, G. Biopharmaceutical benchmarks 2018. *Nat. Biotechnol.* **2018**, *36*, 1136–1145. [CrossRef] [PubMed]
2. Mitragotri, S.; Burke, P.A.; Langer, R. Overcoming the challenges in administering biopharmaceuticals: Formulation and delivery strategies. *Nat. Rev. Drug Discov.* **2014**, *13*, 655–672. [CrossRef] [PubMed]
3. Kraan, H.; Vrieling, H.; Czerkinsky, C.; Jiskoot, W.; Kersten, G.; Amorij, J.-P. Buccal and sublingual vaccine delivery. *J. Control. Release* **2014**, *190*, 580–592. [CrossRef] [PubMed]
4. Rathbone, M.J.; Drummond, B.K.; Tucker, I.G. The oral cavity as a site for systemic drug delivery. *Adv. Drug Deliv. Rev.* **1994**, *13*, 1–22. [CrossRef]
5. Schiele, J.T.; Quinzler, R.; Klimm, H.-D.; Pruszydlo, M.G.; Haefeli, W.E. Difficulties swallowing solid oral dosage forms in a general practice population: Prevalence, causes, and relationship to dosage forms. *Eur. J. Clin. Pharmacol.* **2013**, *69*, 937–948. [CrossRef]
6. Gandhi, R.B.; Robinson, J.R. Oral cavity as a site for bioadhesive drug delivery. *Adv. Drug Deliv. Rev.* **1994**, *13*, 43–74. [CrossRef]
7. Stie, M.B.; Jones, M.; Sørensen, H.O.; Jacobsen, J.; Chronakis, I.S.; Nielsen, H.M. Acids 'generally recognized as safe' affect morphology and biocompatibility of electrospun chitosan/polyethylene oxide nanofibers. *Carbohydr. Polym.* **2019**, *215*, 253–262. [CrossRef]
8. Stie, M.B.; Gätke, J.R.; Wan, F.; Chronakis, I.S.; Jacobsen, J.; Nielsen, H.M. Swelling of mucoadhesive electrospun chitosan/polyethylene oxide nanofibers facilitates adhesion to the sublingual mucosa. *Carbohydr. Polym.* **2020**, *242*, 116428. [CrossRef]
9. Sharma, A.; Gupta, A.; Rath, G.; Goyal, A.; Mathur, R.B.; Dhakate, S.R. Electrospun composite nanofiber-based transmucosal patch for anti-diabetic drug delivery. *J. Mater. Chem. B* **2013**, *1*, 3410–3418. [CrossRef]
10. Xu, L.; Sheybani, N.; Ren, S.; Bowlin, G.L.; Yeudall, W.A.; Yang, H. Semi-Interpenetrating Network (sIPN) Co-Electrospun Gelatin/Insulin Fiber Formulation for Transbuccal Insulin Delivery. *Pharm. Res.* **2014**, *32*, 275–285. [CrossRef]
11. Lancina, M.G.; Shankar, R.K.; Yang, H. Chitosan nanofibers for transbuccal insulin delivery. *J. Biomed. Mater. Res. Part A* **2017**, *105*, 1252–1259. [CrossRef] [PubMed]
12. Berka, P.; Stránská, D.; Semecký, V.; Berka, K.; Doležal, P. In vitro testing of flash-frozen sublingual membranes for storage and reproducible permeability studies of macromolecular drugs from solution or nanofiber mats. *Int. J. Pharm.* **2019**, *572*, 118711. [CrossRef] [PubMed]
13. Edmans, J.G.; Murdoch, C.; Santocildes-Romero, M.E.; Hatton, P.V.; Colley, H.E.; Spain, S.G. Incorporation of lysozyme into a mucoadhesive electrospun patch for rapid protein delivery to the oral mucosa. *Mater. Sci. Eng. C* **2020**, *112*, 110917. [CrossRef] [PubMed]
14. Van Kerrebroeck, P.; Nørgaard, J.P. Desmopressin for the treatment of primary nocturnal enuresis. *Ped. Health* **2009**, *3*, 311–327. [CrossRef]
15. Sharman, A.; Low, J. Vasopressin and its role in critical care. Contin. Educ. *Anaesth. Crit. Care Pain* **2008**, *8*, 134–137. [CrossRef]
16. Iorgulescu, G. Saliva between normal and pathological. Important factors in determining systemic and oral health. *J. Med. Life* **2009**, *2*, 303–3007.
17. Madsen, K.D.; Sander, C.; Baldursdottir, S.; Pedersen, A.M.L.; Jacobsen, J. Development of an ex vivo retention model simulating bioadhesion in the oral cavity using human saliva and physiologically relevant irrigation media. *Int. J. Pharm.* **2013**, *448*, 373–381. [CrossRef]
18. Thirion-Delalande, C.; Gervais, F.; Fisch, C.; Cuiné, J.; Baron-Bodo, V.; Moingeon, P.; Mascarell, L. Comparative analysis of the oral mucosae from rodents and non-rodents: Application to the nonclinical evaluation of sublingual immunotherapy products. *PLoS ONE* **2017**, *12*, e0183398. [CrossRef]
19. Kondo, M.; Yamato, M.; Takagi, R.; Murakami, D.; Namiki, H.; Okano, T. Significantly different proliferative potential of oral mucosal epithelial cells between six animal species. *J. Biomed. Mater. Res. Part A* **2014**, *102*, 1829–1837. [CrossRef]
20. Vass, P.; Démuth, B.; Hirsch, E.; Nagy, B.; Andersen, S.K.; Vigh, T.; Verreck, G.; Csontos, I.; Nagy, Z.K.; Marosi, G. Drying technology strategies for colon-targeted oral delivery of biopharmaceuticals. *J. Control. Release* **2019**, *296*, 162–178. [CrossRef]
21. Vass, P.; Démuth, B.; Farkas, A.; Hirsch, E.; Szabó, E.; Nagy, B.; Andersen, S.K.; Vigh, T.; Verreck, G.; Csontos, I.; et al. Continuous alternative to freeze drying: Manufacturing of cyclodextrin-based reconstitution powder from aqueous solution using scaled-up electrospinning. *J. Control. Release* **2019**, *298*, 120–127. [CrossRef] [PubMed]

International Journal of *Molecular Sciences*

MDPI

Review

Polymeric Nanocomposites for Environmental and Industrial Applications

Mohamed S. A. Darwish [1,*], Mohamed H. Mostafa [1] and Laila M. Al-Harbi [2]

1 Egyptian Petroleum Research Institute, 1 Ahmed El-Zomor Street, El Zohour Region, Nasr City, Cairo 11727, Egypt; m.hassan@epri.sci.eg
2 Chemistry Department, Faculty of Science, King Abdul-Aziz University, P.O. Box 80203, Jeddah 21589, Saudi Arabia; lalhrbi@kau.edu.sa
* Correspondence: msa.darwish@gmail.com

Abstract: Polymeric nanocomposites (PNC) have an outstanding potential for various applications as the integrated structure of the PNCs exhibits properties that none of its component materials individually possess. Moreover, it is possible to fabricate PNCs into desired shapes and sizes, which would enable controlling their properties, such as their surface area, magnetic behavior, optical properties, and catalytic activity. The low cost and light weight of PNCs have further contributed to their potential in various environmental and industrial applications. Stimuli-responsive nanocomposites are a subgroup of PNCs having a minimum of one promising chemical and physical property that may be controlled by or follow a stimulus response. Such outstanding properties and behaviors have extended the scope of application of these nanocomposites. The present review discusses the various methods of preparation available for PNCs, including in situ synthesis, solution mixing, melt blending, and electrospinning. In addition, various environmental and industrial applications of PNCs, including those in the fields of water treatment, electromagnetic shielding in aerospace applications, sensor devices, and food packaging, are outlined.

Keywords: polymeric nanocomposites; sensing; electromagnetic shielding; water treatment; food packaging

Citation: Darwish, M.S.A.; Mostafa, M.H.; Al-Harbi, L.M. Polymeric Nanocomposites for Environmental and Industrial Applications. *Int. J. Mol. Sci.* **2022**, *23*, 1023. https://doi.org/10.3390/ijms23031023

Academic Editor: Ana María Díez-Pascual

Received: 7 December 2021
Accepted: 16 January 2022
Published: 18 January 2022

Publisher's Note: MDPI stays neutral with regard to jurisdictional claims in published maps and institutional affiliations.

Copyright: © 2022 by the authors. Licensee MDPI, Basel, Switzerland. This article is an open access article distributed under the terms and conditions of the Creative Commons Attribution (CC BY) license (https://creativecommons.org/licenses/by/4.0/).

1. Introduction

Polymeric nanocomposites (PNCs) are important materials for industrial as well as research purposes and are used widely in packaging, energy, safety, transportation, electromagnetic shielding, defense systems, sensors, catalysis, and information industry [1–3]. PNCs could resolve several problems and daily challenges of the real world, conferring great future potential to these materials. PNCs are designed based on the principle of size and surface area being associated with much higher reactivity [4]. PNCs are hybrid materials composed of polymers as the matrix and nanomaterials as the nanofillers. PNCs exhibit unparalleled multi functions due to the incorporation of multi-components into an integrated compatible structure, which enables PNCs to be highly applicable in various electronic, magnetic, and optical applications [5,6]. Two types of polymers are available—natural and synthetic. Natural polymers are the ones that occur in nature, from where these may be extracted for use. Natural polymers are often water-based, such as silk, wool, DNA, cellulose, and proteins [7]. Synthetic polymers, on the other hand, include those that are prepared synthetically, such as nylon, polyethylene, polyester, Teflon, and epoxy. Different inorganic nanofillers, including nanoclays, metal-oxide nanoparticles, carbon nanomaterials, and metal nanoparticles, may be incorporated within a polymer matrix to prepare a PNC with improved properties specific to a particular application [8]. The enhanced properties of the fabricated PNC are a consequence of the uniform distribution of nanofillers within the polymer matrix. However, when the nanomaterial fillers aggregate within the polymer matrix due to the Van der Waals forces among the nanoparticles, the effective and desired properties of the fabricated PNC may exhibit a decline [9,10]. This problem could be resolved by the use of nanomaterials with modified/functionalized surfaces, which would result in enhanced dispersion of the nanofillers within the polymer matrix through the enhancement of the reaction and compatibility between the nanofillers and the polymer matrix at their interface [11]. The surface functionalization of the nanofillers may be achieved by fabricating an organic coating through a physical or chemical reaction which would generate a PNC for advanced applications [12,13]. Several applications are reported for such fabricated PNCs in different fields [3,14]. Among these are the stimuli-responsive polymers, or smart polymers as these are alternatively referred to, which exhibit remarkable changes in their properties, in terms of responding to even the slightest changes in the environmental conditions. Stimuli-responsive polymers are sensitive to certain triggers from their external environment, including changes in the temperature, light, electrical field, magnetic field, and chemicals. The present review discusses such polymer nanocomposites, including their preparation methods and applications in the fields of water treatment, electromagnetic shielding in aerospace applications, sensor devices, and food packaging, among others.

2. Preparation Methods

The preparation of PNCs involves the incorporation of different nanofillers into the polymer matrix [15,16]. PNCs may be fabricated using various techniques, such as *in situ* synthesis, solution mixing, melt processing, electrospinning, etc. The selection of the preparation method depends on various parameters, such as the polymeric system used, the target application field, particle distribution, size, etc. [17,18] The various methods of preparation available for PNCs are discussed below.

2.1. In Situ Synthesis

The in situ synthesis of PNCs includes several steps. The first step is the synthesis of the nanomaterial in the presence of polymer. The second step, as illustrated in Figure 1, is the synthesis of the PNC through the polymerization of monomers in the presence of the synthesized nanomaterial. The third step is the simultaneous synthesis of both polymer and nanomaterial. The main advantage of the in situ synthesis of PNCs is that it enables achieving high uniform dispersion of nanofillers throughout the polymer matrix, which improves compatibility and enables high interaction at the interface [19,20]. Several types of PNCs with different characteristics may be fabricated using the in situ method. The characteristics of the fabricated material would depend on the nature of the nanoparticle precursor and the polymer used. In the preparation of high-quality PNCs, two challenges are mainly encountered: (i) the dispersion of the nanoparticles within the polymer matrix and (ii) the interaction between the polymer and the nanoparticles at their interface. These challenges may be overcome by controlling several factors, such as using a suitable solvent, the appropriate functionalization route and selective dispersion techniques [21].

The in situ synthesis method requires several starting materials, such as the polymer, monomer, precursor, nanoparticles, etc. The additional requirements include glass, heat source, long duration, and high energy for conducting the chemical reactions. While in situ synthesis is a promising method for obtaining PNCs with uniform and well-defined structures, it produces only small amounts of the final product and, therefore, is not suitable for large-scale production [22]. The in situ synthesis of PNCs holds better potential when the nanofillers are functionalized. In this context, the synthesis of poly N-isopropylacrylamide (PNIPAM)/magnetic nanoparticles (MNPs) using the in situ synthesis method has been reported. In that study, the MNPs were prepared through co-precipitation and then coated with a silica shell and subsequently modified with γ-methaacryloxpropyl triisoproox-idesilane. The prepared MNPs were then inserted into the NIPAM monomer via the "grafting-through" reaction, producing the desired nanocomposite [23]. In another study, polymethyl methacrylate (PMMA)/superparamagnetic iron oxide NPs (SPIONs)/PEG bis(amine) nanocomposites were applied to water treatment. In that study, the NPs were fabricated through co-precipitation and subsequently modified with PMMA using the emulsion polymerization process in SPION suspension. This was followed by the insertion of PEG bis(amine) onto the PMMA-coated SPIONs [13]. Munnawar et al. prepared a nanocomposite of chitosan (Ct) and ZnO nanoparticles through chemical precipitation. In the chemical reaction, a coordination bond was formed between the functional groups of Ct and Zn^{+2} ions [24]. In another study, graphene oxide was grafted with polylactic acid (PLA) through in situ polycondensation of L-lactic acid in the presence of graphene oxide to produce the nanocomposite [25]. Similarly, a nanocomposite comprising PLA and cloisite was fabricated through ring-opening polymerization of lactide at various loadings of cloisite with the assistance of microwave heating [26].

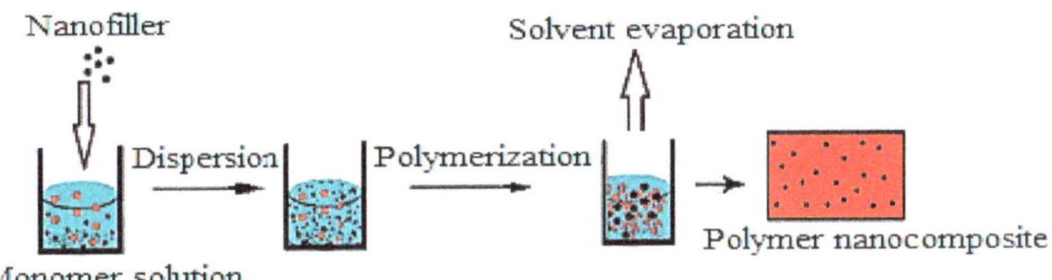

Figure 1. Schematic illustration for the in situ polymerization method (taken from [27]).

2.2. Solution Mixing

The solution mixing method for preparing PNCs relies on evaporating the solvent from the solution, as illustrated in Figure 2. The polymer is first dissolved in a volatile solvent, following which the nanomaterials are dispersed into the polymer solution using sonication. Afterward, the PNCs are produced by performing rapid solvent evaporation [28]. The solution mixing method is a simple and economic approach that does not require a complex design or a large number of chemicals. Moreover, there are no additional energy requirements, and high amounts of PNCs are produced within a short duration. Several studies have demonstrated the preparation of PNCs using the solution mixing method. For instance, the poly (3-hydroxybutyrate-co-3-hydroxyvalerate)/ZnO nanoparticles nanocomposite was prepared using the solution mixing without involving the use of coupling agents, and the process involved the formation of a hydrogen bond between the polymer and the nanoparticles [29]. The solution mixing method has also been used for the preparation of starch nanocomposite films with clay nanolayers, which exhibited enhanced antibacterial activity toward *S. aureus* and *E. coli* [30]. Furthermore, Ct nanocomposite with TiO_2 nanoparticles was prepared using the solution mixing method. The mechanism involved in this preparation could be explained based on the pH at the point of zero charge (PZC), a point at which the sum of all negative surface charges balances the sum of all positive charges. When the pH of the solution is below the pH at PZC, the surface of the TiO_2 nanoparticles becomes positively charged. Conversely, the surface becomes negatively charged when the solution pH is higher than the pH at PZC. In solutions with pH lower than the pH at PZC of TiO_2 nanoparticles and the pKa of Ct, electrostatic repulsion would occur between the charged surface of the TiO_2 nanoparticles and the chains of Ct, which would enable the complete stretching of the flexible Ct chains, thereby resulting in the dispersion of TiO_2 nanoparticles into the Ct matrix [31].

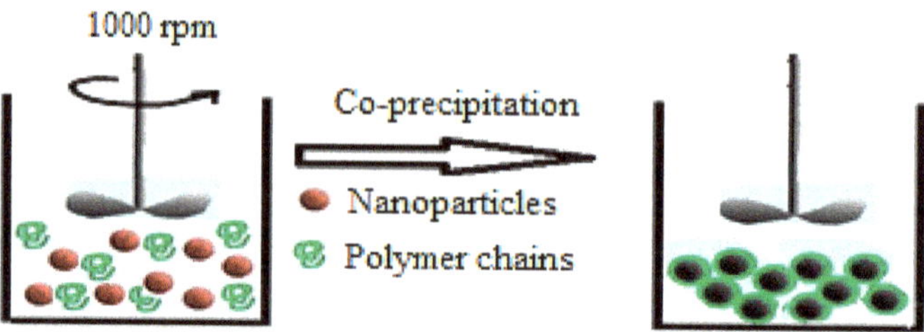

Figure 2. Schematic illustration for the solution mixing method (taken from [27]).

2.3. Melt Blending

The melt blending method for preparing PNCs involves melting the polymer followed by the dropwise addition of the nanomaterials. Figure 3 presents the schematic representation of PNC preparation using the melt blending process. The factors that affect the melt blending process include the type of polymer used, the type of nanoparticles used, the temperature of the process, and the process duration. The production of PNCs through melt blending is achieved using ordinary compounding devices such as mixers or extruders [32]. In comparison to the in situ synthesis and solution mixing method for PNC production, melt blending has the advantage that it does not involve the use of organic solvents. Melt blending is compatible with the existing industrial processes, such as extrusion and injection molding. This process allows continuous, rapid, and simple transformation of the raw ingredients into the desired product. A high temperature during the melt blending process may result in the thermal degradation of the polymers used. Therefore, it is important to adjust the temperature accordingly and use a system for producing PNCs that has appropriate conditions specific to the desired processing efficiency and the desired shape and properties of the final products [33]. Nanocomposites based on PLA and ZnO nanoparticles were prepared using the melt blending process conducted in a HaakeMiniLab II co-rotating twin-screw extruder at 180 °C, 15 min of retention time, and 20 rpm [34]. In another study, poly(-3-hydroxybutyrate-co-3-hydroxyvalerate) (PHBV) nanocomposites with modified montmorillonite and halloysite were prepared through melt blending using a twin screw-rotating extruder at 80 rpm over the temperature range of 150–165 °C [35]. Darwish et al. reported using melt blending for preparing polypropylene nanocomposites using physically prepared Ct/ZnO nanocomposite. Another study involved performing melt blending in a Brabender mixer at 180 °C and 60 rpm as the temperature and rotor speed, respectively [36].

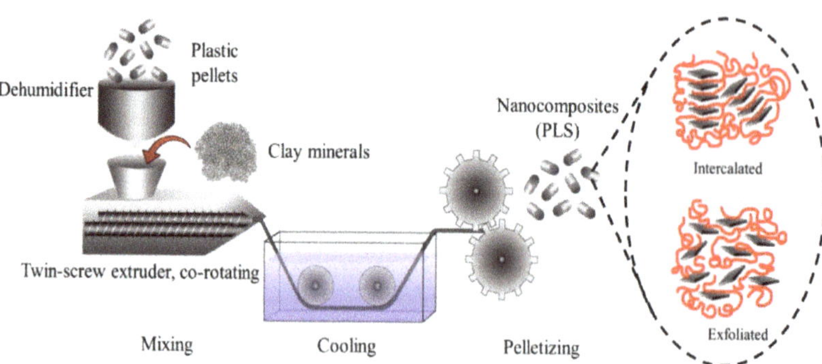

Figure 3. Schematic representation for preparing polymer nanocomposites by melt blending (taken from [37]).

2.4. Electrospinning

Electrospinning is recognized as a successful method for producing nanofibers. The various factors that affect the electrospinning process include the flow rate, polymer concentration, solution viscosity, air humidity, and electric field intensity [38]. The set-up of the electrospinning technique is depicted in Figure 4. Electrospinning is a versatile method used for the preparation of PNCs through the insertion of nanomaterials such as metal nanoparticles, metal-oxide nanoparticles, carbon-based nanomaterials, and clay nanolayers into the polymer matrix [39,40]. For instance, electrospinning has been used for producing electrospun nanofiber mats of cellulose and organically modified montomontrille, which exhibited metal adsorption and removal of Cr^{6+} ions from aqueous solutions [41]. Electrospinning has also been used for preparing Ct nanocomposite fibers with multi-walled carbon nanotubes [42]. Similarly, PHBV nanocomposite fibers with multi-walled carbon nanotubes were produced using electrospinning. The rotating disc collector provides an extra drawing force for the stretching and aligning of the nanomaterial with the electrospun fibers [43–45]. Furthermore, cellulose/multi-walled carbon nanotubes nanocomposite was prepared through electrospinning using 14.25 kV of DC voltage and a horizontally positioned metal needle at the temperature of 27 °C and the relative humidity of 34% [44].

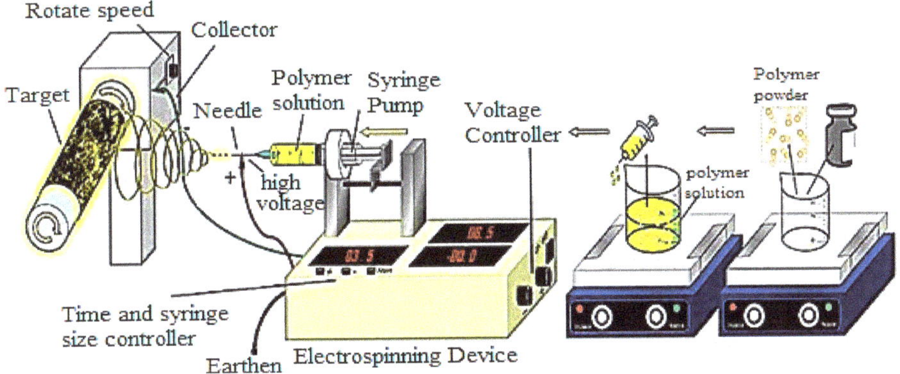

Figure 4. The set-up of electrospinning technique (taken from [45]).

2.5. Other Methods

One of the other methods reported for the preparation of PNCs is microwave heating. In one study, Ct/ZnO nanocomposite was prepared using microwave heating via a complexation reaction between the surface zinc cations of ZnO nanoparticles and the Ct functional groups. The optimum conditions for the preparation process were 800 watts of power and process duration of 10 min. The hydroxyl and amine groups of Ct served as a Lewis base, which formed coordination bonds with the surface zinc ions. As illustrated in Figure 5, the complexation reaction between the Ct functional groups and the surface of the ZnO nanoparticles was achieved via a ligand substitution reaction in which the Ct functional groups substituted water molecules (or the products of protolysis) coordinated to the surface Zn^{+2} ions [46]. In a similar report, a nanocomposite comprising Ct and magnetic nanoparticles was prepared by performing the functionalization of nanoparticles with carboxylic groups, which allowed the covalent bonding of the nanoparticles with the Ct amine groups [47]. In another report, Ct/TiO_2 nanocomposite was prepared via the outer-sphere complexation reaction between the positive charges of the Ct chains and the negative charges present on the surface of TiO_2 nanoparticles [31]. A related study reported the preparation of starch/multi-walled carbon nanotubes via covalent bond formation between the hydroxyl groups of starch and the groups present on the surface of the nanomaterial [48].

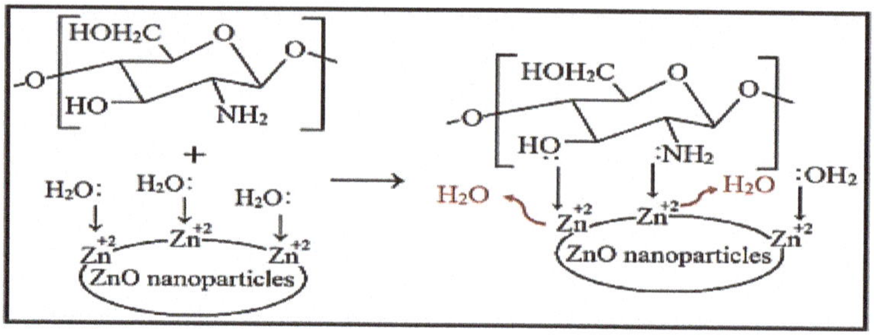

Figure 5. Complexation reaction between chitosan functional groups and Zn^{+2} ions of ZnO nanoparticles (taken from [46]).

When comparing the different methods available for the preparations of PNCs, the physical preparation methods such as solution mixing and melt blending could prove to be advantageous as these are applicable to large-scale production as well. Moreover, these methods require a short duration for attaining the products and involve the use of a limited number of chemicals compared to the in situ synthesis technique. On the other hand, the in situ synthesis method is preferred when uniform and well-defined materials containing strong bonds are desired. However, the in situ synthesis method involves the use of several chemicals and glasses and requires high energy and a long duration for producing even small amounts of materials. On the contrary, the solution mixing method does not require high energy or a long duration for obtaining the products when used in large-scale production. In comparison, the melt blending method requires higher energy, shorter duration, and compounding devices without involving the use of solvents, thereby rendering it further advantageous compared to all other methods in terms of the limited number of chemicals used. In addition, melt blending provides rapid and continuous production of PNCs at an industrial scale. Electrospinning is much more complex compared to any of the other methods as it requires complex equipment and set-up. However, it produces nanocomposite fibers in the nanoscale dimensions with favorable properties.

3. Smart Polymer Nanocomposites

Smart polymer nanocomposites or stimuli-responsive polymer nanocomposites are those that possess a minimum of one chemical and physical property that is controlled by or follows a stimulus-response. These stimuli-responsive properties may be categorized based on their nature as external (physical stimuli) and internal (chemical stimuli). Figure 6 presents the classification of responsive nanocomposites. The chemical stimuli are associated with the pH, biological recognition, solvent type, and chemical recognition, while physical stimuli are related to the magnetic field strength, temperature, electric current, and light [2,9]. These stimuli are discussed in the sections ahead.

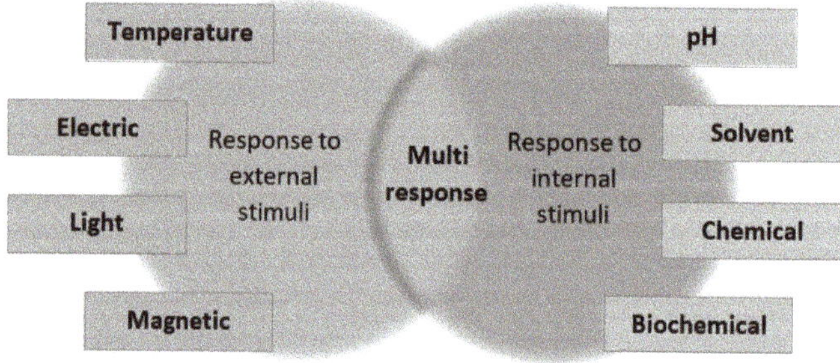

Figure 6. Classification of the responsive nanocomposite (taken from [9]).

3.1. Thermo-Responsive Nanocomposites

Thermo-responsive nanocomposites are commonly used as intelligent materials as these are based on the application of temperature as a facile stimulus. A thermo-responsive polymer exhibits a variation in the hydrogen bonds within the matrix. Hydrogel is a kind of thermo-responsive polymer that exhibits a switch from the hydrolyzed phase to the precipitated phase, which is accompanied by a considerable shift in volume in response to a difference in the temperature [49]. The reversible phase transition of thermo-responsive polymers is illustrated in Figure 7. The switch phase of the polymer occurs when the temperature lowers, and the transition temperature is referred to as the upper critical solution temperature (UCST) (including the copolymers of poly(acrylamide) and poly(acrylic acid)) [50]. On the other hand, the transition phase occurs as the temperature becomes high, and this temperature threshold is referred to as the lower critical solution temperature (LCST) [including poly(N,N-dimethyl acrylamide) and poly(N isopropyl acrylamide)] [51,52]. Nanofillers may also be added to improve the mechanical behavior and the thermo-responsiveness of the polymer matrix, properties that would extend the application of these polymers to be used as a tissue substitute. Various fabricated PNCs have been used as bone substitutes. For instance, Nistor et al. reported the fabrication of collagen/PNIPAAm hydroxyapatite nanocomposite to be used as an artificial extracellular matrix as this nanocomposite exhibited remarkable properties for bone tissue replacement above 33 °C [52]. Oguz et al. prepared methylcellulose-gelatin hydrogels with several calcium phosphate fillers. These hydrogels could be used as injectable nanocomposites that would shift to a solid phase inside the human bone tissue. Another interesting application is the utilization of the dual behavior of thermo-responsive polymers, one of which is the antibacterial effect exhibited by certain metal nanoparticles components of the nanocomposites [53]. Bacteria are capable of rapid growth, which renders their control difficult within the short duration available during the application of drugs. In this context, Arafa et al. fabricated the pluronic 127/gold nanocomposites release system for in vivo application, i.e., to be applied on skin burns [54].

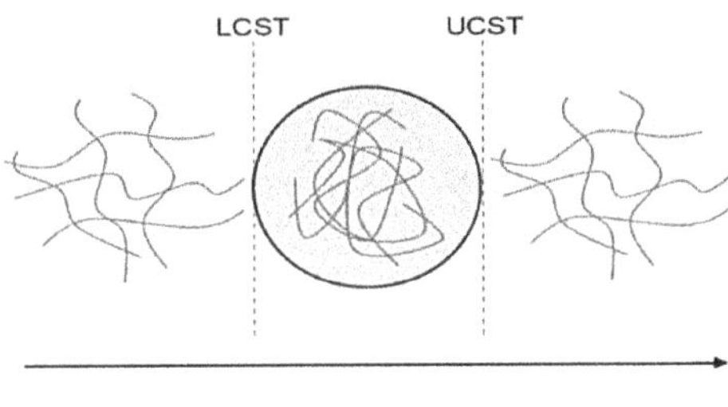

Figure 7. Reversible phase transition of thermo-responsive polymers (taken from [9]).

3.2. Light-Responsive Nanocomposites

Light is simply produced or applied directly from solar energy. Light is not affected by electromagnetic fields as it is a source of remote stimulation. In addition, nanofillers, particularly metal-oxide nanoparticles, exhibit an important property of localized surface plasmon resonances (LSPR) through the interaction between the surface of particles and the light source [55–57]. The shape, nature, and size of the nanomaterial determine LSPR. This property leads to the production of intense electromagnetic fields upon the incidence of light and the occurrence of rapid heat transfer from the nanoparticles to the surrounding environment. It is the diversion to heat release which could produce the response of the polymer matrix, and this type of material is, at certain times, referred to as plasmonic polymer composite. Light-responsive nanocomposites have been studied extensively for application in the field of medicine, including cancer therapy and drug delivery. Among these applications, near-infrared irradiation (NIR) could be a remarkable candidate as it has deep permeation and negligible absorbance in the human tissue, properties that are not possessed by the other electromagnetic irradiation techniques such as those involving UV irradiation. Li et al. performed partial decomposition that was activated through the interaction of NIR with reduced graphene, thereby developing polyethyleneimine functionalized with thiocarbamate as a hydrogen sulfide (H_2S) release platform. The exogenous H_2S was then investigated for application in cancer therapy and tumor generation [58]. Raza et al. performed drug delivery using nanocomposites responsive to NIR for cancer therapy and investigating tissue regeneration [59]. Yang et al. studied the cell-growth promoting and bacteria-inhibiting capabilities of nanocomposites in skin lesions. Moreover, a near-infrared photo-responsive dressing nanocomposite based on a dodecyl-modified and Schiff's base-linked Ct hydrogel, tungsten disulfide nanosheets (photothermal agent), and ciprofloxacin (antimicrobial drug) has been fabricated. This nanocomposite exhibited several advantages, including the ability to mold rapidly, self-adaptability, and injectability. The nanocomposite also exhibited good tissue adherence and excellent biocompatibility [60]. Yue et al. used graphene quantum dots as active NIR nanofillers in the dextran modified with pendant PNIPAAm chains to fabricate a nanocomposite. The nanocomposite comprised buprenorphine and thermo-responsive chains, which played a significant role in drug release [61].

3.3. *Responsive Nanocomposites Based on Electric Current*

The application of electric current in shape memory composites (SMC) has demonstrated promising results. When electrical current is applied, depending on the local heat release, a shift is produced in the shape of materials [62]. The most interesting choice here would be to use nanofillers as the active component, particularly carbon-based nanomaterials such as carbon nanotubes (CNTs), graphene oxide (GO), or graphene. The use of nanofillers would confer the property of responding to the electrical current stimulus due to the good thermal and mechanical behaviors of these nanofillers and their ability to significantly enhance the electronic behavior of the nanocomposite material. However, this property relies on the randomized distribution of the nanofillers within the matrix [63]. Yang et al. performed the in situ polymerization of graphene oxide nanoplatelets to fabricate a nanocomposite comprising 2-acrylamide-2-methyl propane sulfonic acid, acrylamide, and reduced graphene oxide. In comparison to pure polymer, these nanocomposites exhibited enhanced mechanical and electrical response behaviors, which facilitated their application as soft robots [64].

3.4. *Magnetic Responsive Nanocomposites*

Magnetic nanoparticles (MNPs) exhibit magnetic induction heating, which facilitates their application in biomedical fields, including targeted drug delivery and elastomer fabrication [65–67]. The performance of magnetic induction heating relies mainly on the properties of the MNPs and the applied magnetic field conditions. The size of the nanoparticles affects their magnetic domains. While small sizes are constituted of a single domain, the larger sizes are constituted of multiple domains that result in the minimization of the magnetostatic energy. Moreover, in the field of cancer therapy, the conversion of magnetic energy into heat energy within the MNPs has demonstrated outstanding potential [68,69]. The facile separation and controlled placement of functionalized MNPs through the external application of magnetic field enabled the use of these MNPs in various bio-separation and catalytic processes.

MNPs have been applied widely in biomedical fields owing to their superparamagnetic behavior and high biocompatibility. MNPs are suitable candidates for drug delivery agents and heat mediators in hyperthermia cancer treatment. MNPs could also serve as diagnostic and therapeutic agents in magnetic resonance imaging (MRI) [70,71]. The MRI technique relies on the counterbalance between the exceedingly large number of protons in the biological tissue and the small magnetic moment of a proton which undergoes a shift under the effect of a magnetic field. MNPs are influenced by static or alternating magnetic fields (AMF) with a relatively high penetration depth and non-contact stimulation source. The interaction between the magnetic gradient produced by the magnetic field and the magnetic moments in the material results in the development of magnetic behavior. The energy of the AMF could be converted into heat using MNPs via two types of relaxation processes, namely Neel relaxation and Brownian relaxation. Neel relaxation occurs due to the re-orientation of the magnetization that results from the re-orientation occurring inside the magnetic core against the energy barrier. Brownian relaxation occurs due to the rotational diffusion of the whole particle within the carrier liquid [72,73]. In the presence of MNPs, the magnetite poly(dimethylsiloxane) (PDMS) nanocomposites were fabricated using magnetic induction heating. Under the effect of an AC magnetic field, heat was generated by the MNPs that were utilized for accelerating the polymerization process and curing the PDMS. The magnetite PDMS composites with enhanced thermal stability compared to conventional PDMS were produced without the use of a catalyst within a short duration, and these were termed thermally stable elastomers [73]. The unmodified iron oxide nanoparticles exhibit a high surface/volume ratio, which causes the magnetic dipole/dipole attraction, thereby leading to the formation of aggregated particles. The aggregation unstabilizes the colloidal solution, leading to the loss of the size-dependent properties of the nanoparticles (such as their superparamagnetic behavior). Stabilizers comprising polymers and surfactants may be utilized as a coating on the material surface for higher stability through electrostatic repulsions and steric effects. Various polymer-and organic material-shelled iron oxide nanoparticles have been fabricated using this approach [74].

The magnetic response PNCs could exhibit changes in their movement or shape. These PNCs could be utilized as magnetically separable materials for purification systems and localized drug delivery. Moreover, at the nanoscale, hyperthermia might occur in the magnetic material. This may be influenced by the AMF for releasing heat through the oscillation of the nanomagnetic filler associated with the Brownian and Néel relaxation processes. This behavior could be applied to shape memory composites, artificial muscles, killing cancer cells in tumors, and drug release. Soto et al. fabricated polyurethane/Fe_3O_4 nanocomposites and also investigated the influence of the Fe_3O_4 content on shape recovery and heat release when these nanocomposites were exposed to AMF. It was revealed that the time taken to achieve shape recovery was directly correlated with the Fe_3O_4 content [65]. In another work, nanocomposites comprising Kappa carrageenan-g-poly(acrylic acid)/SPION were observed to be effective as in vitro antibacterial agents with the drug efficiency of 105 ± 8 µg/mg exhibited by the drug carrier [75]. SPION/polyvinyl alcohol/PMMA nanocomposites have been used for the delivery of ciprofloxacin. An increased release was achieved by applying a magnetic field, with high PMMA content, and with low PVA content [76]. Nanocomposites comprising PEG/phospholipid-coated iron oxide nanoparticles modified with peptide and a fluorescent dye were used for the confocal imaging of kidney-derived cells and primary human dermal fibroblast cells. These nanocomposites exhibited great potential in tissue imaging [67]. A polyethyleneimine/folic acid-targeted Fe_3O_4 nanocomposite was investigated for application in the in vivo MRI of tumors. This nanocomposite exhibited a high T2 relaxivity of 99.64 mM^{-1} s^{-1} when used as the nanoprobe for conducting the MRI of cancer cells [66].

Hyperthermia may occur on the MNPs, and it could be affected by the AMF for releasing heat through the oscillation of the MNPs associated with the Brownian and Néel relaxation processes. Figure 8 depicts a simple representation of artificially induced hyperthermia. Hyperthermia occurs when the temperature of the living tissue elevates beyond the physiological normal values. Hyperthermia is used in cancer therapy for damaging and killing cancer cells. It is also used for inducing local drug release from thermo-sensitive vehicles. Artificially induced hyperthermia involves locally elevating the temperatures of the influenced cells in the body up to 42 °C. It is used for specifically targeting and destroying the cancer cells without influencing the surrounding healthy cells [77,78]. A study investigated the development of injectable 20 nm dextran coated with iron-oxide nanoparticles and covalently bound antitumor chimeric L6 monoclonal antibody [79,80]. Using the co-precipitation method, modified magnetite nanoparticles exhibiting antibacterial and self-healing properties were prepared. Three sets of these nanoparticles were evaluated for their antibacterial properties and magnetic heating specific absorption rates. The determined concentration of these nanoparticles for 10% growth inhibition (EC10) of *S. aureus* and *E. coli* was 150 mg/L [81].

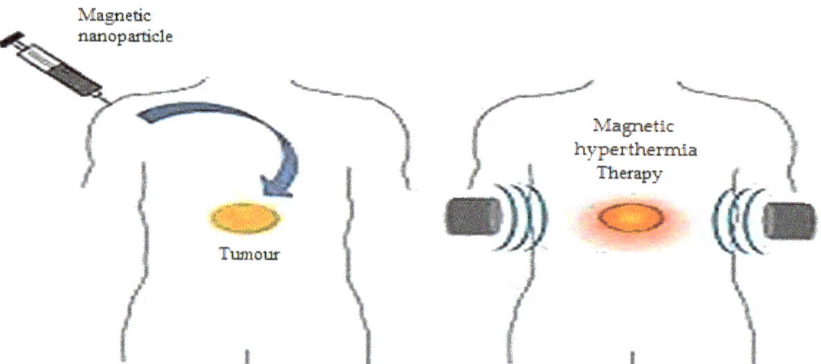

Figure 8. Artificially induced hyperthermia (taken from [77]).

4. Applications of Polymer Nanocomposites

4.1. Water Treatment

Recently, PNCs have been receiving great attention in the field of wastewater treatment and remediation. Freshwater sources have been demonstrating a continuous decrease in both quantity and quality, mainly due to anthropogenic activity. Water is essential for humans and plants and is also a primary resource in various industrial processes. Owing to rapid industrialization, the discharge of wastewater has increased greatly, which has led to diverse pollutants being released into the environment and causing adverse effects on the environment as well as human health. In this regard, PNCs could serve as an efficient and cost-effective material for wastewater treatment. PNCs appear to be promising in resolving the inherent challenges of ordinary particles when used in water treatment [82]. Nanocomposites comprising polymers integrated with nanomaterials exhibit improved properties, such as enhanced resistance to fouling, thermal stability, membrane permeability, good mechanical behavior, higher photocatalytic activity, and higher adsorption [83]. The extent of these properties depends on the type of nanoparticle incorporated, its shape and size, its interaction with the polymer, and its concentration. Moreover, the method used for the modification of PNCs determines their reusability, remediation capacity, and selectivity. PNCs could be used in the field of water treatment to achieve various purposes, including the removal of dyes, metal ions, and microorganisms from water [84]. The roles of PNCs in dye removal, metal removal, and water disinfection are discussed in the sections ahead.

4.1.1. Dye Removal

Dyes are used widely in several industries, such as textiles, paper making, ink, coating, and cosmetics, to induce color into the products. However, severe environmental problems are caused due to the wastewater discharged from these industries, which is polluted with dyes that are carcinogenic, toxic, and hazardous. Therefore, it is essential to remove these dyes for rendering the wastewater suitable for industrial reuse and also to prevent environmental impact. In this regard, several studies have used PNCs as efficient materials for the removal of dyes from polluted water [85–106]. This ability of PNCs is conferred by the synergistic effect of the polymer and the nanomaterial [86]. Table 1 lists a few examples of PNCs that have been utilized for dye removal. PNCs used as adsorbents for dyes exhibit a large surface area, numerous active sites, and chemical and thermal stability. For instance, Zaman et al. fabricated the cellulose/graphene oxide nanocomposite for the enhanced removal of methylene blue dye through adsorption. Approximately 98% of the dye was removed within 135 min, with the nanocomposite achieving an adsorption capacity of 334.19 mg/g [87]. In another work, the methylene blue dye was removed using cellulose/clay nanocomposites and the maximum removal efficiency achieved was 98% [88]. When the polyaniline/TiO_2 nanocomposite was used for the removal of methylene blue dye removal in a different study, the maximum adsorption capacity achieved was 458.10 mg/gL. The adsorption was achieved through membrane diffusion, chemical adsorption, and intraparticle diffusion, and hydrogen bonding, coordination interaction, and electrostatic interaction were revealed as the adsorption mechanisms, as illustrated in Figure 9 [89].

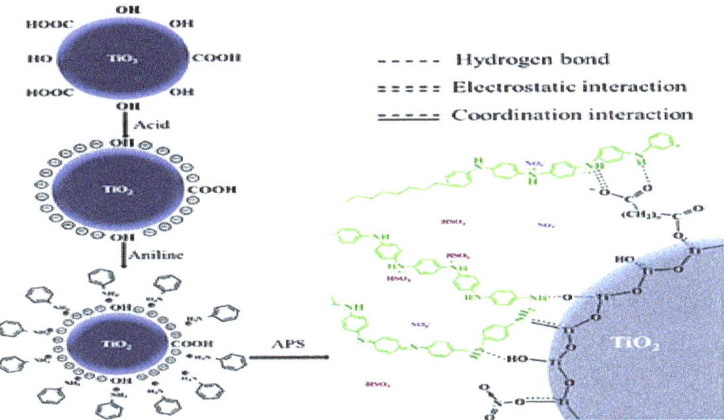

Figure 9. Schematic representation of the adsorption mechanisms by polyaniline/TiO_2 nanocomposite (taken from [89]).

In addition, certain PNCs comprising polymer and nanomaterial photocatalysts such as the nanoparticles of ZnO, TiO_2, and CuO, have been used for dye removal. The presence of these photocatalysts degrades the dyes as these photocatalysts produce hydroxyl free radicals that are capable of degrading organic materials. Therefore, when using PNCs for dye removal, the combination of the adsorption process via the active sites of PNCs and the photocatalytic degradation of the dye due to the action of the photo-catalyst appears to be the most suitable approach [90]. For instance, a study evaluated the efficiency of Ct and Ct/ZnO in methylene blue dye removal and reported the values of 81% and 96.7% of the methylene blue dye removed by Ct and Ct/ZnO, respectively. The enhanced dye removal by the nanocomposites was attributed to the synergistic effect of the adsorption process accomplished by the nanocomposites and the photocatalytic activity of the ZnO nanoparticles [46]. In another study, Kumar et al. fabricated a Ct/CuO film to be used as a simple, portable, recoverable, reusable, and efficient photocatalyst. Using these nanocomposites, up to 99% of Rhodamine B dye was removed within 60 min of irradiation because of the slow recombination rate of the electron–hole pair of the CuO nanoparticles in the chitosan matrix [91]. Shaikh et al. fabricated the PLA/TiO_2 nanocomposites and investigated their efficiency in degrading malachite green and methyl orange under UV light and solar light. In comparison to the UV light conditions, the degradation was faster under the sunlight. In sunlight, a 10^{-4} M solution of each of the two dyes was completely decolorized within 8 min and 20 min, respectively [92]. Karagoza et al. prepared polycaprolactone/Ag/TiO_2 nanocomposites to be used as photocatalysts for the degradation of organic pollutants. The fabricated nanocomposites were able to completely degrade methylene blue and ibuprofen under UV irradiation within approximately 180 min [93]. Meenakshi et al. reported that

Ct/TiO_2 nanocomposites, via photodegradation, could remove Reactive Red 2, methylene blue, and Rhodamine B [94].

PNCs that contain magnetic nanoparticles may be used as reusable magnetic adsorbents for dyes by exploiting the magnetic properties conferred by their component nanoparticles. The PNC magnetic adsorbents exhibit a large surface area, porous structure, and small particle size, and their excellent magnetic properties allow convenient recovery via magnetic separation after the adsorption or regeneration. This enables overcoming the limitations of separation difficulty encountered when using PNC adsorbents while increasing the reusability of the PNCs in dye removal. For instance, the magnetic Ct nanocomposites prepared using the chemical approach were applied as a reusable adsorbent for dyes, and an adsorption capacity of 20.5 mg/g, based on the pseudo-second-order model, was achieved [95]. Esvandi et al. prepared a magnetic nanocomposite using starch, clay, and $MnFe_2O_4$ for the uptake of sunset yellow and Nile blue dyes from water [96]. Guan et al. developed a nanocomposite containing cellulose nanocrystals and zinc oxide for the removal of malachite green and methylene blue dyes. This nanocomposite exhibited rapid removal and high dye removal efficiency, with high dye-removal ratios for malachite green (99.02%) and methylene blue (93.55%) [97].

Table 1. Performance of some polymer nanocomposites in dye removal.

Polymer Nanocomposite	Dye	Results	Ref.
Chitosan/CuO nanocomposites beads	Congo red (CR) Eriochrome black T (EBT)	A total of 97% of dyes were removed within 2 h. Maximum adsorption capacity of CR and EBT were 119.70 and 235.70 mgg^{-1}	[98]
Molecularly imprinted Chitosan/TiO_2 nanocomposite	Rose Bengal (RB)	The adsorption capacity for RB was 79.365 mg/g and enthalpy was 62.279 kJ mol^{-1}	[99]
Chitosan/ZnO nanocomposite	Methylene blue (MB)	96.7% of MB dye was removed	[46]
ZnO/Cellulose nanocrystal nanocomposite	Methylene blue (MB) Malachite green (MG)	93.55% and 99.02% of MB and MG were removed within 5 min. The absorption capacity was 46.77 and 49.51 mg/g for MB and MG	[97]
ZnO/Poly(methyl methacrylate) nanocomposite membrane	Methylene blue (MB)	About 100% of MB was removed within 80 min	[100]
Poly(methyl methacrylate)/Multiwall carbon nanotube nanocomposite	Methyl green (MG)	The Langmuir adsorption capacity for MG was 6.85 mmol/g at 25 °C	[101]
Polyacrylic acid/Fe_3O_4/Carboxylated cellulose nanocrystals nanocomposite	MB	The maximum adsorption capacity for MB was 332 mg g^{-1}	[102]
Fe_3O_4/Starch/Poly (acrylic acid) nanocomposite hydrogel	Methylene violet (MV) Congo red (CR)	A maximum of 93.83% and 99.32% CR and MV dyes with maximum adsorption of 96.7% and 97.5%	[103]
Polylactic acid/Graphene oxide/Chitosan nanocomposite	Crystal violet (CV)	97.8 ± 0.5% of CV was removed	[104]
Polypyrrole/Zeolite nanocomposite	Reactive blue (RB) Reactive red (RR)	A total of 86.2% of RB and 88.3% of RR were adsorbed from synthetic solution	[105]

Dye removal using PNCs may be improved by controlling several factors, such as the types and properties of the polymer and the nanomaterial. Moreover, the methods used for PNC preparation and the design of the fabricated PNC play essential roles in determining the performance of the fabricated PNC in dye removal. The mechanism of dye removal is also an important factor. For instance, selecting a PNC capable of removing dyes through a combination of the adsorption approach and the photocatalytic degradation approach would result in improved performance compared to the PNCs following either of the individual mechanisms. The PNCs possessing remarkable magnetic properties would be further advantageous as these may be recovered conveniently and then reused several times. In addition, the experimental conditions, including dye concentration, PNC concentration, pH, temperature, and contact duration, could affect the dye removal process. Therefore, the integration of all these factors would allow the fabrication of a PNC that would exhibit outstanding performance in dye removal [90]. Generally, it is the laboratories where the PNCs have been applied the most widely for dye removal. However, the scaling up and production in huge amounts for use at the industrial scale remains to be achieved so far and warrants further research and improvement on the production cost and practical application possibility. Several other issues are also required to be dealt with, including the following: (i) the toxicity of the PNCs has to be considered to prevent secondary pollution, (ii) greater efforts are required for the selective removal of a specific dye using the PNC in the presence of other dyes, (iii) PNCs with the ability to be regenerated several times and be used for longer durations should be investigated. In addition, PNCs that are environmentally friendly and highly stable, and also exhibit potential for large-scale production should be produced. The preparation of PNCs at a low cost is an important factor when considering them for large-scale applications. Moreover, efficient methods of separation of the PNCs from solutions after use should be developed to prevent the PNCs from acting as pollutants [106].

4.1.2. Metal Ion Removal

Metal pollution has been increasing rapidly throughout the world due to rapid advancements in urbanization and industrialization, which has led to serious environmental concerns. Metal ion removal has, therefore, become an essential requirement for the protection of the environment and human health. In this regard, PNCs could be useful in metal ion removal. Table 2 lists the performance of a few PNCs in metal ion removal. PNCs may remove metal ions through the adsorption process, which is a facile and efficient mechanism for the removal of heavy metal ions, such as Cu (II), Cd (II), Pb (II), Co (II), Cr (VI), and Ni (II), from solutions. The adsorption of metal ions on a PNC may be affected by several factors, including the concentration of the adsorbent, pH of the solution, the concentration of metal ions in the solution, contact duration, and temperature conditions. In comparison to individual polymer adsorbents, PNCs exhibit a greater number of surface groups available for interaction, a larger number of active sites, better stability, mechanic feasibility, and higher adsorption capacity [107]. Several studies have reported the use of PNCs in metal ion removal. For instance, Cr (VI) was removed using polyaniline β-FeOOH prepared through the blending process (mechanical force) [108]. A related study demonstrated that the magnetite acrylamide amino-amidoxime nanocomposites prepared using the in situ method exhibited excellent sorption properties and could, therefore, be utilized for treating U(VI) present in aqueous solution [109]. In another work, magnetite/poly (1-naphthylamine) nanocomposite was prepared using the in situ method and then used for As (III) removal [110]. Another nanocomposite composed of poly(N-vinylcarbazole) and graphene oxide was fabricated in a study and then used for adsorbing heavy metals from aqueous solutions. The Pb^{2+} adsorption capacity of this nanocomposite could be increased by increasing the graphene oxide content of the nanocomposite due to the resulting increase in the concentration of oxygen-containing groups in the nanocomposite. The highest Pb^{2+} adsorption capacity achieved using this nanocomposite was 887.98 mg g^{-1}, and the adsorption fitted well with the Langmuir model [111]. Another study reported the fabrication of polyethersulfone/Fe-NiO nanocomposites through solution blending, followed by the application of these nanocomposites for salt removal [112]. A novel nanocomposite comprising poly(methyl methacrylate)-grafted alginate/Fe_3O_4 was synthesized through oxidative-free radical-graft copolymerization reaction and then used for the adsorption of Pb^{2+} and Cu^{2+} ions from aqueous media. The pH value of 5 was determined as the optimum condition for the adsorption process in the afore-stated study, and the maximum adsorption capacity achieved for Cu^{2+} and Pb^{2+} ions was 35.71 mg g^{-1} and 62.5 mg g^{-1}, respectively [113]. A. A. Saad reported using ZnO/chitosan/organically nanocomposite for the removal of Cu(II), Cd(II), and Pb(II) ions from polluted water [114]. Figure 10 depicts the schematic representation of the adsorption of metal ions using a nanocomposite.

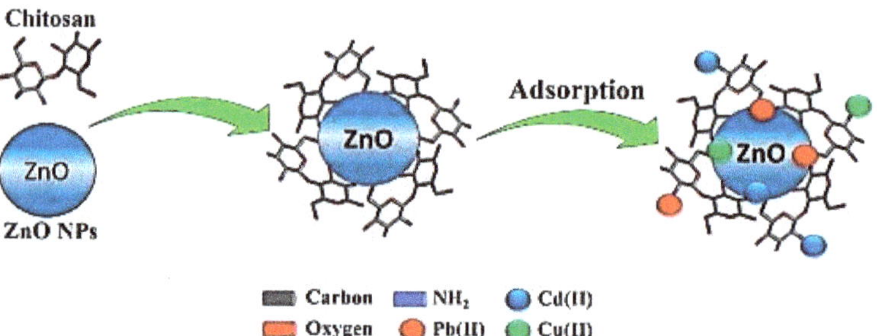

Figure 10. Schematic representation of metal ion adsorption by nanocomposite (Taken from [114]).

Owing to advantageous properties such as high adsorption capacity, great recycling performance, high mechanical strength, and convenient separation from the solution, PNCs have been applied widely for the removal of metal ions from wastewater. The polymer present in a PNC serves as a support for the nanoparticles and also as a chelating material, while the incorporated nanoparticles exhibit effective chelating sites, reducibility, and exceptional magnetic properties. PNCs demonstrate rapid adsorption kinetics, remarkable regeneration capability, and strong chelating abilities for metal ions. Unfortunately, scaling up becomes an issue for these PNCs, and their preparation methods have not been adapted for scaling up from laboratory to industrial level so far. Moreover, the research on production cost and practical application possibilities is scarce. Several other issues also require investigation, such as (i) evaluation of the adsorbent capacity for each metal ion to compare different kinds of adsorbents, (ii) development of reliable approaches for adsorbent regeneration, and (iii) exploring PNCs that exhibit long-term performance and better regeneration capability. In addition, further clarification of the adsorption mechanism using advanced analytical and characterization techniques is required to gain deeper insights into the molecular mechanisms underlying the metal ion removal process and the interaction process [115,116].

Table 2. Performance of some polymer nanocomposites in metal ion removal.

Polymer Nanocomposite	Metal Ion	Results	Ref.
Polyaniline/Reduced graphene oxide nanocomposite	Hg(II)	The adsorption capacity was 1000.00 mg/g	[117]
Fe_3O_4/starch/Poly(acrylic acid) nanocomposite hydrogel	Cu(II) Pb (II)	95.4% of Cu(II) and 88.4% of Pb(II) were removed at pH of 6.0 and 5.5	[103]
Graphene oxide/Chitosan/ Ferrite nanocomposite	Cr (VI)	The adsorption capacity for Cr(VI) was 270.27 mg g^{-1} at pH of 2.0.	[118]
Magnetic chitosan/Functionalized 3D graphene nanocomposite	Pb (II)	The efficiency of Pb(II) removal is 100% at pH of 8.5 within 18 min	[119]
Bacterial cellulose/Amorphous TiO_2 nanocomposite	Pb(II)	A total of 90% of Pb(II) was removed in 120 min at pH 7	[120]

Table 2. *Cont.*

Polymer Nanocomposite	Metal Ion	Results	Ref.
Cellulose/TiO_2 nanocomposite	Zn(II) Cd(II) Pb(II)	Maximum adsorption capacity for Zn(II), Cd(II) and Pb(II) was 102.04, 102.05 and 120.48 mg/g	[121]
Polyacrylamide/Sodium Montmorillonite nanocomposite	Ni (II) Co (II)	A total of 99.3% of Ni(II) and 98.7% of Co (II) was removed at pH 6.	[122]
Polyacrylamide/Bentonite hydrogel nanocomposite	Pb (II) Cd (II)	More than 95% of Pb (II) and Cd (II) were removed within first 20 min. Maximum adsorption capacity for Pb (II) and Cd (II) was 138.33 and 200.41 mg/g.	[123]
Modified mesoporous silica/Poly(methyl methacrylate) nanocomposites	Cu (II)	Maximum adsorption capacity for Cu (II) was 41.5 mg/g at pH 4 and 140 min	[124]
Xanthan gum grafted Polyaniline/ZnO nanocomposite	Cr(VI)	Maximum adsorption capacity was 346.18 mg g^{-1} for Cr(VI)	[125]

4.1.3. Water Disinfection

The pathogenic microorganisms present in drinking water affect human health greatly. Therefore, disinfection of water is necessary, either through the deactivation or by complete removal or killing of the pathogenic microorganisms. In this regard, PNCs exhibiting antimicrobial activity could be utilized for inhibiting the growth of microorganisms present in drinking water [83]. The good antimicrobial activity of PNCs is a consequence of the synergistic effect of its component polymer and nanomaterial. Several studies have reported the use of PNCs as antimicrobial agents for water disinfection. For instance, Chen and Peng prepared a cellulose/silver nanocomposite that exhibited improved antimicrobial activity and high water permeability, which rendered it suitable for application as an antimicrobial agent in the field of water treatment [126]. Sarkandi et al. fabricated a nanocomposite hydrogel comprising cellulose and silver nanoparticles, which exhibited excellent antibacterial activities and 100% reduction in bacterial percentage, while inhibition zones of 2.8 cm and 2.6 cm against *E. coli* and *S. aureus*, respectively, were observed [127]. Munnawar et al. prepared the chitosan/zinc oxide nanocomposite and incorporated it into a polyethersulfone matrix to develop antifouling polyethersulfone membranes, which exhibited outstanding water permeability and prevented microbial fouling [24]. In another study, a polycaprolactone nanocomposite membrane with ZnO nanoparticles was fabricated, which demonstrated enhanced antimicrobial activity against *S. aureus* and *E. coli* [128]. Al-Naamani et al. used a chitosan/ZnO nanocomposite coating to prevent marine biofouling. The nanocomposite used demonstrated antibacterial activity against the marine bacterium *Pseudo alteromonasnigrifaciens* and anti-diatom activity against *Navicula* sp. [129]. In another study, a nanocomposite of polypyrrole/carbon nanotubes/silver was prepared through the in situ oxidative polymerization of pyrrole with $AgNO_3$ containing single-wall carbon nanotubes. The nanocomposite was then used for the inhibition of bacteria in water, and the *E. coli* removal percentage achieved was 87.5–95% [130].

Although the excellent antimicrobial behavior demonstrated by PNCs when used in water disinfection has rendered PNCs effective against different kinds of microorganisms, further investigation is nonetheless warranted in this field of research. For instance, efforts for large-scale PNC production at lower costs have to be increased and the practical application potential of these PNCs in actual-world scenarios has to be ensured. Several other issues have to be addressed as well, such as (i) the preparation of PNCs with excellent antimicrobial activity against a higher number of microorganisms and (ii) the investigation of the antimicrobial activity of PNCs using further advanced techniques to attain deeper insights into the mechanism of their antimicrobial activity.

4.2. Sensor Devices

Sensors have a wide range of applications, including the detection of chemicals and toxic gases for safety purposes, medical diagnosis, and defense applications. In order to be effective, a sensor must have small dimensions, multiple functions, low cost, reliability, rapid response, higher sensitivity, and selectivity. Among these qualities of sensors, rapid response and high sensitivity are achieved with a large specific surface area. In this regard, polymer nanocomposites could be considered promising candidates for fabricating sensors [131]. The use of PNCs in the fabrication of sensors has become common these days. This is because polymeric materials are inexpensive and convenient to fabricate and also exhibit multi-functionality owing to their varied physicochemical and structural behaviors. Moreover, the modification of sensors, such as the addition of side chains and inorganic materials into the bulk matrix, is convenient when using PNCs. The resultant behaviors include enhanced conductive, electrolytic, dielectric, and optical properties, better ion selectivity, and improved molecular recognition capability [132]. The nanofillers used in the preparation of PNCs aimed at sensing applications must possess certain unique behaviors, such as electrochemical, optical, and magnetic properties, for enhancing the sensitivity and detection rate of the sensors.

PNCs have been used widely in various kinds of sensors, including biosensors, gas sensors, and metal ions sensors. Table 3 summarizes the performance of a few PNCs in different sensors. A biosensor is capable of interacting with biological components and thereby detecting biomolecules, including cholesterol, glucose, and DNA. Figure 11 depicts the setup of a biosensor. As visible in the figure, a biosensor comprises a biologically active element that is immobilized on a convenient substrate, a transducer, and a signal processor. The biologically active element may be an enzyme, DNA, or a protein [133]. PNCs have been used as bio-sensing materials for the fabrication of biosensors. For instance, a sensor relying on poly (3,4-ethylenedioxylthiopene)/Au nanocomposite was used for measuring the concentration of 17β-estradiol using square wave voltammetry and cyclic voltammetry. In this biosensor, the transduced signal released was lower because of the interference of bound 17β-estradiol, while the current drop was proportional to the concentration of the contaminant. In addition, the probe exhibited outstanding selectivity as it could distinguish 17β-estradiol from the other structurally similar EDCs [134]. Narang et al. fabricated a NiO–chitosan/ZnO/zinc hexacyanoferrate film to be used as a triglyceride biosensor. When this biosensor was polarized at +0.4 V against Ag/AgCl, its optimum response was obtained within 4 s at 35 °C and pH 6.0. Moreover, a linear relationship existed between the response of the sensor and the concentration of triolein in the concentration range of 50–700 mg/dL, and a sensitivity of 0.05 A/mg/dL was achieved [135]. In another report, cellulose/organic montmorillonite nanocomposites were fabricated for use in bio-macromolecular quorum sensing inhibitors, which were capable of interfering with the quorum-sensing-regulated physiological process of bacteria. This would provide a sustainable and inexpensive approach for dealing with the challenges raised due to microbial infections in numerous products, such as biomedical materials or food packaging [136]. Manno et al. prepared starch/Ag nanocomposites using the green method and utilizing starch as the capping agent. The fabricated nanocomposites exhibited high sensitivity for hydrogen peroxide [137]. Singh et al. reported the fabrication of a nanocomposite electrode based on polypyrrole through the electrochemical deposition of carboxy functionalized multi-walled carbon nanotubes on an indium–tin–oxide (ITO) electrode via p-toluene sulfonic acid (PTS). Subsequently, cholesterol esterase and cholesterol oxidase were immobilized onto this nanocomposite electrode using N-ethyl-N-(3-dimethylaminopropyl) carbodiimide and N-hydroxy succinimide for the detection of cholesterol. Figure 12 presents the schematic representation of this fabricated nanocomposite electrode. The electrode exhibited a response time of 9 s, a linear concentration range of 4×10^{-4} to 6.5×10^{-3} M/L, and thermal stability up to 45 °C [138].

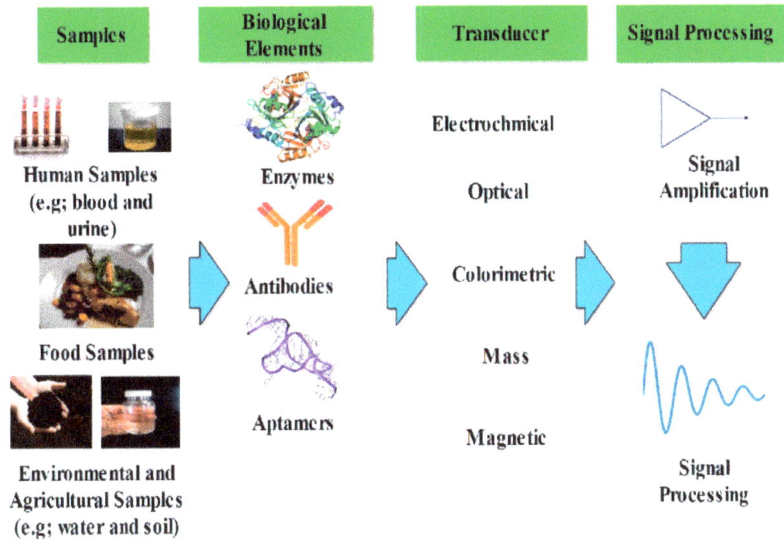

Figure 11. Schematic representation of biosensor (taken from [132]).

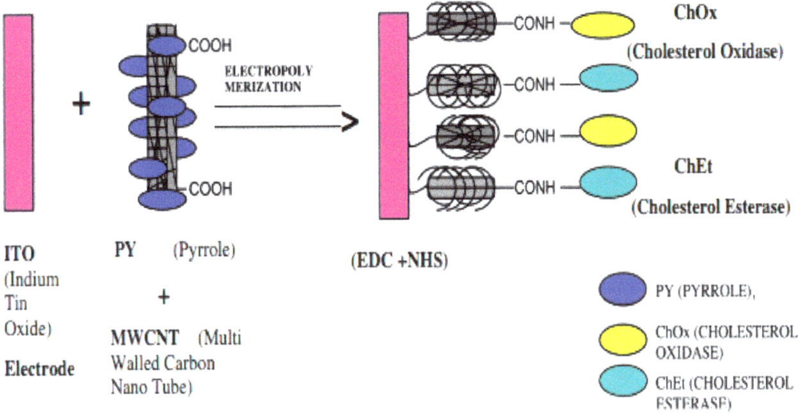

Figure 12. Fabrication of nanocomposite electrode (taken from [138]).

Gas sensors could be used for monitoring and controlling gas emissions from several emitters. A gas sensor comprises a sensing material capable of detecting combustible gases toxic gases, and vapors. PNCs have been utilized in gas sensors due to their properties of large surface area, high sensitivity, high selectivity, and good electrical conductivity. PNCs are able to detect various gases, including ammonia and chloroform vapors. For instance, PLA/multi-wall carbon nanotube-based conductive biopolymer nanocomposites were developed using the spray layer-by-layer technique. The chemo-resistive behavior of these nanocomposites was studied based on exposure to various organic vapors (methanol, chloroform, water, and toluene) exhibiting different physical properties. The results indicated that the largest response from these sensors was obtained for chloroform, demonstrating the ability of nanocomposites to serve as a vapor sensor [139]. Zhu et al. studied a sensor based on the nanocomposites composed of carbon nanotubes and cellulose, which was highly sensitive to humidity. The nanocomposites exhibited superior performance as a humidity sensor, achieving a maximum response value of 69.9% ($\Delta I/I0$) at a relative humidity of 95%.

The sensor also exhibited good long-term stability and bending resistance. In addition, the fabricated humidity sensor could be used for monitoring human breath [140]. Cheng reported a nanocomposite based on polycaprolactone/carbon black, which was used as a sensor for the detection of solvent vapors [141]. Pandey et al. prepared agar gum/silver nanocomposites, which when used as sensors, exhibited a response time of 2–3 s and an ammonia solution detection limit of 1 ppm at room temperature. These nanocomposites appear promising for use as optical sensors for ammonia detection [142]. Dai et al. prepared nanocomposites based on Ct/ZnO/single-walled carbon nanotubes to be used as chemo-resistive humidity sensors. The sensing was achieved owing to the Ct swelling behavior of the surrounding nanotubes, which led to a change in the hopping conduction path between the nanotubes [143]. Chakraborty reported the fabrication of PLA/exfoliated graphene nanocomposites to be used as a sensor for the detection of ethanol vapor [144].

The designing of the sensors for the detection of heavy metal ions is highly desired as heavy metal ions cause a great threat and hazards to the ecosystem. PNCs have been used in heavy metal ion sensors owing to their good conductivity and environmental stability. PNC-based sensors have been used for detecting several heavy metal ions, including cadmium ions (Cd^{2+}), lead ions (Pb^{2+}), and copper ions (Cu^{2+}). For instance, Khachatryan et al. fabricated a sensor based on starch/ZnS quantum dots/L-cysteine nanocomposites for the detection of Pb^{2+} and Cu^{2+} ions [145]. Another sensor based on graphene oxide/carbon nanotubes/poly(O-toluidine) nanocomposite could selectively detect Pb^{2+} ions in aqueous solutions and also exhibited an antimicrobial behavior by inhibiting *E. coli* and *B. subtilis* [146]. Wang et al. used glassy carbon electrodes coated with polyaniline/multi-wall carbon nanotubes for the detection of Pb^{2+} ions in the buffer solution of acetate. The modified electrode exhibited enhanced activity compared to the original glassy carbon electrode [147]. Y. Shao et al. used a screen-printed carbon electrode modified gold nanoparticles/polyaniline/multi-wall carbon nanocomposite for fabricating a highly sensitive sensor for the detection of Cu^{+2} ions with the detection limit of 0.017 μg/L and a linear concentration range of 1–180 μg/L. The sensor also exhibited excellent stability, selectivity, repeatability, and reproducibility [148].

Table 3. Performance of some polymer nanocomposites in sensors.

Polymer Nanocomposite	Type of Sensor	Target	Results	Ref.
NiO– chitosan/ZnO/Zinc hexacyanoferrate nanocomposite film	Biosensor	Triolein	Optimum response: within 4 s linear concentration range: (50–700 mg/dL) Sensitivity: 0.05 A/mg/dL	[135]
GOx/MWCNTs-polyaniline nanocomposite.	Biosensor	Glucose	Electrical conductivity: 3.78×10^{-1} Scm^{-1} Response time: 5 s Linear concentration range: 0.5–22 mM	[149]
Polyaniline/MWCNTs/Au NPs nanocomposite modified glass carbon electrode	Biosensor	Glucose	Detection limit: 0.19 mM Sensitivity: 29.17 mA mM^{-1} cm^{-2} Concentration range: 0.0625–1.19 mM	[150]
Polypyrrole/MWCNTs/GOx nanocomposite modified glassy carbon electrode	Biosensor	Glucose	The linear range: up to 4 mM Sensitivity: 95 $nAmM^{-1}$ Response time: 8 s	[151]
Polypyrrole/MWCNTs/Au NPs/ChOx	Biosensor	Cholesterol	Linear response: (2×10^{-3} to 8×10^{-3} M) Detection limit: 0.1×10^{-3} M Sensitivity: 10.12 mA mM^{-1} cm^{-2}.	[152]

Table 3. *Cont.*

Polymer Nanocomposite	Type of Sensor	Target	Results	Ref.
Polyaniline/ Functionalize MWCNT nanocomposite	Gas Sensor	Ammonia Vapor	High sensitivity (92% for100 ppm) Detection limit: (200 ppb) Response time: (9 s) Recovery time: (30 s)	[153]
Polypyrrole/Nitrogen-doped MWCNTs film fabricated on PI substrate	Gas Sensor	NO_2 gas	The sensor possessed high response of 24.82% (Rg − Ra)/Ra × 100%) under 5 ppm of NO_2. The sensor had outstanding selectivity, repeatability and stability	[154]
Ethylene diamine tetraacetic acid/Polyaniline/MWCNTs. with carbon electrode	Metal ion sensor	Pb^{+2}	Detection limit: 22 pM	[155]
Polypyrrole/MWCNTs deposited on electrode	Metal ion sensor	Pb^{+2} ions	Detection limit: 2.9×10^{-9} mol/L (S/N = 3)	[156]
Polyaniline/MWCNTs -3-aminopropyltriethoxysilane casted on glassy carbon electrode	Metal ion sensor	Cd^{+2} ions	Detection limit: 0.015 μM Linear concentration range:(0.05–50 μM)	[157]
Modified glassy carbon electrode with polythiophene/COOH -MWCNTs/reduced graphene oxide	Metal ion sensor	Hg^{+2} ions	Linear range: (0.1 to 25 μM) Limit of detection: (0.009 μM) Recovery: between 110.7 and 96.79%	[158]

The use of PNCs as sensing materials in sensors is based on several advantages of PNCs in terms of stability, good electrical performance, and chemical properties. PNCs are able to detect various compounds, including biomolecules, gases, and metal ions. However, certain challenges regarding sensitivity, selectivity, and recovery times are encountered. Selectivity remains a major challenge when detecting specific species. The non-repeatability of certain PNCs limits their practical application potential in sensors as it would be economically unaffordable to use the sensor only one time. Factors such as the design of the PNCs, the type of polymer component of the PNC, the type of component nanofillers, and the ratio between the polymer and the nanofillers are important in determining the performance of a PNC as a sensing material. Among the different polymers, conducting polymers could be the most effective due to their remarkable performance as a sensing material. The carbon-based nanomaterials have also been used widely as nanofillers in the preparation of PNCs aimed for sensors as these nanomaterials possess excellent electrical properties. A clear understanding of the PNC properties, the type of interaction between the component polymer and nanomaterial, the effect of the size and shape of the PNC, the surface area of the PNC, its sensing mechanism, porosity, and other factors affecting the sensing capability would contribute immensely to the controlled synthesis of a PNC capable of fulfilling all the desired requirements. Therefore, further research is necessary to develop a PNC to be used as a sensing material that would exhibit improved performance at a lower cost compared to the currently available PNCs, which would, in turn, facilitate the expansion of the scope of application of PNCs in industrial fields [159].

4.3. Electromagnetic Shielding in Aerospace Applications

The advances in the technology and utilization of telecommunication devices and electronics have increased electromagnetic pollution. The resulting electromagnetic interference (EMI) could disrupt equipment, systems, and the electronic devices used in critical fields, such as military, aerospace, and medicine. Long-term exposure to electromagnetic waves could cause adverse effects on human health. Electronic systems and equipment generate waves that exist in the microwave range of the electromagnetic radiation spectrum.

These radiations have to be shielded. In this context, electromagnetic shielding is defined as the practice of electromagnetic field reduction in a space using a blocking field with barriers composed of a magnetic or conductive material. Shielding is achieved within enclosures used for isolating the electrical devices from the "outside world" and isolating the wires from the environment through which the cables run [160]. Figure 13 presents the schematic representation of EMI shielding. The mechanisms that are exploited for shielding could be categorized into three main classes—reflection, absorption, and multiple reflections. Reflection is the primary mechanism exploited in EMI shielding. In order to shield radiation through reflection, the shield has to contain the mobile charge carriers—either electrons or holes—that would interact with the electromagnetic field in the radiation to be shielded. Absorption is the secondary mechanism of EMI shielding. In order to shield radiation mainly through absorption, the shield must have magnetic or/and electric dipoles that would interact with the electromagnetic field in the radiation. The electric dipoles could be provided by BaTiO or other materials with a high value of the dielectric constant, while the magnetic dipoles could be provided by FeO or other materials with a high degree of magnetic permeability. The third mechanism of EMI shielding is multiple reflections, in which reflections occur at various interfaces or surfaces in the shield. This mechanism of shielding requires a large interface or surface area in the shield. The loss owing to multiple reflections is negligible when the distance between the reflecting interfaces or surfaces is large compared to the skin depth. Shielding effectiveness (dB) of a material is defined as the sum of all the losses that have occurred due to reflection, absorption and multiple reflections [161–163].

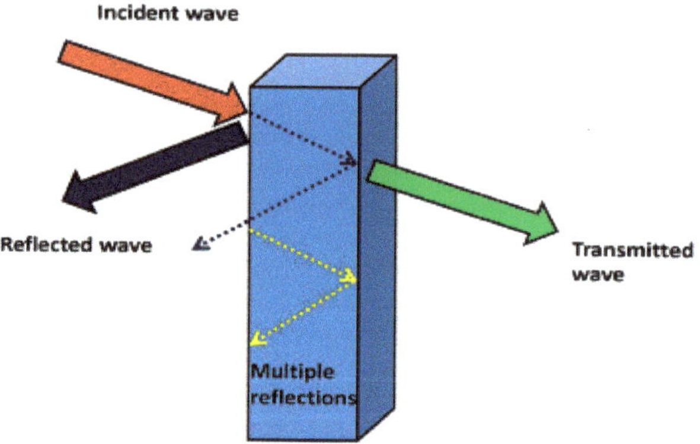

Figure 13. Schematic representation of EMI shielding (taken from [164]).

The effectiveness of shielding relies on conductivity; higher conductivity leads to better shielding efficiency. The development of lightweight materials with high electromagnetic radiation shielding performance to enable the prevention of interference is highly desirable. In this regard, PNCs could serve several purposes, which include providing solutions for aerospace and environmental applications. Various PNCs that comprise conductive nanofillers, such as metal nanoparticles, magnetic nanoparticles, or carbon nanomaterials, could be utilized as EMI shielding materials. Table 4 lists the shielding performance of a few lightweight polymer nanocomposites. In a study, polyvinylpyrrolidone/Fe_3O_4 nanocomposite nanofiber (FCNF) was developed, and its capability as electromagnetic interference (EMI) shielding material was studied using the frequency range of 8.2–12.4 GHz. The EMI shielding efficiency of FCNF increased up to approximately 22 dB, demonstrating that reflection was the secondary shielding mechanism, while the major shielding mechanism was absorption [165]. In another study, electromagnetic shielding composites based on the acrylic resin matrix (AR) were fabricated by inserting up to 30% (wt.%) activated charcoal (AC) loading. It was observed that the 30% (wt.%) AC loading composite exhibited a higher relative permittivity value (~79) compared to the AR (~5). Moreover, the electrical conductivity, porous structure, and permittivity value of the composite contributed to the EMI shielding effectiveness value of −36 dB, revealing the ability of these composites to serve as an efficient coating for EMI shielding [166]. The porous polymer nanocomposites fabricated in a study through the ionic self-assembly of gold nanoparticles on the charged polymer skeleton composed of poly (pyridobisimidazole)-grafted-poly(dimethyl diallyl ammonium chloride) exhibited a shielding effectiveness value of over –64.9 dB in the frequency range of 250 MHz–1.5 GHz with a thickness of only 20 mm [167].

The EMI shielding fabrics based on polyaniline and its composites and fabricated using the in situ polymerization method exhibited the properties of polyaniline and its composites as well as those of the fabrics used (cotton and nylon). The prepared functional fabrics with a thickness of 0.1 mm exhibited the EMI shielding performance of 11–15 dB in the frequency range of 8.2–18 GHz [168]. Lightweight and flexible shielding materials are suitable for a wide variety of millimeter-wave shielding applications. Such materials may be prepared using simple and economical methods. In a study, shielding effectiveness (EMSE) of the nanocomposites composed of heat-treated carbon nanofibers (CNF) in a linear low-density polyethylene matrix was assessed. It was observed that the heat treatment (HT) of carbon nanofibers at 2500 °C had remarkably increased the intrinsic transport and graphitic crystallinity, thereby increasing the electromagnetic shielding efficiency of the nanocomposites. The nanocomposites containing 11% vol. % (20% wt.%) of HT exhibited a DC electrical conductivity of 1.0 6 0.1 3 101 S/m, approximately 10 orders of magnitude better than that of the as-received PR-19 CNF nanocomposites. In the frequency range of 30 MHz–1.5 GHz, the nanocomposite containing HT exhibited average EMSE values of approximately 14 ± 2 dB [169]. Epoxy nanocomposites containing 15% of 2 mm thick single-walled carbon nanotube exhibited an EMI shielding performance of 20–30 dB [170]. A higher shielding was gained through the insertion of highly conductive metal particles, such as that in the case of a 4 µm CNT film with approximately 35% wt.% of iron which exhibited a shielding performance of 61–67 dB [171]. In another study, a five-layer CNT film exhibited extraordinary shielding efficiency of 67–78 dB [172]. When the electromagnetic shielding of carboxymethyl cellulose (CMC) and CMC/metal nanoparticles nanofiber mats was investigated, it was observed that the EMI improved in the presence of metal nanoparticles, and was dependent on the concentration and electrical conductivity of the metal nanoparticles [173].

Table 4. Shielding performance of some lightweight polymer nanocomposites.

Polymer Nanocomposites	Thickness d(mm)	Shielding (dB)	References
Poly (methyl methacrylate)/Multi-walled carbon nanotubes	0.06	27	[174]
Nitrile butadiene rubber/Fe_3O_4	2	80–90	[175]
Poly(vinyl alcohol)/Fe_3O_4	4.5	6	[176]
Polyurethane/Multi-walled Carbon nanotubes	0.1–0.2	20–29	[177,178]
Polyacrylate/Multi-walled carbon nanotubes	1.5	25	[179]
Polypropylene/Carbon black	2.8	40	[180]
Polysulfone/Carbon nofiber	1	45	[181]
Polylactide/Graphene	1.5	15	[182]
Polyaniline/Grahene	2.5	45.1	[183]
Polyetherimide/Graphene	2.3	44	[184]
Poly (methyl methacrylate)/Single-walled carbon nanotubes	4.5	40	[185]

The rapid proliferation of electro/electrical technology has rendered it necessary to suppress unwanted electromagnetic radiation for effective system implementation. In order to meet the ever-increasing demand for EMI suppression, the use of PNCs for shielding has increased. PNCs have a low fabrication cost, lighter weight, and relatively simple processability. The interfaces between the embedded nanofillers and the polymer within the PNCs are crucial. It is possible to optimize PNCs for several bands of frequency to encompass all the electronic/electrical systems used in various fields, including the medical, industrial, and military sectors. PNCs are capable of suppressing EMI through the mechanisms of reflection and absorption. Among the various nanomaterials that are embedded into PNCs, magnetic nanomaterials are the most appropriate for using PNCs in electromagnetic shielding. These magnetic nanomaterials include ferrites, carbon-based materials, and flexible dielectric materials. The concentration, shape, and size of the nanofillers have a significant influence on the electromagnetic properties of PNCs. It is important to refine the existing methods and compositions for developing EMI shielding responses. PNCs with larger interfaces could be utilized for various shielding applications. Moreover, these PNCs would produce remarkable polarization effects and dielectric attributes for better EM energy absorption. It would, therefore, be appropriate to investigate the synergy between PNC phases for understanding the possibilities of modification for obtaining materials exhibiting improved performance [186,187].

Although extensive research has already been conducted regarding the use of PNCs in electromagnetic shielding, several challenges remain to be overcome, including the nanofiller distribution in the PNCs, chemical stability of PNCs at extreme conditions, and thermal stability of PNCs at higher temperatures. Additional PNC preparation methods should be used when applying PNCs in electromagnetic shielding to produce better PNCs with a uniform distribution of nanomaterials. In addition, further research is required to develop radiation shielding materials with higher efficiency, better uniform distribution, and chemical and thermal stability. Moreover, the weight, size, and toxicity of PNCs have to be investigated to obtain lighter and thin shields with the least toxicity. Furthermore, using recycled polymer waste materials from industries for developing PNCs could be a promising solution for reducing costs and addressing environmental issues. This approach would be environmentally friendly and economically viable [188].

4.4. Food Packaging

Food packaging is an icon in the food industry as it plays an important role in modern society. Food packaging facilitates preservation and extension of the shelf life of food products during delivery and until consumption. This is realized through the prevention of unfavorable conditions or factors, such as spoilage microorganisms, chemical contaminants, external forces, moisture, oxygen, light, etc. Figure 14 presents the roles of active packaging in improving the shelf life of food. The food package should prevent microbial contamination, hinder loss or gain of moisture, and act as a barrier against the permeation of oxygen, carbon dioxide, water vapor, and other volatile compounds, including taints and flavors, in addition to the basic properties of packaging materials, such as thermal, optical, and mechanical properties. PNCs represent a novel class of materials that are considered promising candidates as materials for food packaging owing to their outstanding thermal, antimicrobial, mechanical, and barrier properties [189].

Recently, PNCs have been receiving great attention in the food packaging industry. Rahman et al. used a nanocomposite film composed of chitosan and zinc oxide, developed via solution mixing, as the packaging material for extending the shelf life of raw meat [190]. Mahmoodi et al. developed colure biodegradable film nanocomposites composed of PLA, dye, and clay, which exhibited superior light and thermo-mechanical resistance and also provided a barrier to mass transport, rendering these suitable for use as packaging materials [191]. The nanocomposite containing polycaprolactone and zinc oxide nanoparticles exhibited antimicrobial activity against pathogenic *Staphylococcus aureus* and demonstrated great potential as an active food packaging material [192]. Starch nanocomposite films with montmorillonite, prepared through melt processing, are also capable of being used in food packaging. Pirsa et al. developed a novel nanocomposite film based on bacterial cellulose, ZnO NPs, and polypyrrole, which may be useful in antimicrobial active and smart packaging applications [193]. Anothernanocomposite composed of carbon nanotubes and PHBV exhibited enhanced mechanical and thermal properties, facilitating its application in food packaging [194]. Kumar and Gautam reported that starch/ZnO nanocomposite could be used in food packaging applications [195]. Fernández et al. demonstrated that the cellulose/silver nanocomposite exhibited improved antimicrobial activity against spoilage microorganisms during the storage of minimally processed "Piel de Sapo" melon. Moreover, the presence of silver nanoparticles in the nanocomposite resulted in remarkably lower yeast counts and a juicier appearance, in addition to retarding the senescence of melon cuts after ten days of storage [196]. The reinforcement of polycaprolactone with exfoliated graphene oxide produced a nanocomposite that could also be utilized as a packaging material [197]. Cellulose/clay nanocomposite has also demonstrated enhanced mechanical properties, rendering it suitable for use in the packaging of food products [198]. A related study reported the synthesis of a nanocomposite of chitosan with graphene oxide that exhibited outstanding antimicrobial and mechanical behaviors, which render it suitable as a candidate material for the food packaging industry [199].

PNCs have been used in food packaging due to several factors, including their good mechanical and thermal properties, antimicrobial activity, barrier properties, and optical properties. However, certain PNCs may be composed of synthetic polymers or biopolymers incorporated with nanomaterials. It is imperative to limit the use of such PNCs in food packaging, including those that are based on synthetic petroleum-based polymeric films, due to their adverse impacts on human health and also on the environment. On the other hand, the use of PNCs derived from biopolymers could be increased as this exhibit outstanding performance. However, the use of such PNCs for food packaging in industrial fields also remains limited due to several factors, such as their high production cost and unavailability for large-scale production. In order to address these limitations, further research aimed at attaining better mechanical, thermal, and barrier properties along with improved biocompatibility in the PNCs, is required. In addition, the low cost and large scale production of PNC films by avoiding the use of expensive chemicals and rather using economical and natural components would widen the practical application scope of PNCs in various industrial fields [200].

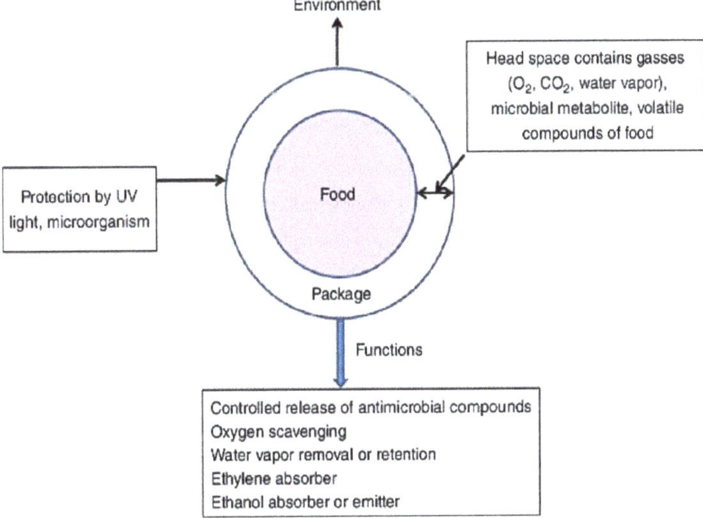

Figure 14. Functions of active packaging to improve self-life of packaged food (taken from [189]).

5. Conclusions

The present review aimed to describe the current scenario regarding the use of polymer nanocomposites in various fields of applications. As discussed, such applications require producing stable nanoparticles that could be functionalized and exhibit biocompatible properties. The successful development of particles exhibiting specific functional properties for industrial and environmental applications is the greatest challenge for the research community in this field. Polymer nanocomposites hold great importance in the progress of material science. The promising physical and chemical properties of the components of PNCs would enable extending the smart behaviors to material applications. However, much remains to be investigated regarding these systems. Indeed, controlling the challenging and critical variables would provide an even greater impetus for developing these materials and ensuring their potential for upscaling. So far, different preparation techniques have been reported for polymer nanocomposites. However, further investigation is warranted in this regard as well because the intrinsic properties of nanofillers and ensuring their uniform dispersion within the polymer remain a great challenge to date. The preparation method and the types of nanofillers are important factors in determining the type of application a nanocomposite would be suitable for. The polymer nanocomposites may be used in water treatment, sensor, electromagnetic shielding and food packaging. This is attributed to the improved properties exhibited by the polymer nanocomposites compared to the original polymer.

Author Contributions: Conceptualization, M.S.A.D., M.H.M. and L.M.A.-H.; writing—original draft preparation, M.S.A.D., M.H.M. and L.M.A.-H.; writing—review and editing, M.S.A.D., M.H.M. and L.M.A.-H. All authors have read and agreed to the published version of the manuscript.

Funding: This research received no external funding.

Institutional Review Board Statement: Not applicable.

Informed Consent Statement: Not applicable.

Data Availability Statement: Not applicable.

Conflicts of Interest: The authors declare no conflict of interest.

References

1. Sun, X.; Huang, C.; Wang, L.; Liang, L.; Cheng, Y.; Fei, W.; Li, Y. Recent Progress in Graphene/Polymer Nanocomposites. *Adv. Mater.* **2021**, *33*, 2001105. [CrossRef]
2. Chow, W.S.; Ishak, Z.A.M. Smart polymer nanocomposites: A review. *Express Polym. Lett.* **2020**, *14*, 416–435. [CrossRef]
3. Alateyah, A.I.; Dhakal, H.; Zhang, Z. Processing, Properties, and Applications of Polymer Nanocomposites Based on Layer Silicates: A Review. *Adv. Polym. Technol.* **2013**, *32*. [CrossRef]
4. Luo, H.; Zhou, X.; Ellingford, C.; Zhang, Y.; Chen, S.; Zhou, K.; Zhang, D.; Bowen, C.R.; Wan, C. Interface design for high energy density polymer nanocomposites. *Chem. Soc. Rev.* **2019**, *48*, 4424–4465. [CrossRef] [PubMed]
5. Wu, H.; Fahy, W.; Kim, S.; Kim, H.; Zhao, N.; Pilato, L.; Kafi, A.; Bateman, S.; Koo, J. Recent developments in polymers/polymer nanocomposites for additive manufacturing. *Prog. Mater. Sci.* **2020**, *111*, 100638. [CrossRef]
6. Mukhopadhyay, R.; Bhaduri, D.; Sarkar, B.; Rusmin, R.; Hou, D.; Khanam, R.; Sarkar, S.; Biswas, J.K.; Vithanage, M.; Bhatnagar, A.; et al. Clay–polymer nanocomposites: Progress and challenges for use in sustainable water treatment. *J. Hazard. Mater.* **2020**, *383*, 121125. [CrossRef]
7. Yuan, M.; Hu, M.; Dai, F.; Fan, Y.; Deng, Z.; Deng, H.; Cheng, Y. Application of synthetic and natural polymers in surgical mesh for pelvic floor reconstruction. *Mater. Des.* **2021**, *209*, 109984. [CrossRef]
8. Jamróz, E.; Kulawik, P.; Kopel, P. The Effect of Nanofillers on the Functional Properties of Biopolymer-Based Films: A Review. *Polymers* **2019**, *11*, 675. [CrossRef]
9. Vera, M.; Mella, C.; Urbano, B.F. Smart Polymer Nanocomposites: Recent Advances and Perspectives. *J. Chil. Chem. Soc.* **2020**, *65*, 4973–4981. [CrossRef]
10. Talaat, A.; Suraj, M.; Byerly, K.; Wang, A.; Wang, Y.; Lee, J.; Ohodnicki, P. Review on soft magnetic metal and inorganic oxide nanocomposites for power applications. *J. Alloy Compd.* **2021**, *870*, 159500. [CrossRef]
11. Mout, R.; Moyano, D.F.; Rana, S.; Rotello, V.M. Surface functionalization of nanoparticles for nanomedicine. *Chem. Soc. Rev.* **2012**, *41*, 2539–2544. [CrossRef]

12. Jawed, A.; Saxena, V.; Pandey, L. Engineered nanomaterials and their surface functionalization for the removal of heavy metals: A review. *J. Water Process Eng.* **2020**, *33*, 101009. [CrossRef]
13. Wanna, Y.; Chindaduang, A.; Tumcharern, G.; Phromyothin, D.; Porntheerapat, S.; Nukeaw, J.; Hofmann, H.; Pratontep, S. Efficiency of SPIONs functionalized with polyethylene glycol bis(amine) for heavy metal removal. *J. Magn. Magn. Mater.* **2016**, *414*, 32–37. [CrossRef]
14. Hashim, A.; Agool, I.R.; Kadhim, K.J. Modern Developments in Polymer Nanocomposites For Antibacterial and Antimicrobial Applications: A Review. *J. Bionanosci.* **2018**, *12*, 608–613. [CrossRef]
15. Luan, J.; Wang, S.; Hu, Z.; Zhang, L. Synthesis Techniques, Properties and Applications of Polymer Nanocomposites. *Curr. Org. Synth.* **2012**, *9*, 114–136. [CrossRef]
16. Saraswathi, M.S.S.A.; Nagendran, A.; Rana, D. Tailored polymer nanocomposite membranes based on carbon, metal oxide and silicon nanomaterials: A review. *J. Mater. Chem. A* **2019**, *7*, 8723–8745. [CrossRef]
17. Müller, K.; Bugnicourt, E.; Latorre, M.; Jorda, M.; Echegoyen Sanz, Y.E.; Lagaron, J.M.; Miesbauer, O.; Bianchin, A.; Hankin, S.; Bölz, U.; et al. Review on the Processing and Properties of Polymer Nanocomposites and Nanocoatings and Their Applications in the Packaging, Automotive and Solar Energy Fields. *Nanomaterials* **2017**, *7*, 74. [CrossRef]
18. Shameem, M.M.; Sasikanth, S.; Annamalai, R.; Raman, R.G. A brief review on polymer nanocomposites and its applications. In *Materials Today: Proceedings*; Elsevier: Amsterdam, The Netherlands, 2021; Volume 45, pp. 2536–2539.
19. Porel, S.; Venkatram, N.; Rao, D.N.; Radhakrishnan, T.P. In situ synthesis of metal nanoparticles in polymer matrix and their optical limiting applications. *J. Nanosci. Nanotechnol.* **2007**, *7*, 1887–1892. [CrossRef]
20. Kashihara, K.; Uto, Y.; Nakajima, T. Rapid in situ synthesis of polymer-metal nanocomposite films in several seconds using a CO_2 laser. *Sci. Rep.* **2018**, *8*, 14719. [CrossRef] [PubMed]
21. Adnan, M.M.; Dalod, A.R.M.; Balci, M.H.; Glaum, J.; Einarsrud, M.-A. In Situ Synthesis of Hybrid Inorganic–Polymer Nanocomposites. *Polymers* **2018**, *10*, 1129. [CrossRef]
22. Akpan, E.; Shen, X.; Wetzel, B.; Friedrich, K. Design and Synthesis of Polymer Nanocomposites. In *Polymer Composites with Functionalized Nanoparticles*; Elsevier: Amsterdam, The Netherlands, 2019; pp. 47–83. [CrossRef]
23. Lü, T.; Zhang, S.; Qi, D.; Zhang, D.; Zhao, H. Thermosensitive poly(N-isopropylacrylamide)-grafted magnetic nanoparticles for efficient treatment of emulsified oily wastewater. *J. Alloy Compd.* **2016**, *688*, 513–520. [CrossRef]
24. Munnawar, I.; Iqbal, S.S.; Anwar, M.N.; Batool, M.; Tariq, S.; Faitma, N.; Khan, A.L.; Khan, A.U.; Nazar, U.; Jamil, T.; et al. Synergistic effect of Chitosan-Zinc Oxide Hybrid Nanoparticles on antibiofouling and water disinfection of mixed matrix polyethersulfone nanocomposite membranes. *Carbohydr. Polym.* **2017**, *175*, 661–670. [CrossRef] [PubMed]
25. Li, W.; Xu, Z.; Chen, L.; Shan, M.; Tian, X.; Yang, C.; Lv, H.; Qian, X. A facile method to produce graphene oxide-g-poly(L-lactic acid) as an promising reinforcement for PLLA nanocomposites. *Chem. Eng. J.* **2014**, *237*, 291–299. [CrossRef]
26. Singla, P.; Mehta, R.; Upadhyay, S. Microwave assisted in situ ring-opening polymerization of polylactide/clay nanocomposites: Effect of clay loading. *Appl. Clay Sci.* **2014**, *95*, 67–73. [CrossRef]
27. de Oliveira, A.D.; Beatrice, C.A.G. Polymer nanocomposites with different types of nanofiller. In *Nanocomposites-Recent Evolutions*; IntechOpen Limited: Rijeka, Croatia, 2018; pp. 103–104. Available online: https://www.intechopen.com/chapters/64843 (accessed on 21 October 2021). [CrossRef]
28. Cheng, S.; Grest, G.S. Dispersing Nanoparticles in a Polymer Film via Solvent Evaporation. *ACS Macro Lett.* **2016**, *5*, 694–698. [CrossRef]
29. Díez-Pascual, A.M.; Diez-Vicente, A.L. ZnO-reinforced poly(3-hydroxybutyrate-co-3-hydroxyvalerate) bionanocomposites with antimicrobial function for food packaging. *ACS Appl. Mater. Interfaces* **2014**, *6*, 9822–9834. [CrossRef] [PubMed]
30. Mehr, R.K.; Peighambardoust, S.J. Preparation and Characterization of Corn Starch/Clay Nanocomposite Films: Effect of Clay Content and Surface Modification. *Starch-Starke* **2018**, *70*, 1700251. [CrossRef]
31. Hussein, L.I.; Abdaleem, A.H.; Darwish, M.S.A.; ElSawy, M.A.; Mostafa, M.H.; Hassan, A. Chitosan/TiO2 nanocomposites: Effect of microwave heating and solution mixing techniques on physical properties. *Egypt J. Chem.* **2020**, *63*, 449–460. [CrossRef]
32. Di, Y.; Iannace, S.; Di Maio, E.; Nicolais, L. Nanocomposites by melt intercalation based on polycaprolactone and organoclay. *J. Polym. Sci. Part B Polym. Phys.* **2003**, *41*, 670–678. [CrossRef]
33. Maniruzzaman, M.; Boateng, J.S.; Snowden, M.J.; Douroumis, D. A Review of Hot-Melt Extrusion: Process Technology to Pharmaceutical Products. *ISRN Pharm.* **2012**, *2012*, 436763. [CrossRef] [PubMed]
34. Anžlovar, A.; Kržan, A.; Žagar, E. Degradation of PLA/ZnO and PHBV/ZnO composites prepared by melt processing. *Arab. J. Chem.* **2018**, *11*, 343–352. [CrossRef]
35. Carli, L.N.; Crespo, J.D.S.; Mauler, R.S. PHBV nanocomposites based on organomodified montmorillonite and halloysite: The effect of clay type on the morphology and thermal and mechanical properties. *Compos. Part A Appl. Sci. Manuf.* **2011**, *42*, 1601–1608. [CrossRef]
36. Darwish, M.S.A.; Mostafa, M.H.; Hussein, L.I.; Abdaleem, A.H.; Elsawy, M.A. Preparation, characterization, mechanical and biodegradation behavior of polypropylene—chitosan/ZnO nanocomposite. *Polym. Technol. Mater.* **2021**, *60*, 1630–1640. [CrossRef]
37. Franco-Urquiza, E.A. Clay-Based Polymer Nanocomposites: Essential Work of Fracture. *Polymers* **2021**, *13*, 2399. [CrossRef] [PubMed]

38. Niu, H.; Wang, X.; Lin, T. Needleless Electrospinning: Developments and Performances. In *Nanofibers—Production, Properties and Functional Applications*; IntechOpen Limited: Rijeka, Croatia, 2011; pp. 17–36. Available online: https://www.intechopen.com/chapters/23290 (accessed on 1 October 2021). [CrossRef]
39. Huang, Z.-M.; Zhang, Y.-Z.; Kotaki, M.; Ramakrishna, S. A review on polymer nanofibers by electrospinning and their applications in nanocomposites. *Compos. Sci. Technol.* **2003**, *63*, 2223–2253. [CrossRef]
40. Yu, W.; Lan, C.-H.; Wang, S.-J.; Fang, P.-F.; Sun, Y.-M. Influence of zinc oxide nanoparticles on the crystallization behavior of electrospun poly(3-hydroxybutyrate-co-3-hydroxyvalerate) nanofibers. *Polymers* **2010**, *51*, 2403–2409. [CrossRef]
41. Cai, J.; Lei, M.; Zhang, Q.; He, J.-R.; Chen, T.; Liu, S.; Fu, S.-H.; Li, T.-T.; Liu, G.; Fei, P. Electrospun composite nanofiber mats of Cellulose@Organically modified montmorillonite for heavy metal ion removal: Design, characterization, evaluation of absorption performance. *Compos. Part A Appl. Sci. Manuf.* **2017**, *92*, 10–16. [CrossRef]
42. Mahdieh, Z.M.; Mottaghitalab, V.; Piri, N.; Haghi, A.K. Conductive chitosan/multi walled carbon nanotubes electrospun nanofiber feasibility. *Korean J. Chem. Eng.* **2011**, *29*, 111–119. [CrossRef]
43. Chan KH, K.; Wong, S.Y.; Tiju, W.C.; Li, X.; Kotaki, M.; He, C.B. Morphologies and electrical properties of electrospun poly [(R)-3-hydroxybutyrate-co-(R)-3-hydroxyvalerate]/multiwalled carbon nanotubes fibers. *J. Appl. Polym. Sci.* **2020**, *116*, 1030–1035. [CrossRef]
44. Lu, P.; Hsieh, Y.-L. Multiwalled Carbon Nanotube (MWCNT) Reinforced Cellulose Fibers by Electrospinning. *ACS Appl. Mater. Interfaces* **2010**, *2*, 2413–2420. [CrossRef] [PubMed]
45. Talal, J.H.; Mohammed, D.B.; Jawad, K.H. Fabrication of Hydrophobic Nanocomposites Coating Using Electrospinning Technique for Various Substrate. *J. Phys. Conf. Ser.* **2018**, *1032*, 012033. [CrossRef]
46. Mostafa, M.H.; Elsawy, M.A.; Darwish, M.S.; Hussein, L.I.; Abdaleem, A.H. Microwave-Assisted preparation of Chitosan/ZnO nanocomposite and its application in dye removal. *Mater. Chem. Phys.* **2020**, *248*, 122914. [CrossRef]
47. López-Cruz, A.; Barrera, C.; Calero-DdelC, V.L.; Rinaldi, C. Water dispersible iron oxide nanoparticles coated with covalently linked chitosan. *J. Mater. Chem.* **2009**, *19*, 6870–6876. [CrossRef]
48. Yan, L.; Chang, P.R.; Zheng, P. Preparation and characterization of starch-grafted multiwall carbon nanotube composites. *Carbohydr. Polym.* **2011**, *84*, 1378–1383. [CrossRef]
49. Hoogenboom, R. Temperature-Responsive Polymers: Properties, Synthesis, and Applications. In *Smart Polymers and their Applications*; Elsevier: Amsterdam, The Netherlands, 2019; pp. 13–44. [CrossRef]
50. Niskanen, J.; Tenhu, H. How to manipulate the upper critical solution temperature (UCST)? *Polym. Chem.* **2017**, *8*, 220–232. [CrossRef]
51. Yaghoubi, Z.; Basiri-Parsa, J. Modification of ultrafiltration membrane by thermo-responsive Bentonite-poly(N-isopropylacrylamide) nanocomposite to improve its antifouling properties. *J. Water Process. Eng.* **2020**, *34*, 101067. [CrossRef]
52. Tatiana, N.M.; Cornelia, V.; Tatia, R.; Aurica, C. Hybrid collagen/pNIPAAM hydrogel nanocomposites for tissue engineering application. *Colloid Polym. Sci.* **2018**, *296*, 1555–1571. [CrossRef]
53. Demir Oğuz, Ö.; Ege, D. Rheological and Mechanical Properties of Thermoresponsive Methylcellulose/Calcium Phosphate-Based Injectable Bone Substitutes. *Materials* **2018**, *11*, 604. [CrossRef]
54. Arafa, M.G.; El-Kased, R.F.; Elmazar, M.M. Thermoresponsive gels containing gold nanoparticles as smart antibacterial and wound healing agents. *Sci. Rep.* **2018**, *8*, 13674. [CrossRef] [PubMed]
55. Duan, Q.; Liu, Y.; Chang, S.; Chen, H.; Chen, J.-H. Surface Plasmonic Sensors: Sensing Mechanism and Recent Applications. *Sensors* **2021**, *21*, 5262. [CrossRef]
56. Pastoriza-Santos, I.; Kinnear, C.; Pérez-Juste, J.; Mulvaney, P.; Liz-Marzán, L.M. Plasmonic polymer nanocomposites. *Nat. Rev. Mater.* **2018**, *3*, 375–391. [CrossRef]
57. Zhu, C.; Lu, Y.; Peng, J.; Chen, J.; Yu, S. Photothermally sensitive poly (N-isopropylacrylamide)/graphene oxide nanocomposite hydrogels as remote light-controlled liquid microvalves. *Adv. Funct. Mater.* **2012**, *22*, 4017–4022. [CrossRef]
58. Li, H.; Yao, Y.; Shi, H.; Lei, Y.; Huang, Y.; Wang, K.; He, X.; Liu, J. A near-infrared light-responsive nanocomposite for photothermal release of H2S and suppression of cell viability. *J. Mater. Chem. B* **2019**, *7*, 5992–5997. [CrossRef] [PubMed]
59. Raza, A.; Hayat, U.; Rasheed, T.; Bilal, M.; Iqbal, H.M. "Smart" materials-based near-infrared light-responsive drug delivery systems for cancer treatment: A review. *J. Mater. Res. Technol.* **2019**, *8*, 1497–1509. [CrossRef]
60. Yang, N.; Zhu, M.; Xu, G.; Liu, N.; Yu, C. A near-infrared light-responsive multifunctional nanocomposite hydrogel for efficient and synergistic antibacterial wound therapy and healing promotion. *J. Mater. Chem. B* **2020**, *8*, 3908–3917. [CrossRef] [PubMed]
61. Yue, J.; He, L.; Tang, Y.; Yang, L.; Wu, B.; Ni, J. Facile design and development of photoluminescent graphene quantum dots grafted dextran/glycol-polymeric hydrogel for thermoresponsive triggered delivery of buprenorphine on pain management in tissue implantation. *J. Photochem. Photobiol. B Biol.* **2019**, *197*, 111530. [CrossRef] [PubMed]
62. Ratna, D.; Karger-Kocsis, J. Recent advances in shape memory polymers and composites: A review. *J. Mater. Sci.* **2008**, *43*, 254–269. [CrossRef]
63. Zhang, B.-T.; Zheng, X.; Li, H.-F.; Lin, J.-M. Application of carbon-based nanomaterials in sample preparation: A review. *Anal. Chim. Acta* **2013**, *784*, 1–17. [CrossRef] [PubMed]
64. Yang, C.; Liu, Z.; Chen, C.; Shi, K.; Zhang, L.; Ju, X.-J.; Wang, W.; Xie, R.; Chu, L.-Y. Reduced Graphene Oxide-Containing Smart Hydrogels with Excellent Electro-Response and Mechanical Properties for Soft Actuators. *ACS Appl. Mater. Interfaces* **2017**, *9*, 15758–15767. [CrossRef] [PubMed]

65. Soto, G.; Meiorin, C.; Actis, D.; Zélis, P.M.; Londoño, O.M.; Muraca, D.; Mosiewicki, M.; Marcovich, N. Magnetic nanocomposites based on shape memory polyurethanes. *Eur. Polym. J.* **2018**, *109*, 8–15. [CrossRef]
66. Li, J.; Zheng, L.; Cai, H.; Sun, W.; Shen, M.; Zhang, G.; Shi, X. Polyethyleneimine-mediated synthesis of folic acid-targeted iron oxide nanoparticles for in vivo tumor MR imaging. *Biomaterials* **2013**, *34*, 8382–8392. [CrossRef] [PubMed]
67. Nitin, N.; LaConte, L.; Zurkiya, O.; Hu, X.; Bao, G. Functionalization and peptide-based delivery of magnetic nanoparticles as an intracellular MRI contrast agent. *JBIC J. Biol. Inorg. Chem.* **2004**, *9*, 706–712. [CrossRef] [PubMed]
68. Maenosono, S.; Saita, S. Theoretical Assessment of FePt Nanoparticles as Heating Elements for Magnetic Hyperthermia. *IEEE Trans. Magn.* **2006**, *42*, 1638–1642. [CrossRef]
69. Stauffer, P.; Cetas, T.C.; Jones, R.C. Magnetic Induction Heating of Ferromagnetic Implants for Inducing Localized Hyperthermia in Deep-Seated Tumors. *IEEE Trans. Biomed. Eng.* **1984**, *31*, 235–251. [CrossRef] [PubMed]
70. Latorre, M.; Rinaldi, C. Applications of magnetic nanoparticles in medicine: Magnetic fluid hyperthermia. *Puerto Rico Health Sci. J.* **2009**, *28*, 227–238.
71. Niemirowicz, K.; Markiewicz, K.; Wilczewska, A.; Car, H. Magnetic nanoparticles as new diagnostic tools in medicine. *Adv. Med. Sci.* **2012**, *57*, 196–207. [CrossRef] [PubMed]
72. Ilg, P.; Kröger, M. Dynamics of interacting magnetic nanoparticles: Effective behavior from competition between Brownian and Néel relaxation. *Phys. Chem. Chem. Phys.* **2020**, *22*, 22244–22259. [CrossRef] [PubMed]
73. Darwish, M.; Stibor, I. Preparation and characterization of magnetite–PDMS composites by magnetic induction heating. *Mater. Chem. Phys.* **2015**, *164*, 163–169. [CrossRef]
74. Yang, J.; Fan, L.; Xu, Y.; Xia, J. Iron oxide nanoparticles with different polymer coatings for photothermal therapy. *J. Nanopart. Res.* **2017**, *19*, 333. [CrossRef]
75. Bardajee, G.R.; Hooshyar, Z.; Rastgo, F. Kappa carrageenan-g-poly (acrylic acid)/SPION nanocomposite as a novel stimuli-sensitive drug delivery system. *Colloid Polym. Sci.* **2013**, *291*, 2791–2803. [CrossRef]
76. Bajpai, A.K.; Gupta, R. Magnetically mediated release of ciprofloxacin from polyvinyl alcohol based superparamagnetic nanocomposites. *J. Mater. Sci. Mater. Electron.* **2011**, *22*, 357–369. [CrossRef] [PubMed]
77. Lemine, O. Magnetic Hyperthermia Therapy Using Hybrid Magnetic Nanostructures. In *Hybrid Nanostructures for Cancer Theranostics*; Elsevier: Amsterdam, The Netherlands, 2018; pp. 125–138. [CrossRef]
78. Kawai, N.; Kobayashi, D.; Yasui, T.; Umemoto, Y.; Mizuno, K.; Okada, A.; Tozawa, K.; Kobayashi, T.; Kohri, K. Evaluation of side effects of radiofrequency capacitive hyperthermia with magnetite on the blood vessel walls of tumor metastatic lesion surrounding the abdominal large vessels: An agar phantom study. *Vasc. Cell* **2014**, *6*, 15. [CrossRef] [PubMed]
79. De Nardo, S.J.; DeNardo, G.L.; Miers, L.A.; Natarajan, A.; Foreman, A.R.; Gruettner, C.; Adamson, G.N.; Ivkov, R. Development of Tumor Targeting Bioprobes (111In-Chimeric L6 Monoclonal Antibody Nanoparticles) for Alternating Magnetic Field Cancer Therapy. *Clin. Cancer Res.* **2005**, *11*, 7087s–7092s. [CrossRef] [PubMed]
80. Ivkov, R.; DeNardo, S.J.; Daum, W.; Foreman, A.R.; Goldstein, R.C.; Nemkov, V.S.; DeNardo, G.L. Application of High Amplitude Alternating Magnetic Fields for Heat Induction of Nanoparticles Localized in Cancer. *Clin. Cancer Res.* **2005**, *11*, 7093s–7103s. [CrossRef]
81. Darwish, M.S.; Nguyen, N.H.; Ševců, A.; Stibor, I.; Smoukov, S.K. Dual-modality self-heating and antibacterial polymer-coated nanoparticles for magnetic hyperthermia. *Mater. Sci. Eng. C* **2016**, *63*, 88–95. [CrossRef] [PubMed]
82. Akharame, M.O.; Fatoki, O.S.; Opeolu, B.O.; Olorunfemi, D.I.; Oputu, O.U. Polymeric Nanocomposites (PNCs) for Wastewater Remediation: An Overview. *Polym. Technol. Eng.* **2018**, *57*, 1801–1827. [CrossRef]
83. Wen, Y.; Yuan, J.; Ma, X.; Wang, S.; Liu, Y. Polymeric nanocomposite membranes for water treatment: A review. *Environ. Chem. Lett.* **2019**, *17*, 1539–1551. [CrossRef]
84. Mohammed, N.; Grishkewich, N.; Tam, K.C. Cellulose nanomaterials: Promising sustainable nanomaterials for application in water/wastewater treatment processes. *Environ. Sci. Nano* **2018**, *5*, 623–658. [CrossRef]
85. Katheresan, V.; Kansedo, J.; Lau, S.Y. Efficiency of various recent wastewater dye removal methods: A review. *J. Environ. Chem. Eng.* **2018**, *6*, 4676–4697. [CrossRef]
86. Khan, M.A.; Govindasamy, R.; Ahmad, A.; Siddiqui, M.; Alshareef, S.; Hakami, A.; Rafatullah, M. Carbon Based Polymeric Nanocomposites for Dye Adsorption: Synthesis, Characterization, and Application. *Polymers* **2021**, *13*, 419. [CrossRef] [PubMed]
87. Zaman, A.; Orasugh, J.T.; Banerjee, P.; Dutta, S.; Ali, M.S.; Das, D.; Bhattacharya, A.; Chattopadhyay, D. Facile one-pot in-situ synthesis of novel graphene oxide-cellulose nanocomposite for enhanced azo dye adsorption at optimized conditions. *Carbohydr. Polym.* **2020**, *246*, 116661. [CrossRef]
88. Peng, N.; Hu, D.; Zeng, J.; Li, Y.; Liang, L.; Chang, C. Superabsorbent Cellulose–Clay Nanocomposite Hydrogels for Highly Efficient Removal of Dye in Water. *ACS Sustain. Chem. Eng.* **2016**, *4*, 7217–7224. [CrossRef]
89. Wang, N.; Chen, J.; Wang, J.; Feng, J.; Yan, W. Removal of methylene blue by Polyaniline/TiO_2 hydrate: Adsorption kinetic, isotherm and mechanism studies. *Powder Technol.* **2019**, *347*, 93–102. [CrossRef]
90. Gelaw, T.B.; Sarojini, B.K.; Kodoth, A.K. Review of the Advancements on Polymer/Metal Oxide Hybrid Nanocomposite-Based Adsorption Assisted Photocatalytic Materials for Dye Removal. *ChemistrySelect* **2021**, *6*, 9300–9310. [CrossRef]
91. Kumar, P.S.; Selvakumar, M.; Babu, S.G.; Jaganathan, S.K.; Karuthapandian, S.; Chattopadhyay, S. Novel CuO/chitosan nanocomposite thin film: Facile hand-picking recoverable, efficient and reusable heterogeneous photocatalyst. *RSC Adv.* **2015**, *5*, 57493–57501. [CrossRef]

92. Shaikh, T.; Rathore, A.; Kaur, H. Poly (Lactic Acid) Grafting of TiO_2 Nanoparticles: A Shift in Dye Degradation Performance of TiO_2 from UV to Solar Light. *ChemistrySelect* **2017**, *2*, 6901–6908. [CrossRef]
93. Karagoz, S.; Kiremitler, N.B.; Sakir, M.; Salem, S.; Onses, M.S.; Sahmetlioglu, E.; Ceylan, A.; Yilmaz, E. Synthesis of Ag and TiO_2 modified polycaprolactone electrospun nanofibers (PCL/TiO_2-Ag NFs) as a multifunctional material for SERS, photocatalysis and antibacterial applications. *Ecotoxicol. Environ. Saf.* **2020**, *188*, 109856. [CrossRef]
94. Farzana, M.H.; Meenakshi, S. Photo-decolorization and detoxification of toxic dyes using titanium dioxide impregnated chitosan beads. *Int. J. Biol. Macromol.* **2014**, *70*, 420–426. [CrossRef] [PubMed]
95. Hosseini, F.; Sadighian, S.; Hosseini-Monfared, H.; Mahmoodi, N.M. Dye removal and kinetics of adsorption by magnetic chitosan nanoparticles. *Desalin. Water Treat.* **2016**, *57*, 24378–24386. [CrossRef]
96. Esvandi, Z.; Foroutan, R.; Peighambardoust, S.J.; Akbari, A.; Ramavandi, B. Uptake of anionic and cationic dyes from water using natural clay and clay/starch/$MnFe_2O_4$ magnetic nanocomposite. *Surfaces Interfaces* **2020**, *21*, 100754. [CrossRef]
97. Guan, Y.; Yu, H.-Y.; Abdalkarim, S.Y.H.; Wang, C.; Tang, F.; Marek, J.; Chen, W.-L.; Militky, J.; Yao, J.-M. Green one-step synthesis of ZnO/cellulose nanocrystal hybrids with modulated morphologies and superfast absorption of cationic dyes. *Int. J. Biol. Macromol.* **2019**, *132*, 51–62. [CrossRef] [PubMed]
98. Srivastava, V.; Choubey, A.K. Investigation of adsorption of organic dyes present in wastewater using chitosan beads immobilized with bio fabricated CuO nanoparticles. *J. Mol. Struct.* **2021**, *1242*, 130749. [CrossRef]
99. Ahmed, M.A.; Abdelbar, N.M.; Mohamed, A.A. Molecular imprinted chitosan-TiO_2 nanocomposite for the selective removal of Rose Bengal from wastewater. *Int. J. Biol. Macromol.* **2018**, *107*, 1046–1053. [CrossRef] [PubMed]
100. Mohammed, M.I.; Khafagy, R.M.; Hussien, M.S.A.; Sakr, G.B.; Ibrahim, M.A.; Yahia, I.S.; Zahran, H.Y. Enhancing the structural, optical, electrical, properties and photocatalytic applications of ZnO/PMMA nanocomposite membranes: Towards multifunctional membranes. *J. Mater. Sci. Mater. Electron.* **2021**, 1–26. [CrossRef]
101. Hussein, M.A.; Albeladi, H.K.; Elsherbiny, A.S.; El-Shishtawy, R.M.; Al-romaizan, A.N. Cross-linked poly (methyl methacrylate)/multiwall carbon nanotube nanocomposites for environmental treatment. *Adv. Polym. Technol.* **2018**, *37*, 3240–3251. [CrossRef]
102. Samadder, R.; Akter, N.; Roy, A.C.; Uddin, M.; Hossen, J.; Azam, S. Magnetic nanocomposite based on polyacrylic acid and carboxylated cellulose nanocrystal for the removal of cationic dye. *RSC Adv.* **2020**, *10*, 11945–11956. [CrossRef]
103. Saberi, A.; Alipour, E.; Sadeghi, M. Superabsorbent magnetic Fe_3O_4-based starch-poly (acrylic acid) nanocomposite hydrogel for efficient removal of dyes and heavy metal ions from water. *J. Polym. Res.* **2019**, *26*, 271. [CrossRef]
104. Zhou, G.; Wang, K.; Liu, H.; Wang, L.; Xiao, X.; Dou, D.; Fan, Y. Three-dimensional polylactic acid@graphene oxide/chitosan sponge bionic filter: Highly efficient adsorption of crystal violet dye. *Int. J. Biol. Macromol.* **2018**, *113*, 792–803. [CrossRef]
105. Senguttuvan, S.; Janaki, V.; Senthilkumar, P.; Kamala-Kannan, S. Polypyrrole/zeolite composite–A nanoadsorbent for reactive dyes removal from synthetic solution. *Chemosphere* **2022**, *287*, 132164. [CrossRef]
106. Agboola, O.; Fayomi, O.; Ayodeji, A.; Ayeni, A.; Alagbe, E.; Sanni, S.; Okoro, E.; Moropeng, L.; Sadiku, R.; Kupolati, K.; et al. A Review on Polymer Nanocomposites and Their Effective Applications in Membranes and Adsorbents for Water Treatment and Gas Separation. *Membranes* **2021**, *11*, 139. [CrossRef] [PubMed]
107. Momina; Ahmad, K. Study of different polymer nanocomposites and their pollutant removal efficiency: Review. *Polymers* **2021**, *217*, 123453. [CrossRef]
108. Ebrahim, S.; Shokry, A.; Ibrahim, H.; Soliman, M. Polyaniline/akaganéite nanocomposite for detoxification of noxious Cr(VI) from aquatic environment. *J. Polym. Res.* **2016**, *23*, 79. [CrossRef]
109. Akl, Z.F.; El-Saeed, S.M.; Atta, A.M. In-situ synthesis of magnetite acrylamide amino-amidoxime nanocomposite adsorbent for highly efficient sorption of U(VI) ions. *J. Ind. Eng. Chem.* **2016**, *34*, 105–116. [CrossRef]
110. Tran, M.T.; Nguyen, T.H.T.; Vu, Q.T.; Nguyen, M.V. Properties of poly (1-naphthylamine)/Fe_3O_4 composites and arsenic adsorption capacity in wastewater. *Front. Mater. Sci.* **2016**, *10*, 56–65. [CrossRef]
111. Musico, Y.L.F.; Santos, C.M.; Dalida, M.L.P.; Rodrigues, D.F. Improved removal of lead(ii) from water using a polymer-based graphene oxide nanocomposite. *J. Mater. Chem. A* **2013**, *1*, 3789–3796. [CrossRef]
112. Bagheripour, E.; Moghadassi, A.; Hosseini, S.; Nemati, M. Fabrication and Characterization of Novel Mixed Matrix Polyethersulfone Nanofiltration Membrane Modified by Iron-Nickel Oxide Nanoparticles. *J. Membr. Sci. Res* **2016**, *2*, 14–19.
113. Mittal, A.; Ahmad, R.; Hasan, I. Poly (methyl methacrylate)-grafted alginate/Fe_3O_4 nanocomposite: Synthesis and its application for the removal of heavy metal ions. *Desalin. Water Treat.* **2015**, *57*, 19820–19833. [CrossRef]
114. Saad, A.H.A.; Azzam, A.M.; El-Wakeel, S.T.; Mostafa, B.B.; El-latif, M.B.A. Removal of toxic metal ions from wastewater using ZnO@ Chitosan core-shell nanocomposite. *Environ. Nanotechnol. Monit. Manag.* **2018**, *9*, 67–75. [CrossRef]
115. Zhao, G.; Huang, X.; Tang, Z.; Huang, Q.; Niu, F.; Wang, X.-K. Polymer-based nanocomposites for heavy metal ions removal from aqueous solution: A review. *Polym. Chem.* **2018**, *9*, 3562–3582. [CrossRef]
116. Samiey, B.; Cheng, C.-H.; Wu, J. Organic-Inorganic Hybrid Polymers as Adsorbents for Removal of Heavy Metal Ions from Solutions: A Review. *Materials* **2014**, *7*, 673–726. [CrossRef] [PubMed]
117. Li, R.; Liu, L.; Yang, F. Preparation of polyaniline/reduced graphene oxide nanocomposite and its application in adsorption of aqueous Hg (II). *Chem. Eng. J.* **2013**, *229*, 460–468. [CrossRef]

118. Samuel, M.S.; Shah, S.S.; Subramaniyan, V.; Qureshi, T.; Bhattacharya, J.; Singh, N.P. Preparation of graphene oxide/chitosan/ferrite nanocomposite for Chromium (VI) removal from aqueous solution. *Int. J. Biol. Macromol.* **2018**, *119*, 540–547. [CrossRef] [PubMed]
119. Nasiri, R.; Arsalani, N.; Panahian, Y. One-pot synthesis of novel magnetic three-dimensional graphene/chitosan/nickel ferrite nanocomposite for lead ions removal from aqueous solution: RSM modelling design. *J. Clean. Prod.* **2018**, *201*, 507–515. [CrossRef]
120. Shoukat, A.; Wahid, F.; Khan, T.; Siddique, M.; Nasreen, S.; Yang, G.; Ullah, M.W.; Khan, R. Titanium oxide-bacterial cellulose bioadsorbent for the removal of lead ions from aqueous solution. *Int. J. Biol. Macromol.* **2019**, *129*, 965–971. [CrossRef] [PubMed]
121. Fallah, Z.; Isfahani, H.N.; Tajbakhsh, M.; Tashakkorian, H.; Amouei, A. TiO_2-grafted cellulose via click reaction: An efficient heavy metal ions bioadsorbent from aqueous solutions. *Cellulose* **2017**, *25*, 639–660. [CrossRef]
122. Moreno-Sader, K.; García-Padilla, A.; Realpe, A.; Acevedo-Morantes, M.; Soares, J.B.P. Removal of heavy metal water pollutants (Co^{2+} and Ni^{2+}) using polyacrylamide/sodium montmorillonite (PAM/Na-MMT) nanocomposites. *ACS Omega* **2019**, *4*, 10834–10844. [CrossRef] [PubMed]
123. Khan, S.A.; Siddiqui, M.F.; Alam Khan, T. Ultrasonic-assisted synthesis of polyacrylamide/bentonite hydrogel nanocomposite for the sequestration of lead and cadmium from aqueous phase: Equilibrium, kinetics and thermodynamic studies. *Ultrason. Sonochem.* **2019**, *60*, 104761. [CrossRef] [PubMed]
124. Mohammadnezhad, G.; Moshiri, P.; Dinari, M.; Steiniger, F. In situ synthesis of nanocomposite materials based on modified-mesoporous silica MCM-41 and methyl methacrylate for copper (II) adsorption from aqueous solution. *J. Iran. Chem. Soc.* **2019**, *16*, 1491–1500. [CrossRef]
125. Ahmad, R.; Hasan, I. Efficient remediation of an aquatic environment contaminated by Cr (VI) and 2, 4-dinitrophenol by XG-g-polyaniline@ ZnO nanocomposite. *J. Chem. Eng. Data* **2017**, *62*, 1594–1607. [CrossRef]
126. Chen, L.; Peng, X. Silver nanoparticle decorated cellulose nanofibrous membrane with good antibacterial ability and high water permeability. *Appl. Mater. Today* **2017**, *9*, 130–135. [CrossRef]
127. Sarkandi, A.F.; Montazer, M.; Harifi, T.; Rad, M.M. Innovative preparation of bacterial cellulose/silver nanocomposite hydrogels: In situ green synthesis, characterization, and antibacterial properties. *J. Appl. Polym. Sci.* **2021**, *138*, 49824. [CrossRef]
128. Liu, L.; Zhang, Y.; Li, C.; Cao, J.; He, E.; Wu, X.; Wang, F.; Wang, L. Facile preparation PCL/ modified nano ZnO organic-inorganic composite and its application in antibacterial materials. *J. Polym. Res.* **2020**, *27*, 78. [CrossRef]
129. Al-Naamani, L.; Dobretsov, S.; Dutta, J.; Burgess, J.G. Chitosan-zinc oxide nanocomposite coatings for the prevention of marine biofouling. *Chemosphere* **2017**, *168*, 408–417. [CrossRef] [PubMed]
130. Salam, M.A.; Obaid, A.Y.; El-Shishtawy, R.M.; Mohamed, S.A. Synthesis of nanocomposites of polypyrrole/carbon nanotubes/silver nano particles and their application in water disinfection. *RSC Adv.* **2017**, *7*, 16878–16884. [CrossRef]
131. Shrivastava, S.; Jadon, N.; Jain, R. Next-generation polymer nanocomposite-based electrochemical sensors and biosensors: A review. *TrAC Trends Anal. Chem.* **2016**, *82*, 55–67. [CrossRef]
132. Luka, G.; Ahmadi, A.; Najjaran, H.; Alocilja, E.; DeRosa, M.; Wolthers, K.; Malki, A.; Aziz, H.; Althani, A.; Hoorfar, M. Microfluidics Integrated Biosensors: A Leading Technology towards Lab-on-a-Chip and Sensing Applications. *Sensors* **2015**, *15*, 30011–30031. [CrossRef]
133. Gupta, S.; Sharma, A.; Verma, R.S. Polymers in biosensor devices for cardiovascular applications. *Curr. Opin. Biomed. Eng.* **2020**, *13*, 69–75. [CrossRef]
134. Olowu, R.A.; Arotiba, O.; Mailu, S.N.; Waryo, T.T.; Baker, P.; Iwuoha, E. Electrochemical Aptasensor for Endocrine Disrupting 17β-Estradiol Based on a Poly(3,4-ethylenedioxylthiopene)-Gold Nanocomposite Platform. *Sensors* **2010**, *10*, 9872–9890. [CrossRef] [PubMed]
135. Narang, J.; Chauhan, N.; Pundir, C. Construction of triglyceride biosensor based on nickel oxide–chitosan/zinc oxide/zinc hexacyanoferrate film. *Int. J. Biol. Macromol.* **2013**, *60*, 45–51. [CrossRef] [PubMed]
136. Demircan, D.; Ilk, S.; Zhang, B. Cellulose-Organic Montmorillonite Nanocomposites as Biomacromolecular Quorum-Sensing Inhibitor. *Biomacromolecules* **2017**, *18*, 3439–3446. [CrossRef] [PubMed]
137. Manno, D.; Filippo, E.; Di Giulio, M.; Serra, A. Synthesis and characterization of starch-stabilized Ag nanostructures for sensors applications. *J. Non-Cryst. Solids* **2008**, *354*, 5515–5520. [CrossRef]
138. Singh, K.K.; Solanki, P.R.; Basu, T.; Malhotra, B.D. Polypyrrole/multiwalled carbon nanotubes-based biosensor for cholesterol estimation. *Polym. Adv. Technol.* **2012**, *23*, 1084–1091. [CrossRef]
139. Kumar, B.; Castro, M.; Feller, J. Poly(lactic acid)–multi-wall carbon nanotube conductive biopolymer nanocomposite vapour sensors. *Sens. Actuators B Chem.* **2012**, *161*, 621–628. [CrossRef]
140. Zhu, P.; Liu, Y.; Fang, Z.; Kuang, Y.; Zhang, Y.; Peng, C.; Chen, G. Flexible and Highly Sensitive Humidity Sensor Based on Cellulose Nanofibers and Carbon Nanotube Composite Film. *Langmuir* **2019**, *35*, 4834–4842. [CrossRef]
141. Cheng, J.; Wang, L.; Huo, J.; Yu, H. A novel polycaprolactone-grafted-carbon black nanocomposite-based sensor for detecting solvent vapors. *J. Appl. Polym. Sci.* **2011**, *121*, 3277–3282. [CrossRef]
142. Pandey, S.; Goswami, G.; Nanda, K.K. Green synthesis of biopolymer–silver nanoparticle nanocomposite: An optical sensor for ammonia detection. *Int. J. Biol. Macromol.* **2012**, *51*, 583–589. [CrossRef]
143. Dai, H.; Feng, N.; Li, J.; Zhang, J.; Li, W. Chemiresistive humidity sensor based on chitosan/zinc oxide/single-walled carbon nanotube composite film. *Sens. Actuators B Chem.* **2019**, *283*, 786–792. [CrossRef]

144. Chakraborty, G.; Pugazhenthi, G.; Katiyar, V. Exfoliated graphene-dispersed poly (lactic acid)-based nanocomposite sensors for ethanol detection. *Polym. Bull.* **2018**, *76*, 2367–2386. [CrossRef]
145. Khachatryan, G.; Khachatryan, K. Starch based nanocomposites as sensors for heavy metals—detection of Cu^{2+} and Pb^{2+} ions. *Int. Agrophys.* **2019**, *33*, 121–126. [CrossRef]
146. Nie, D.; Han, Z.; Yu, Y.; Shi, G. Composites of multiwalled carbon nanotubes/polyethyleneimine (MWCNTs/PEI) and molecularly imprinted polymers for dinitrotoluene recognition. *Sens. Actuators B Chem.* **2016**, *224*, 584–591. [CrossRef]
147. Wang, Z.; Liu, E.; Gu, D.; Wang, Y. Glassy carbon electrode coated with polyaniline-functionalized carbon nanotubes for detection of trace lead in acetate solution. *Thin Solid Films* **2011**, *519*, 5280–5284. [CrossRef]
148. Shao, Y.; Dong, Y.; Bin, L.; Fan, L.; Wang, L.; Yuan, X.; Li, D.; Liu, X.; Zhao, S. Application of gold nanoparticles/polyaniline-multi-walled carbon nanotubes modified screen-printed carbon electrode for electrochemical sensing of zinc, lead, and copper. *Microchem. J.* **2021**, *170*, 106726. [CrossRef]
149. Sharma, A.L.; Kumar, P.; Deep, A. Highly Sensitive Glucose Sensing With Multi-Walled Carbon Nanotubes—Polyaniline Composite. *Polym. Technol. Eng.* **2012**, *51*, 1382–1387. [CrossRef]
150. Zeng, X.; Zhang, Y.; Du, X.; Li, Y.; Tang, W. A highly sensitive glucose sensor based on a gold nanoparticles/polyaniline/multi-walled carbon nanotubes composite modified glassy carbon electrode. *New J. Chem.* **2018**, *42*, 11944–11953. [CrossRef]
151. Tsai, Y.-C.; Li, S.-C.; Liao, S.-W. Electrodeposition of polypyrrole–multiwalled carbon nanotube–glucose oxidase nanobiocomposite film for the detection of glucose. *Biosens. Bioelectron.* **2006**, *22*, 495–500. [CrossRef] [PubMed]
152. Alagappan, M.; Immanuel, S.; Sivasubramanian, R.; Kandaswamy, A. Development of cholesterol biosensor using Au nanoparticles decorated f-MWCNT covered with polypyrrole network. *Arab. J. Chem.* **2020**, *13*, 2001–2010. [CrossRef]
153. Maity, D.; Kumar, R.T.R. Polyaniline Anchored MWCNTs on Fabric for High Performance Wearable Ammonia Sensor. *ACS Sens.* **2018**, *3*, 1822–1830. [CrossRef] [PubMed]
154. Liu, B.; Liu, X.; Yuan, Z.; Jiang, Y.; Su, Y.; Ma, J.; Tai, H. A flexible NO2 gas sensor based on polypyrrole/nitrogen-doped multiwall carbon nanotube operating at room temperature. *Sens. Actuators B Chem.* **2019**, *295*, 86–92. [CrossRef]
155. Zhao, Y.; Yang, X.; Pan, P.; Liu, J.; Yang, Z.; Wei, J.; Xu, W.; Bao, Q.; Zhang, H.; Liao, Z. All-Printed Flexible Electrochemical Sensor Based on Polyaniline Electronic Ink for Copper (II), Lead (II) and Mercury (II) Ion Determination. *J. Electron. Mater.* **2020**, *49*, 6695–6705. [CrossRef]
156. Lo, M.; Seydou, M.; Bensghaïer, A.; Pires, R.; Gningue-Sall, D.; Aaron, J.-J.; Mekhalif, Z.; Delhalle, J.; Chehimi, M.M. Polypyrrole-Wrapped Carbon Nanotube Composite Films Coated on Diazonium-Modified Flexible ITO Sheets for the Electroanalysis of Heavy Metal Ions. *Sensors* **2020**, *20*, 580. [CrossRef]
157. Alruwais, R.S.; Adeosun, W.A.; Alsafrani, A.E.; Marwani, H.M.; Asiri, A.M.; Khan, I.; Jawaid, M.; Khan, A. Development of Cd (II) Ion Probe Based on Novel Polyaniline-Multiwalled Carbon Nanotube-3-aminopropyltriethoxylsilane Composite. *Membranes* **2021**, *11*, 853. [CrossRef] [PubMed]
158. Al-Refai, H.H.; Ganash, A.A.; Hussein, M.A. Polythiophene-based MWCNTCOOH@RGO nanocomposites as a modified glassy carbon electrode for the electrochemical detection of Hg (II) ions. *Chem. Pap.* **2021**, 1–16. [CrossRef]
159. Shukla, P.; Saxena, P. Polymer Nanocomposites in Sensor Applications: A Review on Present Trends and Future Scope. *Chin. J. Polym. Sci.* **2021**, *39*, 665–691. [CrossRef]
160. He, P.; Cao, M.-S.; Cao, W.-Q.; Yuan, J. Developing MXenes from Wireless Communication to Electromagnetic Attenuation. *Nano-Micro Lett.* **2021**, *13*, 115. [CrossRef] [PubMed]
161. Geetha, S.; Kumar, K.K.S.; Rao, C.R.; Vijayan, M.; Trivedi, D.C.K. EMI shielding: Methods and materials—A review. *J. Appl. Polym. Sci.* **2009**, *112*, 2073–2086. [CrossRef]
162. Thomassin, J.-M.; Jérôme, C.; Pardoen, T.; Bailly, C.; Huynen, I.; Detrembleur, C. Polymer/carbon based composites as electromagnetic interference (EMI) shielding materials. *Mater. Sci. Eng. R Rep.* **2013**, *74*, 211–232. [CrossRef]
163. Chung, D.D.L. Materials for electromagnetic interference shielding. *Mater. Chem. Phys.* **2020**, *255*, 123587. [CrossRef]
164. Deeraj, B.D.S.; Mathew, M.S.; Parameswaranpillai, J.; Joseph, K. EMI shielding materials based on thermosetting polymers. In *Materials for Potential EMI Shielding Applications*; Elsevier: Amsterdam, The Netherlands, 2020; pp. 101–110. [CrossRef]
165. Nasouri, K.; Shoushtari, A.M. Fabrication of magnetite nanoparticles/polyvinylpyrrolidone composite nanofibers and their application as electromagnetic interference shielding material. *J. Thermoplast. Compos. Mater.* **2017**, *31*, 431–446. [CrossRef]
166. Khan, S.U.D.; Arora, M.; Wahab, M.A.; Saini, P. Permittivity and Electromagnetic Interference Shielding Investigations of Activated Charcoal Loaded Acrylic Coating Compositions. *J. Polym.* **2014**, *2014*, 193058. [CrossRef]
167. Li, J.; Liu, H.; Guo, J.; Hu, Z.; Wang, Z.; Wang, B.; Liu, L.; Huang, Y.; Guo, Z. Flexible, conductive, porous, fibrillar polymer–gold nanocomposites with enhanced electromagnetic interference shielding and mechanical properties. *J. Mater. Chem. C* **2017**, *5*, 1095–1105. [CrossRef]
168. Joseph, N.; Varghese, J.; Sebastian, M.T. In situ polymerized polyaniline nanofiber-based functional cotton and nylon fabrics as millimeter-wave absorbers. *Polym. J.* **2017**, *49*, 391–399. [CrossRef]
169. Villacorta, B.S.; Ogale, A.A.; Hubing, T.H. Effect of heat treatment of carbon nanofibers on the electromagnetic shielding effectiveness of linear low density polyethylene nanocomposites. *Polym. Eng. Sci.* **2013**, *53*, 417–423. [CrossRef]
170. Huang, Y.; Li, N.; Ma, Y.; Du, F.; Li, F.; He, X.; Lin, X.; Gao, H.; Chen, Y. The influence of single-walled carbon nanotube structure on the electromagnetic interference shielding efficiency of its epoxy composites. *Carbon* **2007**, *45*, 1614–1621. [CrossRef]

171. Wu, Z.P.; Li, M.M.; Hu, Y.Y.; Li, Y.S.; Wang, Z.X.; Yin, Y.H.; Chen, Y.S.; Zhou, X. Electromagnetic interference shielding of carbon nanotube macrofilms. *Scr. Mater.* **2011**, *64*, 809–812. [CrossRef]
172. Zhang, W.; Xiong, H.; Wang, S.; Li, M.; Gu, Y. Electromagnetic characteristics of carbon nanotube film materials. *Chin. J. Aeronaut.* **2015**, *28*, 1245–1254. [CrossRef]
173. Gouda, M.; Hebeish, A.; Aljaafari, A. New route for development of electromagnetic shielding based on cellulosic nanofibers. *J. Ind. Text.* **2017**, *46*, 1598–1615. [CrossRef]
174. Pande, S.; Singh, B.; Mathur, R.; Dhami, T.; Saini, P.; Dhawan, S. Improved Electromagnetic Interference Shielding Properties of MWCNT–PMMA Composites Using Layered Structures. *Nanoscale Res. Lett.* **2009**, *4*, 327–334. [CrossRef]
175. Al-Ghamdi, A.A.; Al-Hartomy, O.A.; Al-Salamy, F.; El-Mossalamy, E.H.; Daiem, A.M.A.; El-Tantawy, F. Novel electromagnetic interference shielding effectiveness in the microwave band of magnetic nitrile butadiene rubber/magnetite nanocomposites. *J. Appl. Polym. Sci.* **2012**, *125*, 2604–2613. [CrossRef]
176. Chiscan, O.; Dumitru, I.; Postolache, P.; Tura, V.; Stancu, A. Electrospun PVC/Fe3O4 composite nanofibers for microwave absorption applications. *Mater. Lett.* **2012**, *68*, 251–254. [CrossRef]
177. Gupta, T.K.; Singh, B.P.; Dhakate, S.R.; Singh, V.N.; Mathur, R.B. Improved nanoindentation and microwave shielding properties of modified MWCNT reinforced polyurethane composites. *J. Mater. Chem. A* **2013**, *1*, 9138–9149. [CrossRef]
178. Hoang, A.S. Electrical conductivity and electromagnetic interference shielding characteristics of multiwalled carbon nanotube filled polyurethane composite films. *Adv. Nat. Sci. Nanosci. Nanotechnol.* **2011**, *2*, 25007. [CrossRef]
179. Li, Y.; Chen, C.; Zhang, S.; Ni, Y.; Huang, J. Electrical conductivity and electromagnetic interference shielding characteristics of multiwalled carbon nanotube filled polyacrylate composite films. *Appl. Surf. Sci.* **2008**, *254*, 5766–5771. [CrossRef]
180. Al-Saleh, M.H.; Sundararaj, U. X-band EMI shielding mechanisms and shielding effectiveness of high structure carbon black/polypropylene composites. *J. Phys. D. Appl. Phys.* **2012**, *46*, 35304. [CrossRef]
181. Nayak, L.; Khastgir, D.; Chaki, T.K. A mechanistic study on electromagnetic shielding effectiveness of polysulfone/carbon nanofibers nanocomposites. *J. Mater. Sci.* **2013**, *48*, 1492–1502. [CrossRef]
182. Kashi, S.; Gupta, R.K.; Baum, T.; Kao, N.; Bhattacharya, S.N. Morphology, electromagnetic properties and electromagnetic interference shielding performance of poly lactide/graphene nanoplatelet nanocomposites. *Mater. Des.* **2016**, *95*, 119–126. [CrossRef]
183. Yu, H.; Wang, T.; Wen, B.; Lu, M.; Xu, Z.; Zhu, C.; Chen, Y.; Xue, X.; Sun, C.; Cao, M. Graphene/polyaniline nanorod arrays: Synthesis and excellent electromagnetic absorption properties. *J. Mater. Chem.* **2012**, *22*, 21679–21685. [CrossRef]
184. Ling, J.; Zhai, W.; Feng, W.; Shen, B.; Zhang, J.; Zheng, W.G. Facile Preparation of Lightweight Microcellular Polyetherimide/Graphene Composite Foams for Electromagnetic Interference Shielding. *ACS Appl. Mater. Interfaces* **2013**, *5*, 2677–2684. [CrossRef]
185. Das, N.C.; Liu, Y.; Yang, K.; Peng, W.; Maiti, S.; Wang, H. Single-walled carbon nanotube/poly(methyl methacrylate) composites for electromagnetic interference shielding. *Polym. Eng. Sci.* **2009**, *49*, 1627–1634. [CrossRef]
186. Ganguly, S.; Bhawal, P.; Ravindren, R.; Das, N.C. Polymer Nanocomposites for Electromagnetic Interference Shielding: A Review. *J. Nanosci. Nanotechnol.* **2018**, *18*, 7641–7669. [CrossRef]
187. Lyu, L.; Liu, J.; Liu, H.; Liu, C.; Lu, Y.; Sun, K.; Fan, R.; Wang, N.; Lu, N.; Guo, Z.; et al. An Overview of Electrically Conductive Polymer Nanocomposites toward Electromagnetic Interference Shielding. *Eng. Sci. 2018 2*, 26–42. [CrossRef]
188. Vas, J.V.; Thomas, M.J. Electromagnetic Shielding Effectiveness of Layered Polymer Nanocomposites. *IEEE Trans. Electromagn. Compat.* **2017**, *60*, 376–384. [CrossRef]
189. Shankar, S.; Rhim, J.-W. Polymer Nanocomposites for Food Packaging Applications. In *Functional and Physical Properties of Polymer Nanocomposites*; Dasari, A., Njuguna, J., Eds.; John Wiley and Sons: Hoboken, NJ, USA, 2016; Volume 29, pp. 29–55. [CrossRef]
190. Rahman, P.M.; Mujeeb, V.A.; Muraleedharan, K. Flexible chitosan-nano ZnO antimicrobial pouches as a new material for extending the shelf life of raw meat. *Int. J. Biol. Macromol.* **2017**, *97*, 382–391. [CrossRef] [PubMed]
191. Mahmoodi, A.; Ghodrati, S.; Khorasani, M. High-Strength, Low-Permeable, and Light-Protective Nanocomposite Films Based on a Hybrid Nanopigment and Biodegradable PLA for Food Packaging Applications. *ACS Omega* **2019**, *4*, 14947–14954. [CrossRef]
192. Pina, H.D.V.; De Farias, A.J.A.; Barbosa, F.C.; Souza, J.W.D.L.; Barros, A.B.D.S.; Cardoso, M.J.B.; Fook, M.V.L.; Wellen, R.M.R. Microbiological and cytotoxic perspectives of active PCL/ZnO film for food packaging. *Mater. Res. Express* **2020**, *7*, 025312. [CrossRef]
193. Pirsa, S.; Shamusi, T.; Kia, E.M. Smart films based on bacterial cellulose nanofibers modified by conductive polypyrrole and zinc oxide nanoparticles. *J. Appl. Polym. Sci.* **2018**, *135*, 46617. [CrossRef]
194. Yu, H.-Y.; Qin, Z.-Y.; Sun, B.; Yang, X.-G.; Yao, J.-M. Reinforcement of transparent poly(3-hydroxybutyrate-co-3-hydroxyvalerate) by incorporation of functionalized carbon nanotubes as a novel bionanocomposite for food packaging. *Compos. Sci. Technol.* **2014**, *94*, 96–104. [CrossRef]
195. Kumar, P.; Gautam, S. Developing ZnO Nanoparticle embedded antimicrobial starch biofilm for food packaging. *arXiv* **2019**, arXiv:1909.05083.
196. Fernández, A.; Picouet, P.; Lloret, E. Cellulose-silver nanoparticle hybrid materials to control spoilage-related microflora in absorbent pads located in trays of fresh-cut melon. *Int. J. Food Microbiol.* **2010**, *142*, 222–228. [CrossRef]
197. Malik, N. Thermally exfoliated graphene oxide reinforced polycaprolactone-based bactericidal nanocomposites for food packaging applications. *Mater. Technol.* **2020**, 1–10. [CrossRef]

198. Ahmadzadeh, S.; Keramat, J.; Nasirpour, A.; Hamdami, N.; Behzad, T.; Aranda, L.; Vilasi, M.; Desobry, S. Structural and mechanical properties of clay nanocomposite foams based on cellulose for the food-packaging industry. *J. Appl. Polym. Sci.* **2016**, *133*, 1–10. [CrossRef]
199. Grande, C.D.; Mangadlao, J.; Fan, J.; De Leon, A.; Delgado-Ospina, J.; Rojas, J.G.; Rodrigues, D.F.; Advincula, R. Chitosan Cross-Linked Graphene Oxide Nanocomposite Films with Antimicrobial Activity for Application in Food Industry. *Macromol. Symp.* **2017**, *374*, 1600114. [CrossRef]
200. Mujtaba, M.; Morsi, R.E.; Kerch, G.; Elsabee, M.Z.; Kaya, M.; Labidi, J.; Khawar, K.M. Current advancements in chitosan-based film production for food technology; A review. *Int. J. Biol. Macromol.* **2019**, *121*, 889–904. [CrossRef]

International Journal of *Molecular Sciences*

MDPI

Article

Fabrication of $CaCO_3$-Coated Vesicles by Biomineralization and Their Application as Carriers of Drug Delivery Systems

Chiho Miyamaru, Mao Koide, Nana Kato, Shogo Matsubara and Masahiro Higuchi *

Department of Life Science and Applied Chemistry, Graduate School of Engineering, Nagoya Institute of Technology, Gokiso-cho, Show-ku, Nagoya 4668-555, Japan; c.miyamaru.378@nitech.jp (C.M.); m.koide.636@nitech.jp (M.K.); n.kato.575@stn.nitech.ac.jp (N.K.); matsubara.shogo@nitech.ac.jp (S.M.)

* Correspondence: higuchi.masahiro@nitech.ac.jp; Tel.: +81-52-735-7156

Abstract: We fabricated $CaCO_3$-coated vesicles as drug carriers that release their cargo under a weakly acidic condition. We designed and synthesized a peptide lipid containing the Val-His-Val-Glu-Val-Ser sequence as the hydrophilic part, and with two palmitoyl groups at the *N*-terminal as the anchor groups of the lipid bilayer membrane. Vesicles embedded with the peptide lipids were prepared. The $CaCO_3$ coating of the vesicle surface was performed by the mineralization induced by the embedded peptide lipid. The peptide lipid produced the mineral source, CO_3^{2-}, for $CaCO_3$ mineralization through the hydrolysis of urea. We investigated the structure of the obtained $CaCO_3$-coated vesicles using transmission electron microscopy (TEM). The vesicles retained the spherical shapes, even in vacuo. Furthermore, the vesicles had inner spaces that acted as the drug cargo, as observed by the TEM tomographic analysis. The thickness of the $CaCO_3$ shell was estimated as ca. 20 nm. $CaCO_3$-coated vesicles containing hydrophobic or hydrophilic drugs were prepared, and the drug release properties were examined under various pH conditions. The mineralized $CaCO_3$ shell of the vesicle surface was dissolved under a weakly acidic condition, pH 6.0, such as in the neighborhood of cancer tissues. The degradation of the $CaCO_3$ shell induced an effective release of the drugs. Such behavior suggests potential of the $CaCO_3$-coated vesicles as carriers for cancer therapies.

Keywords: $CaCO_3$-coated vesicle; DDS carrier; peptide lipid; mineralization; drug release; under weakly acidic condition

Citation: Miyamaru, C.; Koide, M.; Kato, N.; Matsubara, S.; Higuchi, M. Fabrication of $CaCO_3$-Coated Vesicles by Biomineralization and Their Application as Carriers of Drug Delivery Systems. *Int. J. Mol. Sci.* **2022**, 23, 789. https://doi.org/10.3390/ijms23020789

Academic Editors: Axel T. Neffe and Raghvendra Singh Yadav

Received: 20 December 2021
Accepted: 10 January 2022
Published: 12 January 2022

Publisher's Note: MDPI stays neutral with regard to jurisdictional claims in published maps and institutional affiliations.

Copyright: © 2022 by the authors. Licensee MDPI, Basel, Switzerland. This article is an open access article distributed under the terms and conditions of the Creative Commons Attribution (CC BY) license (https://creativecommons.org/licenses/by/4.0/).

1. Introduction

Cancer is the leading cause of death for humankind. Therapies for cancer are surgery, drug, radiation, etc. Among these, drug therapies are attracting attention as a method that does not damage the "quality of life" of patients, because it suppresses physical invasion. In recent years, molecular-targeted drugs [1,2] and immune checkpoint inhibitors [3,4] have been developed and are being used in drug therapies. These drugs suppress side effects and achieve high anti-cancer effects; however, these drugs are very expensive. On the other hand, many drugs are abandoned in the development stage due to their significant side effects despite their high efficacy. The drug delivery system (DDS) has been refocused to enable the use of these drugs with prominent side-effects. DDS is a system that delivers the drugs "in a minimum amount", "at the right time", and "to the right place". Polymer micelles [5,6], dendrimers [7–9], hydrogels [10–13], and mesoporous nanoparticles [14–16] have been reported as DDS carriers. In addition, vesicles have been used for DDS carriers. Vesicles have the following advantages as DDS carriers. They have a low toxicity and antigenicity because the main component is a lipid. The vesicles can be encapsulated hydrophilic drugs in the internal aqueous phase, and incorporated hydrophobic them in the interior of the bilayer membrane, respectively. It is possible to add a specific recognition ability by embedding proteins and/or peptides [17,18] in the bilayer membrane. That is, vesicles can be obtained with a targeting ability for specific tissues [19–21]. Although vesicles have many of the above advantages as DDS carriers, they have the drawback of being unstable in vivo because of metabolically disruption. We have to consider the difference between cancer and normal tissues in the molecular design of the carriers for drug therapies. The new blood vessels around the cancer tissue are incomplete, and carriers that do not erupt from normal blood vessels permeate from the blood vesicle wall and accumulated in the cancer tissue. The effect of this carrier accumulating in the cancer tissue is called the "enhanced permeability and retention (EPR) effect" [22], and is found in carriers with a diameter from 50 to 200 nm [23,24]. In addition, cancer cells grow rapidly and are actively metabolized. Therefore, as lactic acid accumulates, the CO_2 concentration also increases. Furthermore, because the proton pump operates actively, the pH value around the cancer tissue is relatively lower than that of the normal tissue [25–27]. From the above, it is necessary that the DDS carrier in drug therapy for cancer diseases has a size capable of exhibiting the EPR effect and that can release the drug under weakly acidic conditions.

In this paper, we attempted the fabrication of stable $CaCO_3$-cated vesicles as DDS carriers for drug therapies in cancer. $CaCO_3$ is the main inorganic component of the shells and is not toxic. The important point is that $CaCO_3$ is insoluble under physiological pH conditions around normal tissue, but dissolves under weakly acidic conditions in the neighborhood of cancer tissue. We hypothesize that the $CaCO_3$-coated vesicles are able to effectively release the drugs, owing to the collapse of the shell as it is dissolved in the vicinity of the cancer tissues. $CaCO_3$-coated vesicles were obtained by self-supplied $CaCO_3$ mineralization [28,29] using the embedded peptide lipid with the Val-His-Val-Glu-Val-Ser sequence as the catalyst of urea hydrolysis. The vesicles maintained their spherical shape even under a high vacuum; however, the vesicles collapsed easily under the weakly acidic condition owing to the dissolution of the $CaCO_3$ shells. Both the encapsulated hydrophobic and hydrophilic drugs were released by dissolving the $CaCO_3$ shell under weakly acidic conditions, such as in the neighborhood of cancer tissues.

2. Results and Discussion

2.1. Structure of $CaCO_3$-Coated Vesicles Fabricated by the Peptide Lipids Induced Minerarlzation

2.1.1. $CaCO_3$-Coating of the Vesicle Surface

In our previous studies [28,29], we have been reported that the Val-His-Val-Glu-Val-Ser peptide acts as mineral source supplier for $CaCO_3$ mineralization through the hydrolysis of urea. We have proposed the mechanism of urea hydrolysis by the Val-His-Val-Glu-Val-Ser peptide as follows. The imidazole group of the His residue activates the hydroxyl group of the Ser residue of the peptides by taking proton [29]. The interaction between His and Ser residues among the Val-His-Val-Glu-Val-Ser peptides is well-known as the "charge relay effect" that is seen in the serine protease [30]. The activated hydroxyl group of the Ser residue hydrolyzes one urea molecule to two ammonium cations, and has one carbonate anion as the mineral source of the $CaCO_3$ mineralization. We designed and synthesized the peptide lipid with a hydrophilic Val-His-Val-Glu-Val-Ser sequence as a mineral source supply site, and with two hydrophobic palmitoyl groups as the anchors for embedding in a vesicle membrane. Figure 1 shows the chemical structure of the peptide lipid. In this figure, we show the reaction mechanism of urea hydrolysis by the peptide lipids.

Azolectin vesicles embedded the peptide lipids were prepared. The obtained vesicle with an interior containing 150 mM NaCl, and an exterior containing 100 mM NaCl, 25 mM Urea, and 25 mM $Ca(OAc)_2$ in an isotonic condition. Under this condition, we considered that the vesicle remained spherical and the Val-His-Val-Glu-Val-Ser sequence of the peptide lipid presented on the outer surface of the vesicle hydrolyzed urea to form the $CaCO_3$ shell. Figure 2 shows the schematic picture of the fabrication of the $CaCO_3$-coated vesicle by mineralization under the isotonic condition. In addition, we confirmed that the thermal pyrolysis of urea in an aqueous solution occurred only above 30 °C [29]. Therefore, $CaCO_3$ coating on the vesicle outer surface, owing to the mineralization by peptide lipid induced urea hydrolysis, was performed at 20 °C without bulk mineralization.

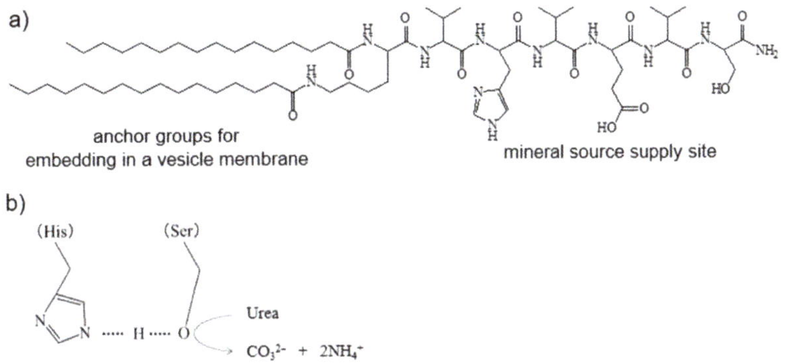

a)

anchor groups for embedding in a vesicle membrane

mineral source supply site

b)

(His) (Ser)

Urea

CO_3^{2-} + $2NH_4^+$

Figure 1. (**a**) Chemical structure of the peptide lipid. (**b**) Hydrolysis mechanism of urea by charge relay between His and Ser.

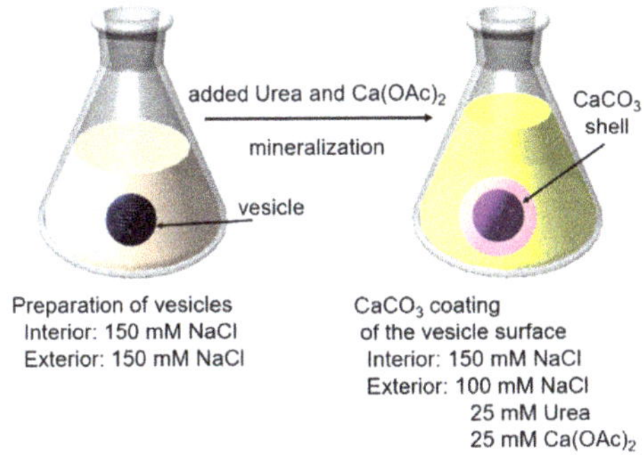

Figure 2. Schematic picture of the fabrication of the $CaCO_3$-coated vesicle by the self-supplied mineralization under the isotonic condition.

2.1.2. Structure of $CaCO_3$-Coated Vesicles

We observed the morphological changes of $CaCO_3$-coated vesicles obtained by the mineralization reaction using transmission electron microscopic (TEM) observations. Figure 3 shows the TEM images of the vesicles after $CaCO_3$ mineralization for 8, 11, 18, and 25 days, respectively. For comparison, the TEM image of the vesicle before $CaCO_3$ mineralization is shown in Figure 3e. The vesicle before mineralization did not have a clear structure. This implies that the non-coated vesicle was crushed under the high vacuum during TEM observation. From the TEM images after 8 (Figure 3a) and 11 days (Figure 3b), when the mineralization reactions were short, the vesicles were crushed and deformed. Furthermore, many white spots were observed on the vesicle surface after 8 days of mineralization (Figure 3a). This suggests that if the mineralization period is short, uncoated areas with $CaCO_3$ occur. On the other hand, when increasing of the mineralization period, the crushed structure disappeared and spherical structures that could maintain those shapes even in vacuo were clearly observed (Figure 3c,d). This suggests that $CaCO_3$-coating by mineralization for 18 days fabricates stable vesicles a maintaining spherical structure.

The hydrodynamic diameter of the $CaCO_3$-coated vesicles obtained by 18-day mineralization was estimated by dynamic light scattering (DLS) measurements. The diameter of the $CaCO_3$-coated vesicle was 193.6 ± 84.4 nm (Figure A1a). This value corresponded to the diameter of the spherical particles observed by TEM, and the size was within the range where the EPR effect could be expected. However, the standard deviation of the radius of the $CaCO_3$-coated vesicle was high, and there were coated vesicles of 200 nm or more. The size of the $CaCO_3$-coated vesicles could be controlled by adjusting the period of mineralization. In addition, the diameter of the vesicle before mineralization was 122.7 ± 22.7 nm (Figure A1b). This implies that the diameter increased owing to the surface coating by $CaCO_3$ mineralization.

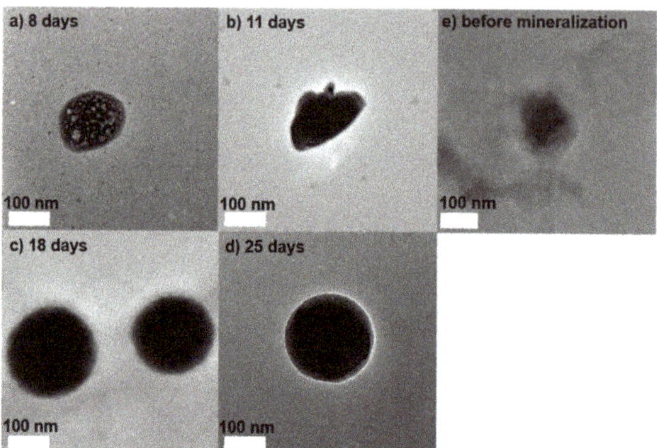

Figure 3. TEM images of vesicles after $CaCO_3$ mineralization for (**a**) 8 days, (**b**) 11 days, (**c**) 18 days, and (**d**) 25 days. (**e**) TEM image of the vesicle before mineralization.

For comparison, $CaCO_3$ mineralization was performed on the surface of the azolectin vesicle (azolectin/peptide lipid ratio of 200:1) with a lower peptide lipid content. Furthermore, mineralization was also performed on the surface of the dipalmitoylphosphatidylcholine (DPPC) vesicle containing the peptide lipid (DPPC/peptide lipid ratio of 100:1) which has a gel state at 20 °C [31]. In either system, after 18 days of mineralization, the vesicles collapsed under TEM observation, and the spherical structure could not be maintained in vacuo. These results suggest that the Val-His-Val-Glu-Val-Ser sequence in the peptide lipids needs to collide in the vesicle membrane and interact between His and Ser in order to hydrolyze urea. That is, the peptide lipids must exist at a collisional concentration in the fluid vesicle membrane.

The composition distribution of $CaCO_3$-coated vesicles obtained after 18-day mineralization was evaluated using energy dispersive X-ray spectrometry (EDX) mapping. Figure 4 shows the elemental mappings of phosphorus (Figure 4a) assigned to azolectin, which is the major component of the vesicle membrane, and calcium (Figure 4b) corresponded to the mineralized $CaCO_3$. $CaCO_3$ was found only in the vicinity of the vesicle, and formation in the bulk was not observed.

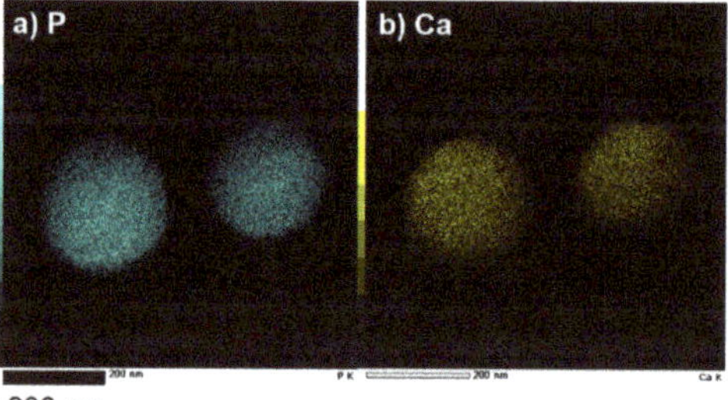

Figure 4. EDX mapping images of (**a**) phosphorous and (**b**) calcium for $CaCO_3$-coated vesicles after 18-day mineralization. "P" and "Ca" indicate phosphorus of azolectin and calcium of $CaCO_3$, respectively.

We investigated the crystal phase of the mineralized $CaCO_3$ shell on the vesicle using X-ray diffraction (XRD) measurements. $CaCO_3$ takes on different crystal phases as most stable calcite phase, semi-stable aragonite phase, and unstable vaterite phase [32]. Figure 5 shows the XRD profile of the $CaCO_3$-coated vesicles obtained after the 18-day mineralization. In this figure, the standard profiles of calcite [33], aragonite [34], and vaterite [35] are shown. The XRD profile of the $CaCO_3$-coated vesicle was very similar to the standard profile of the calcite phase. We assigned crystal faces of the diffraction peaks shown in Figure 5 from those of 2θ and a relative intensity, along with those of the standard calcite XRD profile (Table A1). The 2θ and relative intensities for the $CaCO_3$-coated vesicle were in good agreement with those of calcite. This implies that the mineralized $CaCO_3$ on the vesicle surface took the most stable calcite phase. In this system, $CaCO_3$ was formed on the vesicle surface through the reaction between Ca^{2+} and $CO_3{}^{2-}$. $CO_3{}^{2-}$ was supplied by the peptide lipid owing to the urea hydrolysis. On the other hand, Ca^{2+} was captured on the vesicle surface as a result of the electrostatic interaction with the phosphate group of azolectin, and/or the carboxy group of the Glu side chain of the peptide lipid. It is considered that $CaCO_3$ took the most stable calcite phase, as there was no clear regularity in the spatial arrangement of phosphate and carboxyl groups on the vesicle surface.

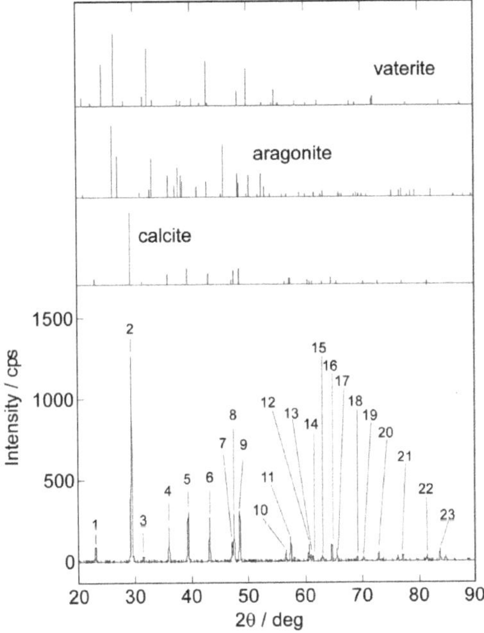

Figure 5. XRD profile of $CaCO_3$-coated vesicles after mineralization for 18 days. Standard profiles of calcite, aragonite, and vaterite are shown in this figure, respectively. The assignment of each diffraction peak is shown in Table A1.

To evaluate the internal structure of the $CaCO_3$-coated vesicle obtained after the 18-day mineralization, three-dimensional TEM (3D-TEM) observations and a tomographic analysis were performed. Figure 6 shows the 3D-TEM and tomography image of the $CaCO_3$-coated vesicle. We can see that there was a cavity inside the $CaCO_3$-coated vesicle. Furthermore, the thickness of the $CaCO_3$ shell was ca. 20 nm. These results suggest that the vesicle obtained for the 18-day mineralization had a stable spherical structure whose surface was coated with calcite, while maintaining the intrnal aqueous phase. The $CaCO_3$-coated vesicle was useful as the DDS carrier because it had a particle size capable of exhibiting the EPR effect and could encapsulate both hydrophilic and hydrophobic drugs.

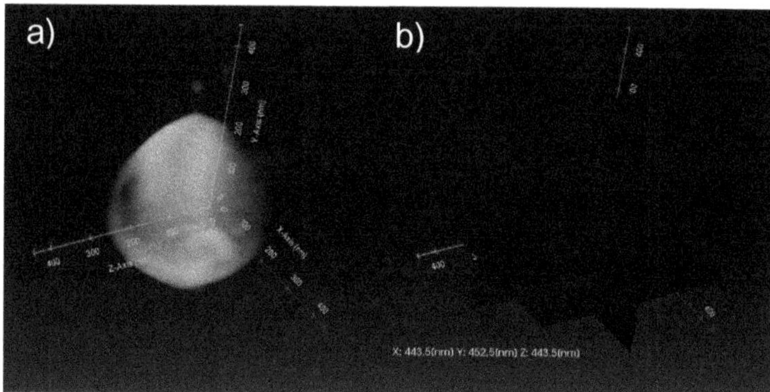

Figure 6. (**a**) 3D-TEM and (**b**) tomography images of $CaCO_3$-coated vesicle for 18-day mineralization. Axis length information; *X* axis: 443.5 nm; *Y* axis: 452.5 nm; *Z* axis: 443.5 nm.

2.2. *Drug Release Properties of $CaCO_3$-Coated Vesicles*

2.2.1. Dissolution Behaviors of $CaCO_3$ Shells

We successfully fabricated the $CaCO_3$-coated vesicle as a stable DDS carrier with a particle size capable of exerting an EPR effect. We considered that the $CaCO_3$-coated vesicles act as environmentally responsive DDS carriers, with the advantage that the $CaCO_3$ shells dissolve under acidic conditions. The strategy for using $CaCO_3$-coated vesicles as DDS carriers is follows. First, the $CaCO_3$ shell on the vesicle dissolves under the weakly acidic conditions, such as in the neighborhood of cancer tissues. Next, the vesicle lost the $CaCO_3$ shell collapses and releases the drugs. We first investigated the dissolubility of $CaCO_3$ shell on the vesicle under various pH conditions. The quantification of free Ca^{2+} generated by the dissolution of the $CaCO_3$ shell was performed by fluorescence measurements using 8-amino-2-[(2-amino-5-methylphenoxy)methyl]-6-methoxyquinoline-*N,N,N′,N′*-tetraacetic acid, tetrapotassium salt (Quin 2) [36]. We used the $CaCO_3$-coated vesicle obtained for the 18-day mineralization. Figure 7 shows the concentration changes in the free Ca^{2+} produced by the dissolution of the $CaCO_3$ shell in the solution mimicking the pH environment in the neighborhood of cancer (pH 6.0) and normal (pH 7.4) tissues, respectively. For comparison, the concentration changes in free Ca^{2+} under weakly basic conditions at pH 8.0 are also shown in this figure. The $CaCO_3$ shells of the vesicles were gradually dissolved and there was a slight increase in the free Ca^{2+} concentration in the external aqueous phase when the $CaCO_3$-coated vesicles were dispersed in the pH 8.0 buffer. Under pH 7.4, mimicking the normal tissue neighborhood, the free Ca^{2+} concentration increased monotonically, although the dissolution rate of the Ca shells was faster than at pH 8.0. On the other hand, the dissolution phenomena of Ca shells in the pH environment at pH 6.0, which mimicked the neighborhood of the cancer tissue, was clearly different from that at pH 8.0 and pH 7.5. Under this condition, the $CaCO_3$ shells dissolved rapidly in 2 days, and then the concentration of free Ca^{2+} gradually increased. This result indicates that the shells of $CaCO_3$-coated vesicles showed effective dissolubility under the weakly acidic condition as compared with the neutral and weakly basic conditions. That is, the $CaCO_3$-coated vesicle had the ability to recognize environmental pH conditions.

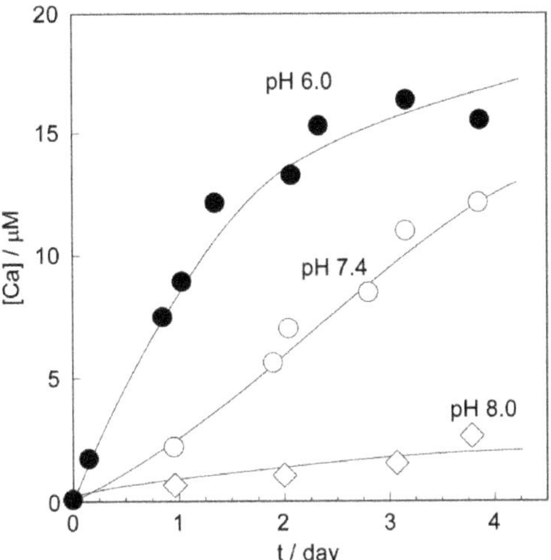

Figure 7. The concentration changes in free Ca^{2+} produced by the dissolution of the $CaCO_3$ shell under various pH conditions of pH 8.0, pH 7.4, and pH 6.0, respectively.

2.2.2. Hydrophilic and Hydrophobic Drug Release Properties

Next, we investigated the drug release behavior from the $CaCO_3$-coated vesicles under various pH conditions. We used the $CaCO_3$-coated vesicles obtained from the 18-day mineralization, which were used for the shell dissolubility measurements. However, the vesicles used in the drug release experiments encapsulated hydrophilic Rhodamine 6G (Rh6G) in the internal aqueous phases or incorporated the hydrophobic Pyrene (Py) in the lipid bilayer membranes. The release of hydrophilic Rh6G was evaluated by permeation from the vesicle interior to the outer aqueous phase as the amount of Rh6G per 1 g of $CaCO_3$-coated vesicles (Figure 8a). On the other hand, the release of hydrophobic Py was evaluated by the transfer amount from the bilayer membrane of $CaCO_3$-coated vesicles to that of the pure azolectin vesicle as Py amount per 1 g of $CaCO_3$-coated vesicles (Figure 8b). The amount of drug release increased with decreasing the pH of the medium in both hydrophilic Rh6G and hydrophobic Py. This propensity was consisted with that of the $CaCO_3$ shell dissolubility shown in Figure 7. This means that the drug release from the $CaCO_3$-coated vesicle was caused by the dissolution of the shell. On the other hand, the profiles between the transfer of hydrophobic Py and the release of hydrophilic Rh6G were observed to be clearly different under the weakly acidic condition of pH 6.0, mimicking the neighborhood of the cancer tissue. The transfer profile of Py showed the saturated curve (Figure 8b; pH 6.0). However, the release profile of Rh6G at pH 6.0 showed a slower release of up to 3 days. Then, the rapid increase in the amount of Rh6G released was observed, and the release amount reached the equilibrium value after 8 days.

This difference in drug release behavior can be explained as follows. Here, we discuss the initial process of the drug release phenomena as a result of the dissolution behavior of the $CaCO_3$ shell. The release of hydrophilic Rh6G from the $CaCO_3$-coated vesicle is thought to be caused by the dissolution of the $CaCO_3$ shell and the subsequent collapse of the vesicle (Figure 9a). On the other hand, it is thought that the release of hydrophobic Py occurred after contact between the $CaCO_3$-coated vesicle and the pure azolectin vesicle (Figure 9b). That is, the rapid transfer of Py was caused in the initial state owing to the immediate contact of the $CaCO_3$-coated vesicle to the azolectin vesicle as soon as partial dissolution of the $CaCO_3$ shell occurred. This is implied from the rapid rising in the Py transfer profile (Figure 8b; pH 6.0) during the period, when $CaCO_3$ shell dissolution occurred rapidly in the first 2 days (Figure 7; pH 6.0). The amount of hydrophilic Rh6G released showed a monotonous increasing tendency (Figure 8a; pH 6.0) during the rapid dissolution of the $CaCO_3$ shell, but the release amount increased rapidly after 3 days, during which the concentration of free Ca^{2+} was gradually increased (Figure 7; pH 6.0). The Py transfer occurred through contact of the exposed bilayer membrane of the $CaCO_3$-coated vesicle with the azolectin vesicle (Figure 9b). However, the release of hydrophilic Rh6G did not occur until the vesicle disintegrated after the dissolution of the $CaCO_3$ shell (Figure 9a). The hydrophilic Rh6G could not permeate the hydrophobic bilayer membrane, so the release of Rh6G required the vesicle disintegration. The hydrophilic drugs were encapsulated in the inner aqueous phase of the vesicle, while the hydrophobic drugs were incorporated in the bilayer membrane. It is thought that drug releases were performed by different mechanisms due to the difference in the location of the drugs.

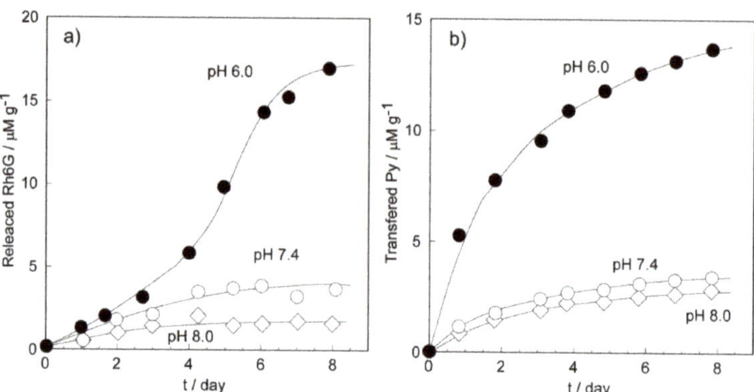

Figure 8. (**a**) The release profile of the hydrophilic Rh6G from the vesicle interior to the outer aqueous phase, and (**b**) the transfer of hydrophobic Py from the bilayer membrane of $CaCO_3$-coated vesicles to that of the pure azolectin vesicle under the various pH conditions.

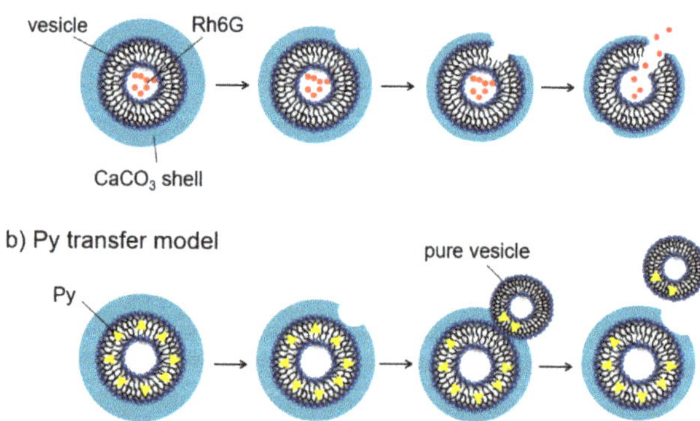

Figure 9. Schematic pictures of the drug release mechanism. (**a**) Release of hydrophilic Ph6G as a result of the collapse of the vesicle. (**b**) Transfer of hydrophobic Py as a result of contact with the vesicle.

3. Materials and Methods

3.1. Materials

3.1.1. Peptide Lipid

The amino acid sequence of the peptide lipid (Figure 1), Val-His-Val-Glu-Val-Ser, was chosen as the hydrolysis site of the urea to produce the $CO_3{}^{2-}$, which is the mineral source of $CaCO_3$ [28,29]. Lysine was introduced at the *N*-terminal of the peptide. Two palmitoyl groups were attached as the anchor to the vesicle through a condensation reaction between the amino group of *N*-terminal lysine and the carboxyl group of palmitic acid (Nacalai Tesque Inc., Kyoto, Japan). Peptide lipid synthesis was carried out using 9-fluorenylmethoxycarbonyl (Fmoc) chemistry on 4-(2,4-dimethoxyphenyl-Fmoc-aminomethoxyl)phenoxyacetyl-norleucine loaded cross-linked ethoxylate acrylate (CLEAR-amide) resin [37]. Fmoc-amino acids (Fmoc-Val, Fmoc-His(Trt), Fmoc-Glu(OBut)-H_2O, and Fmoc-Ser(But)) and clear-amide resin were purchased from Peptide Institute, Inc. (Osaka, Japan). The Fmoc-amino acid for *N*-terminal lysine residue was Fmoc-Lys(Fmoc), which was purchased from Watanabe Chemical Industrials, Ltd. (Hiroshima, Japan). Two palmitoyl groups were attached at the *N*-terminal amino group and side chain of the lysine residue by the same chemistry. Peptide lipid cleavage and deprotection of the side chain protecting groups were performed at the same time to obtain the peptide lipid. The obtained peptide lipid was identified by matrix-assisted laser desorption/ionization time-of-flight mass spectroscopy (MALDI-TOF-MS) on a JEM-S3000 system (JEOL Ltd., Tokyo, Japan). Molecular weights of the synthesized peptide lipid were obtained to be 1272.9 and 1294.9 from the MALDI-TOF-MS measurement (Figure A2). The calculated molecular weight of the peptide lipid was 1272.7. The molecular weights obtained by the MALDI-TOF-MS measurement were in fair agreement with the calculated values as $[M+H]^+$ ($m/z = 1273.7$) and $[M+Na]^+$ ($m/z = 1295.7$), respectively. From the MALDI-TOF-MS study, we obtained evidence indicating the successful synthesis of the designed peptide lipid.

3.1.2. Vesicles

The peptide lipid was dissolved in 2,2,2-trifluoroethanol (TFE; Nacalai Tesque Inc., Kyoto, Japan). Azolectin (Nacalai Tesque Inc., Kyoto, Japan) was dissolved in chloroform (Nacalai Tesque Inc., Kyoto, Japan). These solutions were mixed and poured into a glass flask and then a thin film was formed on the interior surface of the flask from the evaporation of the solvents. The molar ration of the peptide lipid to azolectin was fixed at 0.01. The molecular weight of azolectin used was that of dioleoylphosphatidylcholine, 786. An aqueous solution containing 150 mM NaCl (Nacalai Tesque Inc., Kyoto, Japan) was added to this flask, and was sonicated by Branson Sonifier 250 (Danbury, CT, USA) for 10 min, under a nitrogen atmosphere at 0 °C. The pH of the vesicle dispersion was adjusted to pH 7.4. For comparison, a dipalmitoylphosphatidylcholine (DPPC; FUJIFILM Wako Pure Chemical Co., Osaka, Japan) vesicle, which is in a gel state at room temperature, was also prepared in the same manner.

We used model drugs rhodamine 6G (Rh6G; Nacalai Tesque Inc., Kyoto, Japan), as a hydrophilic drug, and pyrene (Py; FUJIFILM Wako Pure Chemical Co., Osaka, Japan), as a hydrophobic drug. Encapsulation of these drug models in the vesicles was performed as follows: For the encapsulation of the hydrophilic drug model, Rh6G was added to the aqueous solution containing 150 mM NaCl. The Rh6G encapsulated vesicle was obtained by sonication in the aqueous solution containing Rh6G. The concentration of Rh6G in the 150 mM NaCl aqueous solution was 1 mM. On the other hand, to introduce the hydrophobic model drug, Py, into the bilayer membrane of the vesicle, Py was mixed into the thin film consisting of the peptide lipid and azolectin. Py and azolectin were dissolved in chloroform. The molar ration of the Py to azolectin was 0.12. The TFE solution containing the peptide lipid was added to the chloroform solution in the glass flask and then the thin film containing Py was formed through the evaporation of the solvents. An aqueous solution containing 150 mM NaCl was added to the flask and was sonicated to prepare the Py incorporated vesicle.

3.1.3. $CaCO_3$ Coating on the Vesicle Surface by Mineralization

$CaCO_3$ coating on the vesicle surface was performed by mineralization using CO_3^{2-} generated by urea hydrolysis induced by the Val-His-Val-Glu-Val-Ser sequence of the peptide lipid [28]. Aqueous solutions containing 150 mM urea (Nacalai Tesque Inc., Kyoto,

Japan) and 150 mM calcium acetate (Nacalai Tesque Inc., Kyoto, Japan) were added to the vesicle dispersions either with or without the model drugs obtained above. The volume ration of the vesicle dispersion, aqueous solutions of urea, and calcium acetate was 4:1:1. Under this condition, the inner aqueous phase of the vesicle contained 150 mM NaCl, and the outer consisted of 100 mM NaCl, 25 mM urea, and 25 mM calcium acetate. The obtained vesicle dispersion was an isotonic system in which no osmotic pressure was generated between the inside and outside of the bilayer membrane. The vesicle dispersions were gently shaken at 20 °C, after which the pyrolysis of urea did not occur [29]. After 18 days of mineralization, the vesicle dispersions were centrifuged at 10,000 rpm for 20 min. The precipitates were re-dispersed in water adjusted to pH 8.0, and were centrifuged to obtain the precipitates again. This centrifugation and redispersion cycle were performed twice more to remove the unreacted calcium acetate, urea, and the unincluded model drugs from the vesicle outer aqueous phase. The vesicle dispersions were lyophilized, and the vesicles with or without model drugs were stored frozen until they were used in the experiments.

3.2. Methods

3.2.1. Transmission Electron Microscopic Observations

The morphologies of the $CaCO_3$-coated vesicles obtained by mineralization were determined using a scanning transmission electron microscope (STEM; JEM-z2500, JEOL Ltd., Tokyo, Japan) equipped with an Ultra Scan CCD camera (US1000; Gatan Inc., Pleasanton, CA, USA) in TEM mode. Elemental analysis and mappings of the $CaCO_3$-coated vesicles were performed in STEM mode equipped with an energy dispersive X-ray spectrometer (EDX; EX-37001, JEOL Ltd., Tokyo, Japan). An aliquot of the vesicle dispersion obtained after various mineralization periods was placed on an elastic carbon-coated STEM grid, and the $CaCO_3$-coated vesicles were allowed time to adsorb onto its surface. After the adsorption, the excess solution was removed by absorption onto filter paper, and the grid was rinsed with water to remove the unreacted urea and calcium acetate.

Three-dimensional TEM tomography was performed to analyze the internal structure of the $CaCO_3$-coated vesicle. Projection images under sample rotation angles from $-60°$ to $+60°$ (at $1°$ increments) were automatically acquired. The projection images were reconstructed into 3D images using a TEM tomography system (TEMography, SYSTEM IN FRONTIER Inc., Tokyo, Japan). TEM and STEM observations were performed using unstained samples at an acceleration voltage of 200 kV.

3.2.2. Dynamic Light Scattering Measurements

The hydrodynamic diameters of the $CaCO_3$-caoted vesicles were evaluated by dynamic light scattering (DLS) measurement. The lyophilized $CaCO_3$-coated vesicle obtained after 18 days of mineralization was re-dispersed in water adjusted to pH 7.4 to prepare a measurement sample. The concentration of the $CaCO_3$-coated vesicle was 0.1 mg/mL. The DLS measurements were performed at 25 °C using a Zetersizer ZS (Malvern Panalytical Ltd., Cambridge, UK). For comparison, DLS measurements were also performed on $CaCO_3$-uncoated vesicles before mineralization.

3.2.3. X-ray Diffraction Measurements

The crystal structure of the $CaCO_3$ shell formed on the vesicle surface was investigated by X-ray diffraction (XRD) measurement. The lyophilized $CaCO_3$-coated vesicle obtained after 18 days of mineralization was used. XRD measurements were carried out using a SmartLabSE (Shimazu Co., Kyoto, Japan) equipped with a 1.8 kW CuKα ceramic X-ray tube operating at 40 kV and 30 mA.

3.2.4. Dissolution Behaviors of Shells on the $CaCO_3$-Coated Vesicles

We investigated the dissolubility of the $CaCO_3$ shell on the vesicle surface under various pH conditions. The lyophilized $CaCO_3$-coated vesicle obtained after 18 days of mineralization was re-dispersed in 100 mM 2-[4-2-(hydroxyethyl)-1-piperazinyl]ethanesulfonic acid (HEPES, Nacalai Tesque Inc., Kyoto, Japan)–Tris(hydroxylmethyl)aminomethane (Tris, Nacalai Tesque Inc., Kyoto, Japan) buffers adjusted to pH 6.0, 7.4, and 8.0, respectively. The concentration of the $CaCO_3$-coated vesicle was 0.01 g/L. The $CaCO_3$-coated vesicle dispersion at each pH condition was gently shaken at 36 °C. We measured the concentration changes of the Ca^{2+} generated by the dissolution of the $CaCO_3$ shell. The quantification of free Ca^{2+} was performed by fluorescence measurements using 8-amino-2-[(2-amino-5-methylphenoxy)methyl]-6-methoxyquinoline-*N,N,N′,N′*-tetraacetic acid, tetrapotassium salt (Quin 2, Dojindo Co. Ltd., Kumamoto, Japan) as a quantification reagent for Ca^{2+}. At regular intervals, 1 mL of the vesicle dispersion was collected from the dispersion of each pH condition, and 20 μL of 1.44 mM Quin 2 aqueous solution was added, and then fluorescence measurements were carried out using a spectrofluorophotometer (RF-5300, Shimadzu Co., Kyoto, Japan). The excitation wavelength of Quin 2 was 339 nm, and the fluorescence was observed at 492 nm. We created the calibration curve of the relationship between the Ca^{2+} concentration and Quin 2 fluorescence intensity in each buffer solution. The Ca^{2+} concentrations were calculated from the obtained fluorescence intensities, based on the calibration curves.

3.2.5. Drug Release Experiments

We performed drug release measurements of the hydrophilic model drug, Rh6G, from the $CaCO_3$-coated vesicle. The encapsulated Rh6G in the vesicle interior was quenched, and only the Rh6G released from the vesicle emitted fluorescence. The amounts of drug released were determined from the fluorescence intensity of Rh6G in the external aqueous phase. The $CaCO_3$-coated vesicle encapsulated Rh6G obtained after 18 days of mineralization was re-dispersed in 100 mM HEPES–Tris buffers adjusted to pH 6.0, 7.4, and 8.0, respectively. The concentration of the $CaCO_3$-coated vesicle was 0.01 g/L. The $CaCO_3$-coated vesicle dispersion at each pH condition was gently shaken at 36 °C, and the intensity changes of Rh6G fluorescence were monitored on a spectrofluorophotometer. The excitation and emission wavelength of Rh6G were 527 and 551 nm, respectively. The amounts of drug release under each pH condition were calculated based on the calibration curve obtained in each buffer solution.

Furthermore, we investigated the release of the hydrophobic model drug, Py, from the $CaCO_3$-coated vesicle. We evaluated the transfer amounts of Py from the $CaCO_3$-coated vesicle to the pure azolectin vesicle by the fluorescence measurements. The pure azolectin vesicles, which were reservoirs of released Py from the $CaCO_3$-coated vesicles, were prepared by sonication in buffer solutions adjusted to pH 6.0, 7.4, and 8.0, respectively. The concentration of pure azolectin vesicles was 0.1 wt%. Then, 1 mg of the $CaCO_3$-coated vesicle containing Py obtained after 18 days of mineralization was weighted into a microtube. The pure azolectin vesicle dispersions adjusted at pH 6.0, 7.4, and 8.0 were added to the microtubes, and then the microtubes were each shaken at 36 °C. At regular intervals, the dispersions were centrifuged at 10,000 rpm for 10 min, and the azolectin vesicles with transferred Py were collected as a supernatant. The amounts of transferred Py under each pH condition were determined by the fluorescence measurement of the collected supernatant, based on the calibration curve of the vesicle containing 1 wt% Py obtained in each buffer solution. The excitation and emission wavelength of Py were 342 and 375 nm, respectively.

4. Conclusions

In this study, we attempted to fabricate a new DDS carrier especially useful for cancer therapy. The vesicle surface was coated with $CaCO_3$ owing to the self-supplied mineralization induced by the embedded peptide lipid. The $CaCO_3$-coated vesicle maintained its spherical shape even under a high vacuum; however, the vesicle collapsed easily under the weakly acidic condition, mimicking the neighborhood of the cancer tissue, owing to the dissolution of the $CaCO_3$ shell. It was also confirmed that the $CaCO_3$-coated vesicle could encapsulate both the hydrophilic and hydrophobic drugs as the DDS carrier. The encapsulated drugs, hydrophilic Rh6G and hydrophobic Py, were released effectively through the dissolution of the $CaCO_3$ shell under weakly acidic conditions such as the neighborhood of cancer tissues. The release of these drugs behaved differently depending on the location of the drug in the vesicle. The obtained $CaCO_3$-coated vesicle is expected to be an effective DDS carrier in cancer drug therapy.

Author Contributions: Conceptualization, M.H.; methodology, S.M. and M.H.; validation, C.M., M.K., N.K., S.M. and M.H.; formal analysis, C.M., M.K., N.K., S.M. and M.H.; investigation, M.H.; writing—original draft preparation, M.H.; writing—review and editing, M.H.; visualization, M.H.; supervision, M.H.; project administration, M.H.; funding acquisition, M.H. All authors have read and agreed to the published version of the manuscript.

Funding: This research was funded by JSPS KAKENHI Grant-in-Aid for Scientific Research (C), grant number JP20K05264.

Institutional Review Board Statement: Not applicable.

Informed Consent Statement: Not applicable.

Data Availability Statement: Not applicable.

Conflicts of Interest: The authors declare no conflict of interest.

Appendix A

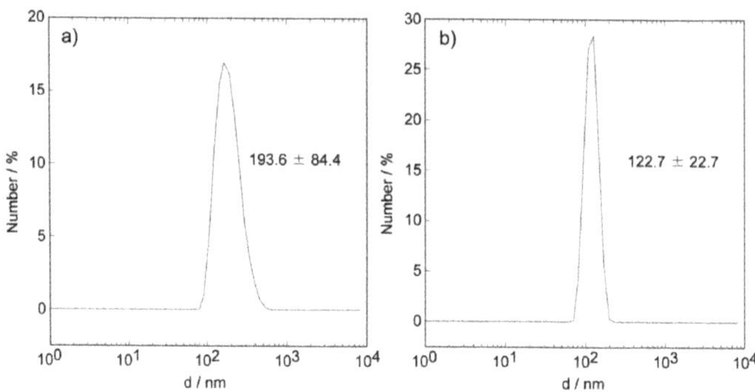

Figure A1. Histograms for hydrodynamic diameter of (**a**) $CaCO_3$-coated vesicles obtained by 18-day mineralization, and (**b**) that of vesicles before mineralization.

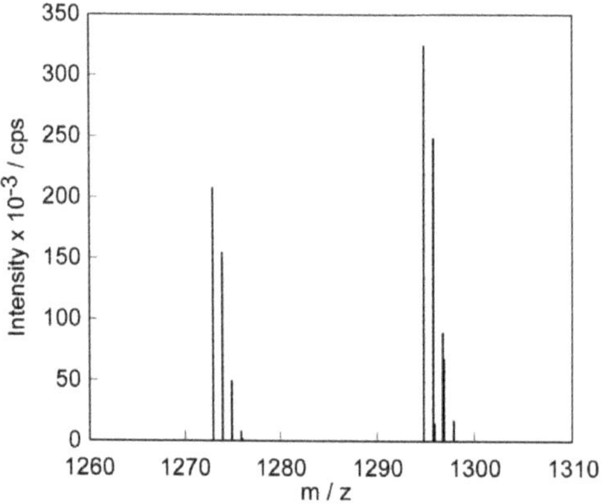

Figure A2. MALDI-TOF-MS spectrum of the peptide lipid.

Table A1. The assignment of crystal faces from diffraction peak and relative intensity shown in Figure 5, and standard XRD data of calcite phase of $CaCO_3$.

		Experiment		Standard (Calcite)	
No.	Crystal Face	2θ/deg	Intensity	2θ/deg	Intensity
1	(0 1 2)	23.02	7.25	23.07	8.10
2	(1 0 4)	29.36	100	29.42	100
3	(0 0 6)	31.42	2.19	31.46	2.43
4	(1 1 0)	35.96	16.9	36.00	13.7
5	(1 1 3)	39.6	0.915	39.44	20.2
6	(2 0 2)	43.14	21.1	43.19	14.3
7	(0 2 4)	47.44	22.3	47.15	6.16
8	(0 1 8)	47.54	7.99	47.54	19.7
9	(1 1 6)	48.46	24.1	48.54	20.0
10	(2 1 1)	56.56	4.93	56.61	3.73
11	(1 2 2)	57.38	11.4	57.44	9.18
12	(2 1 4)	60.64	7.63	60.72	5.28
13	(2 0 8)	60.94	3.08	61.05	2.41
14	(1 1 9)	61.28	2.41	61.43	3.18
15	(1 2 5)	63.00	2.38	63.10	2.29
16	(3 0 0)	64.64	7.71	64.71	6.84
17	(0 0 12)	65.54	5.33	65.67	3.91
18	(2 1 7)	69.12	1.95	69.24	1.51
19	(0 2 10)	70.20	1.79	70.30	2.13
20	(1 2 8)	72.84	3.60	72.95	2.69
21	(1 1 12)	77.02	2.72	77.23	1.93
22	(2 1 10)	81.44	2.57	81.60	2.24
23	(1 3 4)	83.74	5.25	83.84	1.57

References

1. Lee, Y.T.; Tan, Y.J.; Oon, C.E. Molecular targeted therapy: Treating cancer with specificity drugs. *Eur. J. Pharmacol.* **2018**, *834*, 188–196. [CrossRef] [PubMed]
2. Kimura, M.; Yasui, F.; Usami, E.; Kawachi, S.; Iwai, M.; Go, M.; Ikeda, Y.; Yoshimura, T. Cost-effectiveness and safety of the molecular targeted drugs afatinib, gefitinib and erlotinib as first-line treatments for patients with advanced EGFR mutation-postive non-small cell lung cancer. *Mol. Clin. Oncol.* **2018**, *9*, 201–206. [CrossRef] [PubMed]
3. Delaunay, M.; Cadranel, J.; Lusque, A.; Meyer, N.; Gounant, V.; Moro-Sibilot, D.; Michot, J.-M.; Raimboug, J.; Girard, N.; Guisier, F.; et al. Immune-checkpoint inhibitors associated with interstitial lung disease in cancer patients. *Eur. Respir. J.* **2017**, *50*, 1700050. [CrossRef] [PubMed]
4. Yamagata, A.; Yokoyama, T.; Fukuda, Y.; Ishida, T. Impact of interstitial lung disease associated with immune checkpoint inhibitors on prognosis in patients with non-small-cell lung cancer. *Cancer Chemother. Pharamacol.* **2021**, *87*, 251–258. [CrossRef] [PubMed]
5. Misha, D.; Kang, H.C.; Bae, Y.H. Reconstitutable charged polymeric (PLGA)(2)-b-PEI micelles for gene therapeutics delivery. *Biomaterials* **2011**, *32*, 3845–3854. [CrossRef] [PubMed]
6. Yokoyama, M. Polymeric micelles as drug carriers: Their light and shadows. *J. Drug Target.* **2014**, *22*, 576–583. [CrossRef] [PubMed]
7. Morgan, M.T.; Carnahan, M.A.; Finkelstein, S.; Prata, C.A.H.; Degoricijia, L.; Lee, S.J.; Grinstaff, M.W. Dendric supramolecular assemblies for drug delivery. *Chem. Commun.* **2005**, *34*, 4309–4311. [CrossRef]
8. Tomas, T.P.; Joice, M.; Sumit, M.; Silpe, J.E.; Kotlyar, A.; Bharathi, S.; Kukowska-Latallo, J.; Baker, J.R.; Choi, S.K. Design and in vitro validation of multivalent dendrimer methotrexates as a folate-targeting anticancer therapeutic. *Curr. Pharm. Des.* **2013**, *19*, 6594–6605. [CrossRef]
9. Perli, G.; Wang, Q.; Braga, C.B.; Bertuzzi, D.L.; Fontana, L.A.; Soares, M.C.P.; Ruiz, J.; Megiatt, J.D., Jr.; Astruc, D.; Ornelas, C. Self-assembly of a triazolylferrocenyl dendrimer in water yields nontraditional intrinsic green fluorescent vesosomes for nanotheranostic applications. *J. Am. Chem. Soc.* **2021**, *143*, 12948–12954. [CrossRef] [PubMed]
10. Shang, S.; Huang, S.; Zhang, A.; Feng, J.; Yang, S. The binary complex of poly(PEGMA-co-MAA)hydrogel and PLGA nanoparticles as a novel oral drug delivery system for ibuprofen delivery. *J. Biomater. Sci. Polym. Ed.* **2017**, *28*, 1874–1887. [CrossRef]
11. Murai, K.; Kurumisawa, K.; Nomura, Y.; Matsumoto, M. Regulated drug release abilities of calcium carbonate-gelatin hybrid nanocarriers fabricated via a self-organizational process. *ChemMedChem* **2017**, *12*, 1595–1599. [CrossRef]
12. Chander, S.; Kulkarni, G.; Dhiman, N.; Kharkwai, H. Protein-based nanohydrogels for bioactive delivery. *Front. Chem.* **2021**, *9*, 573748. [CrossRef] [PubMed]

13. Lee, J.H.; Tachibana, T.; Yamana, K.; Kawasaki, R.; Yabuki, A. Simple formation of cancer drug-containing self-assembled hydrogels with temperature and pH-responsive release. *Langmuir* **2021**, *37*, 11269–11275. [CrossRef]
14. Murai, K.; Higuchi, M.; Kinoshita, T.; Nagata, K.; Kato, K. Design of the nanocarrier having the regulated durug release ability utilizing a reversible conformational transition of peptide responsed to slight pH changes. *Phys. Chem. Chem. Phys.* **2013**, *15*, 11454–11460. [CrossRef] [PubMed]
15. Beagan, A.M.; Alghamdi, A.A.; Lahmadi, S.S.; Halwani, M.A.; Almeataq, M.S.; Alhazaa, A.N.; Alotaibi, K.M.; Alswieleh, A.M. Folic acid-terminated poly(2-diethyl amino ethyl methacrylate) brush-gated magnetic mesoporous nanoparticles as a smart drug delivery system. *Polymers* **2021**, *13*, 59. [CrossRef] [PubMed]
16. Martínez-Edo, G.; Xue, E.Y.; Ha, S.Y.Y.; Pontón, I.; González-Delgado, J.A.; Borrós, S.; Torres, T.; Ng, D.K.P.; Sanchez-Garcia, D. Nanoparticles for triple drug release for combined chemo- and photodynamic therapy. *Chem.-A Eur. J.* **2021**, *27*, 14610–14618. [CrossRef]
17. Arai, M.; Miura, T.; Ito, Y.; Kinoshita, T.; Higuchi, M. Protein sensing device with multi-recognition ability composed of self-organized glycopeptide bundle. *Int. J. Mol. Sci.* **2021**, *22*, 366. [CrossRef]
18. Sato, Y.; Asawa, K.; Hung, T.; Noiri, M.; Nakamura, N.; Ekdahl, K.N.; Nilson, B.; Ishihara, K.; Teramura, Y. Induction of spontaneous liposome adsorption by exogenous surface modification with cell-penetrating peptide-conjugated lipids. *Langmuir* **2021**, *37*, 9711–9723. [CrossRef]
19. Maruyama, K.; Ishida, O.; Kasaoka, S.; Takizawa, T.; Utoguchi, N.; Shinohara, A.; Chiba, M.; Kobayashi, H.; Eriguchi, M.; Yanagie, H. Intercellular targeting of sodium mercaptoundecahydrododecaborate (BSH) to solid tumors by transferrin-PEG liposomes, for boron neutron-capture therapy (BNCT). *J. Control. Release* **2004**, *98*, 195–207. [CrossRef]
20. Joanitti, G.A.; Sawant, R.S.; Torchilin, V.P.; Freitas, S.M.; Azevedo, R.B. Optimizing liposome for delivery of Bowman-Birk protease inhibitors—Platforms for multiple biomedical applications. *Colloids Surf. B Biointerfaces* **2018**, *167*, 474–482. [CrossRef]
21. Norouzi, M.; Amerian, M.; Amerian, M.; Atyabi, F. Clinical applications of nanomedicine in cancer therapy. *Drug Discov. Today* **2020**, *25*, 107–125. [CrossRef] [PubMed]
22. Matsumura, Y.; Maeda, H. A new concept for macromolecular therapeutics in cancer chemotherapy: Mechanism of tumoritropic accumulation of proteins and the antitumor agent smancs. *Cancer Res.* **1986**, *46 Pt 1*, 6387–6392.
23. Allen, T.M.; Cullis, P.R. Drug delivery systems: Entering the mainstream. *Science* **2004**, *303*, 1818–1822. [CrossRef] [PubMed]
24. Kang, H.; Rho, S.; Stiles, W.R.; Hu, S.; Baek, Y.; Hwang, D.W.; Kashiwagi, S.; Kim, M.S.; Choi, H.S. Size-dependent EPR effect of polymeric nanoparticles on tumor targeting. *Adv. Healthc. Mater.* **2020**, *9*, 1901203. [CrossRef]
25. Swietach, P.; Vaughan-Jones, R.D.; Harris, A.L.; Hulikova, A. The chemistry, physiology and pathology of pH in cancer. *Philos. Trans. R. Soc. Lond. B Biol. Sci.* **2014**, *369*, 20130099. [CrossRef] [PubMed]
26. Ward, C.; Meehan, J.; Gray, M.E.; Murray, A.F.; Argyle, D.J.; Kunkler, I.H.; Langdon, S.P. The impact of tumor pH on cancer progression: Strategies for clinical intervention. *Explor. Target. Anti-tumor Ther.* **2020**, *1*, 71–100. [CrossRef]
27. Koltai, T. The pH paradigm in cancer. *Eur. J. Clin. Nutr.* **2020**, *74*, 14–19. [CrossRef]
28. Murai, K.; Higuchi, M.; Kinoshita, T.; Nagata, K.; Kato, K. Calcium carbonate biomineralization utilizing a multifunctional β-sheet peptide template. *Chem. Commun.* **2013**, *49*, 9947–9949. [CrossRef]
29. Murai, K.; Kinoshita, T.; Nagata, K.; Higuchi, M. Mineralization of calcium carbonate on multifunctional peptide assembly actiong as mineral source supplier and template. *Langmuir* **2016**, *32*, 9351–9359. [CrossRef]
30. Bringer, W.S.; Chao, T.L. Is the serine protease "charge relay system" a charge relay system? *Biochem. Biophys. Res. Commun.* **1975**, *63*, 78–83. [CrossRef]
31. Matsuki, H.; Kaneshina, S. Thermodynamics of bilayer phase transitions of phospholipids. *Netsu Sokutei* **2006**, *33*, 74–82.
32. Plummer, L.N.; Busenberg, E. The solubilites of calcite, aragonite and vaterite in CO_2-H_2O solutions between 0 and 90°C, and an evaluation of the aqueous model for the system $CaCO_3$-CO_2-H_2O. *Geochim. Cosmochim. Acta* **1982**, *46*, 1011–1040. [CrossRef]
33. Graft, D.L. Crystallographic tables for the rhombohedral carbonates. *Am. Mineral.* **1961**, *46*, 1283–1316.
34. De Villiers, J.R.P. Crystal structure of aragonite, strontianite, and witherite Am. *Mineral.* **1971**, *56*, 758–767.
35. Wang, J.; Becker, U. Structure and carbonate orientation of vaterite ($CaCO_3$). *Am. Mineral.* **2009**, *94*, 380–386. [CrossRef]
36. Miyoshi, N.; Hara, K.; Kimura, S.; Nakanishi, K.; Fukuda, M. A New Method of Determining Intercellular Free Ca^{2+} Concentration Using Quin 2-fluorescence, Photochem. *Photobiol.* **1991**, *53*, 415–418. [CrossRef] [PubMed]
37. Carpino, L.A.; Han, G.Y. 9-Fluorenymethoxycarbonyl amino-protecting group. *J. Org. Chem.* **1972**, *37*, 3404–3409. [CrossRef]

International Journal of *Molecular Sciences*

MDPI

Article

Engineering of Vaginal Lactobacilli to Express Fluorescent Proteins Enables the Analysis of Their Mixture in Nanofibers

Spase Stojanov [1,2], Tina Vida Plavec [1,2], Julijana Kristl [2], Špela Zupančič [2] and Aleš Berlec [1,2,*]

1 Department of Biotechnology, Jožef Stefan Institute, SI-1000 Ljubljana, Slovenia; spase.stojanov@ijs.si (S.S.); tina.plavec@ijs.si (T.V.P.)
2 Faculty of Pharmacy, University of Ljubljana, SI-1000 Ljubljana, Slovenia; julijana.kristl@ffa.uni-lj.si (J.K.); spela.zupancic@ffa.uni-lj.si (Š.Z.)
* Correspondence: ales.berlec@ijs.si

Abstract: Lactobacilli are a promising natural tool against vaginal dysbiosis and infections. However, new local delivery systems and additional knowledge about their distribution and mechanism of action would contribute to the development of effective medicine. This will be facilitated by the introduction of the techniques for effective, inexpensive, and real-time tracking of these probiotics following their release. Here, we engineered three model vaginal lactobacilli (*Lactobacillus crispatus* ATCC 33820, *Lactobacillus gasseri* ATCC 33323, and *Lactobacillus jensenii* ATCC 25258) and a control *Lactobacillus plantarum* ATCC 8014 to express fluorescent proteins with different spectral properties, including infrared fluorescent protein (IRFP), green fluorescent protein (GFP), red fluorescent protein (mCherry), and blue fluorescent protein (mTagBFP2). The expression of these fluorescent proteins differed between the *Lactobacillus* species and enabled quantification and discrimination between lactobacilli, with the longer wavelength fluorescent proteins showing superior resolving power. Each *Lactobacillus* strain was labeled with an individual fluorescent protein and incorporated into poly (ethylene oxide) nanofibers using electrospinning, as confirmed by fluorescence and scanning electron microscopy. The lactobacilli retained their fluorescence in nanofibers, as well as after nanofiber dissolution. To summarize, vaginal lactobacilli were incorporated into electrospun nanofibers to provide a potential solid vaginal delivery system, and the fluorescent proteins were introduced to distinguish between them and allow their tracking in the future probiotic-delivery studies.

Keywords: lactobacilli; vaginal probiotics; fluorescent proteins; electrospinning; nanofibers; probiotic analysis; probiotic delivery

Citation: Stojanov, S.; Plavec, T.V.; Kristl, J.; Zupančič, Š.; Berlec, A. Engineering of Vaginal Lactobacilli to Express Fluorescent Proteins Enables the Analysis of Their Mixture in Nanofibers. *Int. J. Mol. Sci.* **2021**, *22*, 13631. https://doi.org/10.3390/ijms222413631

Academic Editor: Raghvendra Singh Yadav

Received: 1 December 2021
Accepted: 17 December 2021
Published: 20 December 2021

Publisher's Note: MDPI stays neutral with regard to jurisdictional claims in published maps and institutional affiliations.

Copyright: © 2021 by the authors. Licensee MDPI, Basel, Switzerland. This article is an open access article distributed under the terms and conditions of the Creative Commons Attribution (CC BY) license (https://creativecommons.org/licenses/by/4.0/).

1. Introduction

The healthy human vagina is home to around 50 microbial species, of which the most dominating are bacteria from the genus *Lactobacillus* [1]. The main ones found are *Lactobacillus crispatus*, *Lactobacillus jensenii*, *Lactobacillus gasseri*, and *Lactobacillus iners*. Each of these is dominant in its community type, and *L. crispatus* is the most abundant [2]. Vaginal infections are favored by dysbiosis of the normal vaginal microbiota, whereby the numbers of lactobacilli decrease. This can allow the overgrowth of several opportunistic pathogens, including *Gardnerella vaginalis*, *Atopobium vaginae*, and *Candida albicans* [3].

Antimicrobial drugs are the first line of defense against bacterial infections. However, their frequent use can lead to antimicrobial resistance and also to high infection recurrence rates (i.e., ~50%) [3]. These high recurrence rates appear to be associated with the nonselective mechanisms of the antimicrobial drugs, where as well as the pathogens, the normal lactobacilli are also reduced [4,5]. Therefore, new therapeutic strategies are required to normalize vaginal dysbiosis. Re-establishing the vaginal microbial balance with *Lactobacillus* bacteria as probiotics can then prevent the overgrowth of vaginal pathogens and thus prevent recurrence of the vaginal infection. The dominant vaginal species, *L. crispatus*, has been shown to be active against vaginal pathogens, both alone and in combination with other lactobacilli [6–11]. Antimicrobial properties of other vaginal lactobacilli against vaginal and uropathogens have also been reported [12–15].

Despite the beneficial properties of vaginal lactobacilli, their application as probiotics is limited by the lack of an appropriate delivery system and their low viability and high sensitivity to environmental factors [16]. The currently used liquid, semi-solid, and solid dosage forms for vaginal drug delivery have several limitations, including short residence time, discomfort, leakage, imprecise dosing, and variable drug distribution [17]. On the other hand, nanofibers have a high surface-to-volume ratio and can provide high drug loading, controlled release, cell binding, good bioavailability, and cost-effectiveness [18].

Nanofibers are produced by electrospinning, which is based on the drying of a thin liquid jet that is formed from a drop of polymer solution in a strong electric field [19] and are used in numerous applications [20,21]. They also represent an effective material for delivering different types of drugs to the nasal, oral, and vaginal mucosa [22], while at the same time, they protect drugs from environmental factors [23]. Different compounds can be incorporated into nanofibers, including small drug molecules, proteins, nucleic acids, and cells. *Lactobacillus* bacteria have also been incorporated into electrospun nanofibers, with *L. plantarum* being the most frequently used [24–28]. Other *Lactobacillus* species have also been electrospun using different polymers, namely, agrowaste amended with poly (vinyl alcohol) for *Lactobacillus acidophilus* [29], poly (vinyl alcohol) and sodium alginate for *Lactobacillus rhamnosus* [30], poly (vinyl alcohol) for *Lactobacillus gasseri* [31], and Eudragit L100 and sodium alginate for *Lactobacillus paracasei* [32]. In a recent study, we successfully incorporated nine different *Lactobacillus* species into polyethylene oxide (PEO) nanofibers (i.e., *L. acidophilus, L. delbrueckii* ssp. *bulgaricus, L. casei, L. gasseri, L. paracasei, L. plantarum, L. reuteri, L. rhamnosus, L. salivarius*) with high viability after electrospinning process [33]. However, few studies have reported incorporation of vaginal lactobacilli. One example was the incorporation of *L. gasseri* CRL1320 and *L. rhamnosus* CRL1332 into polyvinyl-alcohol nanofibers, with skimmed milk lactose and glycerol serving as bioprotective agents [34]. In another study, vaginal *L. acidophilus* was incorporated into polyvinyl alcohol and polyvinylpyrrolidone nanofibers [35].

Apart from the lack of delivery systems, vaginal probiotics use is hampered by the lack of research into their distribution and mechanism of action [36]. The spatial identification of individual strains in mixtures is particularly challenging, as stains (such as Syto 9) cannot be used to distinguish them, and custom antibodies against specific surface antigens or fluorescent in-situ hybridization have to be used instead [37]. Genetic engineering of lactobacilli for the production of fluorescent proteins is a rapid and effective method for tracking and distinguishing lactobacilli. Fluorescent proteins with different spectral properties have already been used to study the distribution and properties of lactobacilli [38–42]. However, few studies have reported the incorporation of fluorescent lactobacilli into nanofibers. In our recent study, *L. plantarum* expressing red fluorescent protein mCherry was incorporated into PEO nanofibers to evaluate nanofiber dissolution and lactobacilli release [27].

The aim and novelty of the present study was to engineer vaginal lactobacilli to express fluorescent proteins and incorporate them into nanofibers. Four different species of lactobacilli (*L. crispatus, L. gasseri, L. jensenii,* and *L. plantarum*) were genetically modified to express compatible fluorescent proteins with different spectral properties: infrared fluorescent protein (IRFP); green fluorescent protein (GFP); red fluorescent protein (mCherry); and blue fluorescent protein (mTagBFP2). Quantification of their fluorescence, the overlap between the fluorescences of the different fluorescent proteins, and the differentiation between the fluorescent species were evaluated. The four engineered species were mixed with PEO solution and electrospun into nanofibers as the potential delivery system. By using genetic engineering, we have introduced a new technique for effective, inexpensive, and real-time tracking of probiotics following their incorporation into nanofibers and nanofiber dissolution.

2. Results

2.1. Genetic Constructs for Expression of Fluorescent Proteins in Vaginal Lactobacilli

To complement the existing pNZ-ldh-GFP plasmid that encodes GFP under the control of the *ldh* promoter from *L. plantarum*, genes that encode the fluorescent proteins IRFP, mCherry, and mTagBFP2 were scarlessly fused with the *ldh* promoter using overlap-extension PCR (Figure 1). The gene fusions were cloned into the pNZ1848 plasmid, thereby replacing the inducible nisin promoter and yielding plasmids pNZ-ldh-IRFP, pNZ-ldh-mCherry, and pNZ-ldh-mTagBFP2. All four plasmids were separately transformed in all four *Lactobacillus* species, yielding 16 combinations (Table S1) and thus providing a wide range of possibilities to distinguish between them when used simultaneously. They were characterized in the following experiment with respect to the intensity of the expressed fluorescent proteins.

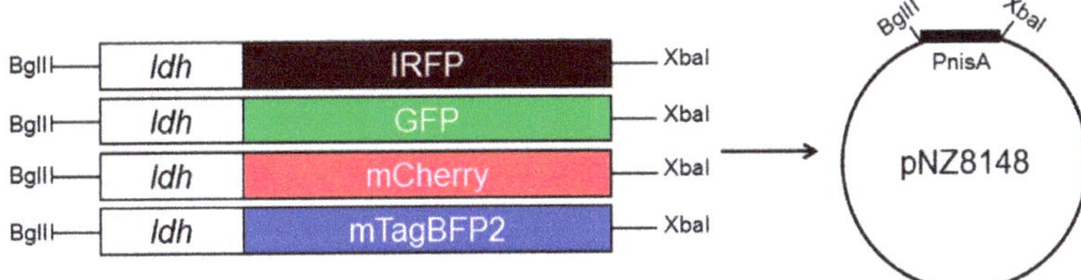

Figure 1. Assemblies of the *ldh* promoter and the four genes that encode the fluorescent proteins (IRFP, GFP, mCherry, mTagBFP2) in pNZ8148 plasmid. BglII, XbaI—restriction recognition sites that were used for cloning; PnisA—nisin promoter.

2.2. Analysis of Fluorescent Protein Expression in Vaginal Lactobacilli

Fluorescent protein expression in lactobacilli was confirmed by measuring the fluorescence of all 16 bacterial suspensions and was shown to be appropriate for the detection of bacteria (Figure 2). The expression of the individual fluorescent proteins depended on the culture conditions and differed between the species. Growing the bacteria with shaking (aeration) resulted in higher fluorescence per unit OD_{600} in comparison to the samples grown without shaking (Figure 2) and significantly steeper slopes of regression lines (Table S1). This is in accordance with previous observations of oxygen-mediated post-translational activation of fluorescent proteins [43]. Fluorescent proteins GFP and TagBFP2 were expressed in all four strains at aerobic conditions, with 42,000 $\pm$ 2000 fluorescence units (FU) in suspensions at OD_{600} 3.0. By contrast, the expression of IRFP and mCherry was strain-dependent and was highest in *L. plantarum* (1100 FU for IRFP and 42,500 FU for mCherry). In comparison, the fluorescence of mCherry in other lactobacilli was lower by a factor of 2.4 for *L. gasseri*, a factor of 20.5 for *L. jensenii*, and a factor of 424 for *L. crispatus*. The fluorescence of these bacteria expressing the fluorescent proteins correlated linearly with a bacterial concentration in the OD_{600} range between 0.25 and 3.00 (Figure 2). The coefficients of determination (R^2) for the engineered bacteria were above 0.96 with the exception of *L. plantarum* expressing mTagBFP2 and *L. crispatus* expressing IRFP, where R^2 was lower (Supplementary Material: Table S1).

The nontransformed lactobacilli were used as controls because of the possibility of their autofluorescence. No or little concentration-dependent autofluorescence was seen for IRFP, GFP, and mCherry. Although bacteria showed relatively strong and concentration-dependent autofluorescence when measured using settings corresponding to mTagBFP2, the absolute fluorescence of nontransformed bacteria was significantly lower in comparison to the transformed bacteria.

The expression of the fluorescent proteins in these vaginal lactobacilli was also detected under a confocal microscope (Figure 3). Fluorescence was detected for all the engineered lactobacilli, while no fluorescence was detected for the nontransformed lactobacilli when using the settings for IRFP and mCherry. Some autofluorescence was seen for the nontransformed bacteria when using the settings for GFP and mTagBFP2; however, the signals were lower in comparison to those of the engineered bacteria.

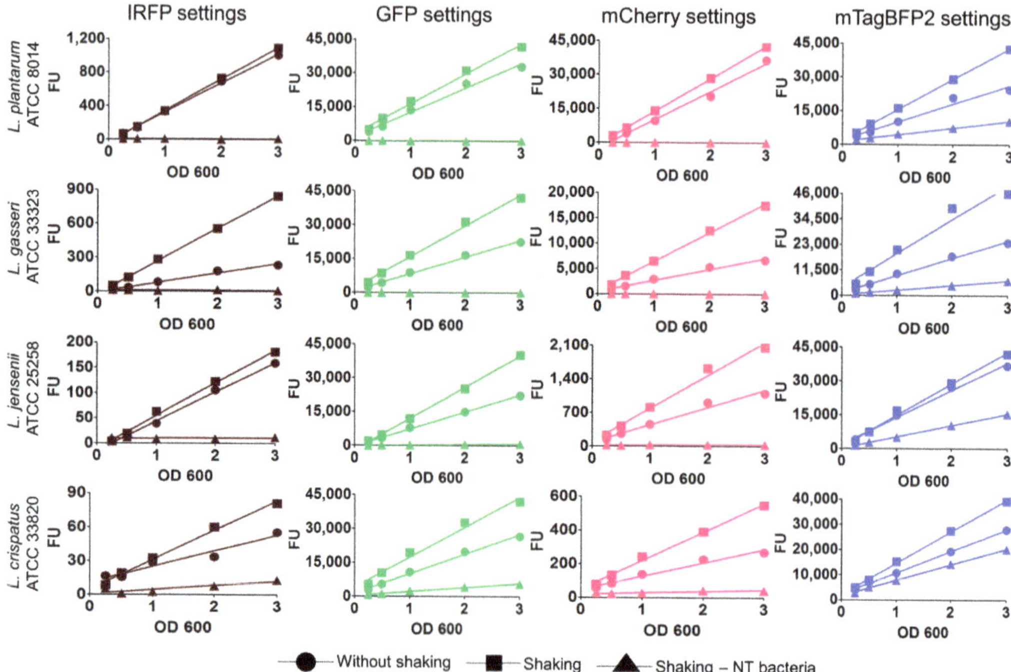

Figure 2. Concentration-dependent fluorescence of lactobacilli expressing different fluorescent proteins. Linear regression analyses of fluorescence and OD_{600} of lactobacilli grown at 37 °C without and with shaking (aeration) are shown. Nontransformed (NT) bacteria were used as controls. FU—fluorescence unit.

2.3. Fluorescence-Based Distinction between Lactobacilli in Mixture and Assessment of the Influence of Spectral Overlap

The individual *Lactobacillus* species were transformed with all four of the plasmids that encoded these fluorescent proteins with different spectral properties. Engineered strains of the same species of lactobacilli were mixed in different proportions, and their fluorescence was measured in a suspension (Figure 4). For the majority of the mixtures, the individual fluorescent strain was clearly distinguished by significantly higher fluorescence when using the settings corresponding to the fluorescent protein it expressed (i.e., the relevant excitation and emission wavelengths). Additionally, the fluorescence signals correlated with the content of the individual strain in the mixtures. The lowest specificity was observed with the mTagBFP2 settings, especially in the case of *L. plantarum*, where considerable fluorescence was also seen for other fluorescent proteins. This was attributed to shorter wavelengths and bacterial autofluorescence.

2.4. Selection and Analysis of Optimal Lactobacilli-Fluorescent Protein Combinations

To enable the distinction between the *Lactobacillus* species in complex mixtures or probiotic products, and thereby facilitate characterization in further studies, each species was defined with an individual fluorescent protein on the basis of the fluorescence properties of all 16 lactobacilli and fluorescent protein combinations, as follows: *L. plantarum*, IRFP; *L. crispatus*, GFP; *L. gasseri*, mCherry; and *L. jensenii*, mTagBFP2 (Figure 4). The fluorescence of the individual strains and their mixtures in the different proportions was measured, and the nontransformed lactobacilli were used as controls (Figure 5). Again, the individual fluorescent species were clearly distinguished by significantly higher fluorescence when using the settings corresponding to the fluorescent protein they expressed (i.e., the relevant excitation and emission wavelengths), and the fluorescence signals correlated with the contents of the individual strain in the mixtures (Figure 5). When using the settings for IRFP and mCherry (i.e., for determination of *L. plantarum* and *L. gasseri*, respectively), no fluorescence was observed for the nontransformed bacteria (i.e., no autofluorescence) or for the other fluorescent lactobacilli. However, stronger autofluorescence of the nontransformed lactobacilli was seen for the measures with the GFP settings (i.e., for *L. crispatus*), with the strongest seen for the measures with the mTagBFP2 settings (i.e., for *L. jensenii*). The autofluorescence was also species specific, whereby *L. plantarum* showed the lowest autofluorescence using the GFP and mTagBFP2 settings. Considerable overlap was observed between *L. crispatus* that expressed GFP and *L. jensenii* that expressed mTagBFP2, where individual bacteria were detected in both of the fluorescence channels.

However, with confocal microscopy, the majority of the lactobacilli that expressed the different fluorescent proteins were distinguished on the basis of their fluorescence when in mixtures of equal ratios (Figure 6).

2.5. Incorporation of Fluorescent Lactobacilli into PEO Electrospun Nanofibers

Fluorescent species of vaginal lactobacilli were successfully incorporated into PEO nanofibers as a potential vaginal delivery system, both individually and as mixtures, including the smallest species (*L. plantarum*: length, 1.28 ± 0.32 μm; width, 0.52 ± 0.04 μm) and the largest species (*L. crispatus*: length, 7.73 ± 1.89 μm; width, 0.81 ± 1.31 μm). The mean diameter of the PEO nanofibers without the bacteria was 170 ±40 nm, and the bacteria incorporation was seen as characteristic thickenings along the nanofibers as reported previously [27,33]. The incorporation of *L. plantarum*, *L. gasseri*, and *L. crispatus* resulted in the increase of the mean nanofiber diameter by app. 100 nm, while the incorporation of *L. jensenii* caused no increase (Figure 7). This may be due to the bacterial release of molecules, such as exopolysaccharides or ions, that can influence the conductivity or viscosity of the polymer suspension.

Effective incorporation of these bacteria into the nanofibers was also confirmed using confocal microscopy (Figure 8). All of the recombinant species retained their fluorescence following their incorporation in the nanofibers, while no fluorescence was observed for the nontransformed lactobacilli, which were incorporated as the controls. The different lactobacilli could be distinguished in the mixtures on the basis of their fluorescence, although some overlap was observed for *L. crispatus* expressing GFP and *L. jensenii* expressing mTagBFP2. This was similar to the data obtained for the bacterial suspensions (Figure 6).

2.6. Release of Lactobacilli from Nanofibers

The lactobacilli from the nanofibers retained their fluorescence after the dissolution of nanofibers. The lower intensity of the fluorescence observed with the dissolved nanofibers was in line with the lower concentration of the lactobacilli in the dissolved nanofibers in comparison to the original dispersions. Namely, the concentration of bacteria per PEO mass was estimated to be on average 3.6-fold higher in suspension than in nanofibers, preventing direct comparison of absolute fluorescence values. Nevertheless, similar to the data above, the fluorescence intensities correlated with the bacterial concentrations in the dissolved nanofibers, as well as in the control 4% (w/v) PEO bacterial dispersions (Figure 9a). Here, a 4% (w/v) PEO solution and the dissolved PEO nanofibers without bacteria were used as the negative controls with significantly lower fluorescence. With the exception of *L. jensenii*, the bacteria-containing polymer dispersion and dispersion from nanofibers showed no autofluorescence, regardless of their concentration.

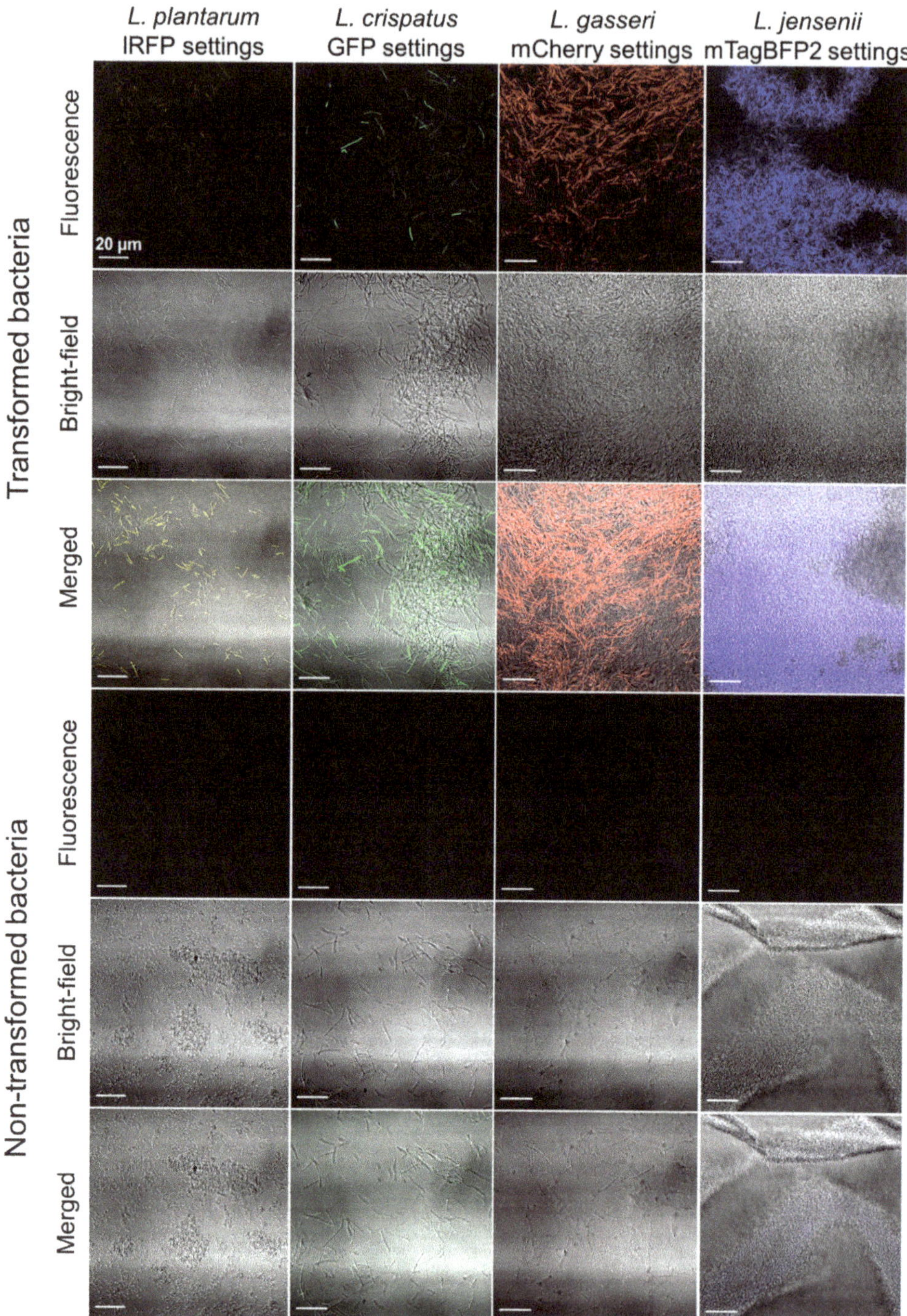

Figure 3. Representative confocal microscopy images of the lactobacilli expressing the different fluorescent proteins, as *L. plantarum* expressing IRFP, *L. gasseri* expressing mCherry, *L. crispatus* expressing GFP, and *L. jensenii* expressing mTagBFP2, in comparison to the nontransformed bacteria.

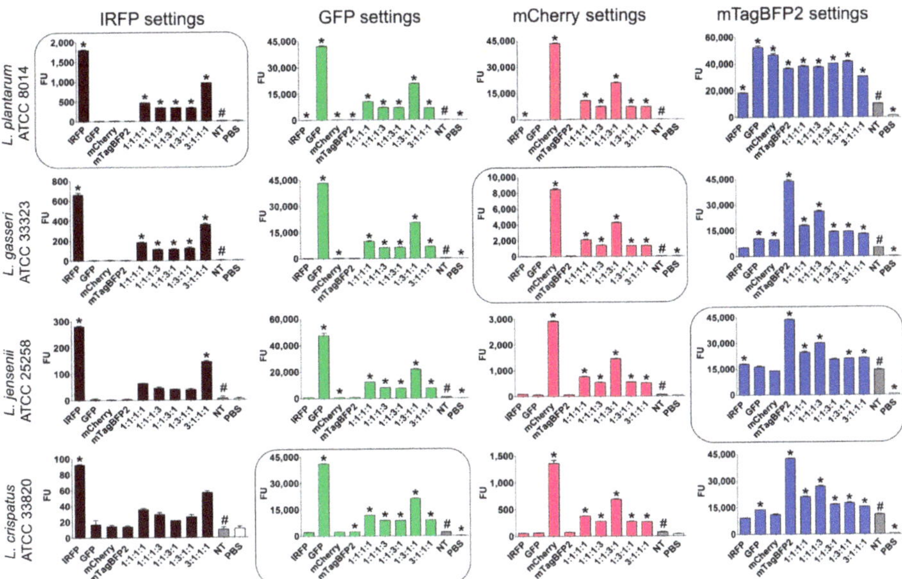

Figure 4. Distinction between the same species of *Lactobacillus* expressing different fluorescent proteins. Fluorescence was measured for the individual fluorescent strains and their mixtures, using settings corresponding to all four of the fluorescent proteins. The ratios indicate the proportions of the species expressing the fluorescent proteins in the following order: IRFP:GFP:mCherry:mTagBFP2. The encircled graphs represent the combination of fluorescent proteins and lactobacilli that were selected for further studies. * $p < 0.05$ (Student's t tests) relative to nontransformed strain (NT, high-lighted with # for clarity). FU—fluorescence units; PBS—phosphate-buffered saline.

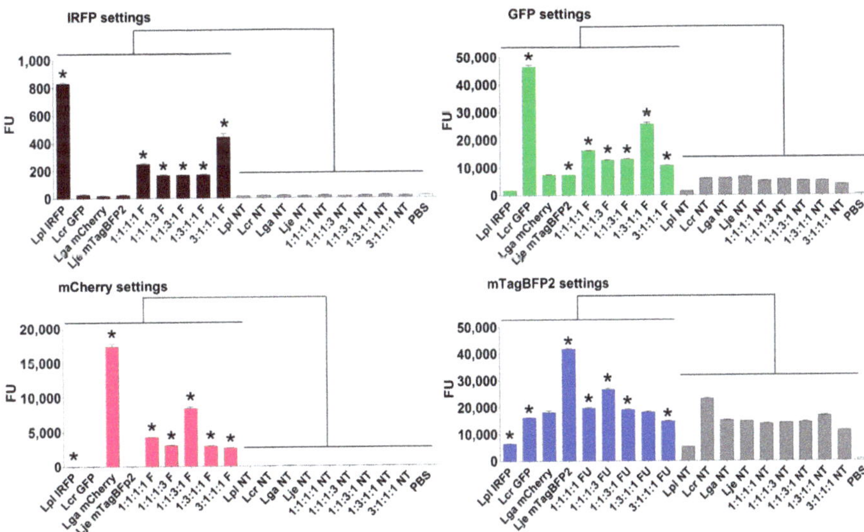

Figure 5. Fluorescence-based distinction of the different fluorescent lactobacilli and the nontransformed (NT) lactobacilli. Fluorescence was measured for the individual fluorescent species or their mixtures using the settings for all four of the fluorescent proteins. The ratios indicate the proportions of species in the mixtures in the following order: *L. plantarum* expressing IRFP; *L. crispatus* expressing GFP; *L. gasseri* expressing mCherry; and *L. jensenii* expressing mTagBFP2. * $p < 0.05$ (Student's t tests), obtained by comparing fluorescent strain (F) to its nontransformed counterpart (NT). Lpl—*L. plantarum*; Lga—*L. gasseri*; Lcr—*L. crispatus*; Lje—*L. jensenii*; PBS—phosphate-buffered saline.

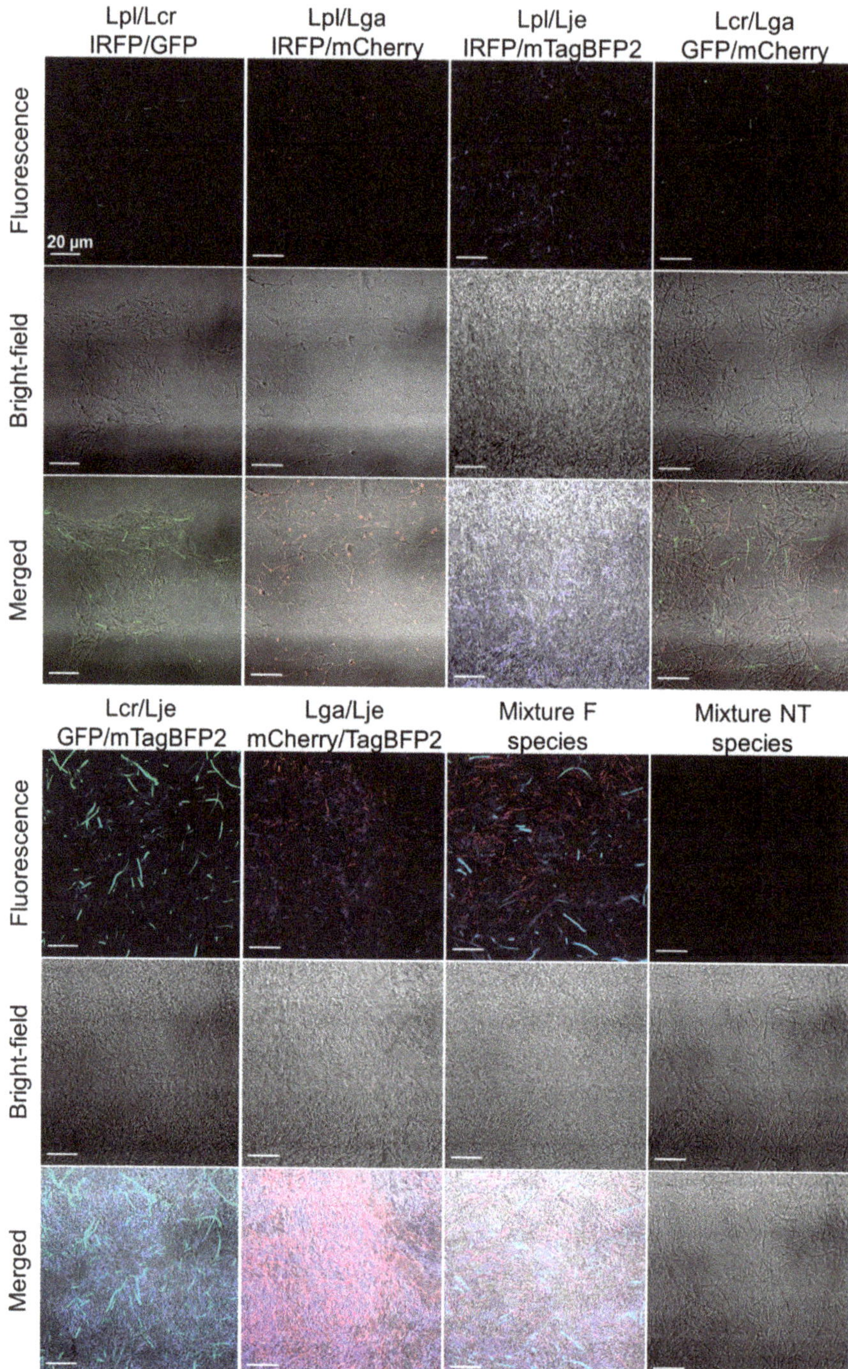

Figure 6. Representative confocal microscopy images of the mixtures of the lactobacilli expressing the different fluorescent proteins (F). Lpl—*L. plantarum*; Lga—*L. gasseri*; Lcr—*L. crispatus*; Lje—*L. jensenii*; NT—nontransformed species. Fluorescence images were obtained by using settings for denoted fluorescent proteins and merging thus obtained images, whereby settings for all four fluorescent proteins were used for the mixture.

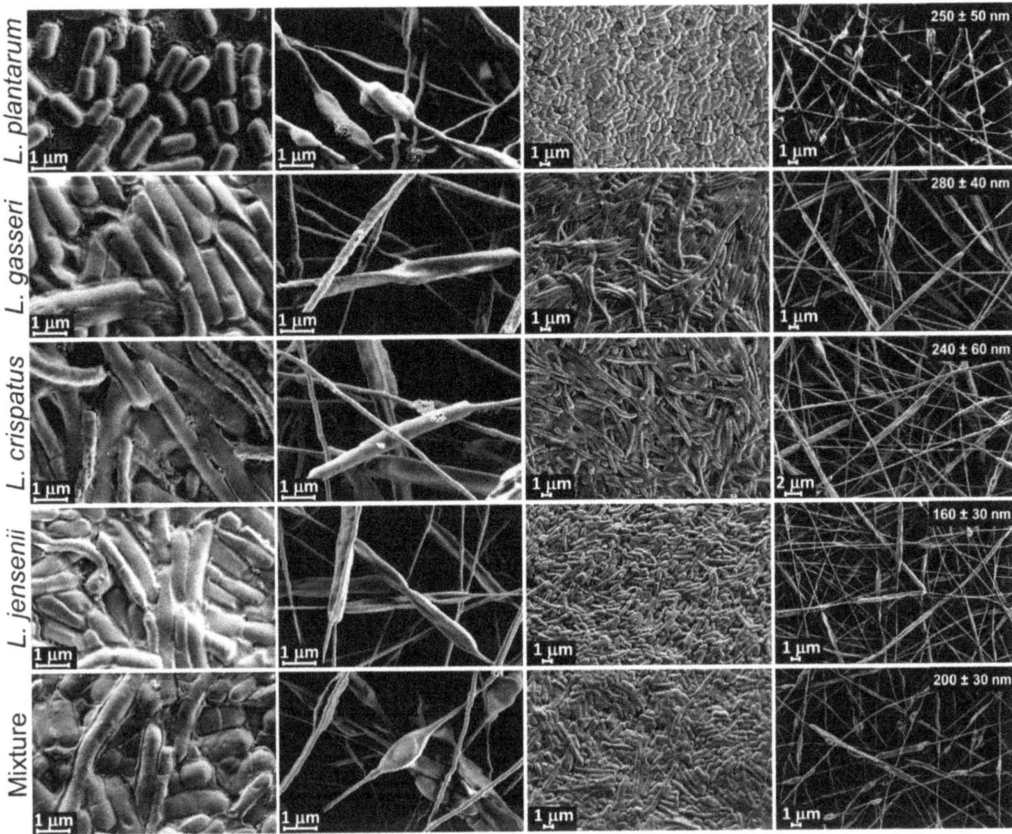

Figure 7. Scanning electron microscopy images of individual lactobacilli and as mixtures under high and low magnification. Columns 1 and 3, air-dried bacteria from water suspensions; Columns 2 and 4, bacteria incorporated into nanofibers. Average nanofiber diameters are specified in the last column.

As well as the individual species (Figure 9a), mixtures of all four of these fluorescent lactobacilli were incorporated into PEO nanofibers, with the fluorescence evaluated following the dissolution (Figure 9b). The IRFP-expressing and mCherry-expressing bacteria produced no fluorescence when the settings for the other fluorescent proteins were used. This was not the case for the GFP-expressing and mTagBFP2-expressing bacteria, for which significant fluorescence was observed also when using the settings for the other fluorescent proteins. Additionally, fluorescence overlap was seen, in terms of the fluorescence detected for mCherry-expressing and GFP-expressing bacteria when using the GFP and mTagBFP2 settings, respectively.

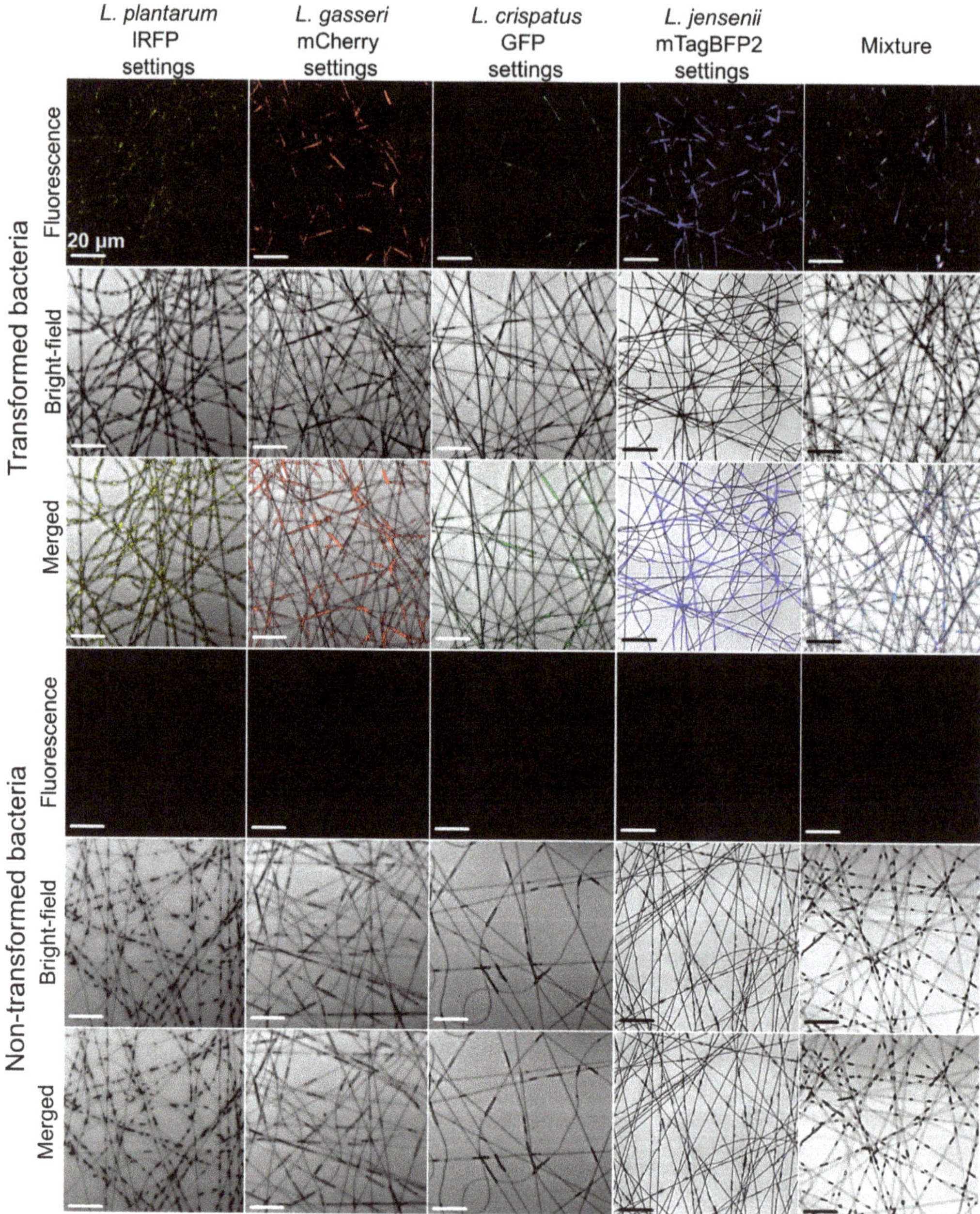

Figure 8. Representative confocal microscopy images of the fluorescent lactobacilli and the nontransformed lactobacilli when incorporated in PEO electrospun nanofibers individually or as mixture of all 4 fluorescent lactobacilli.

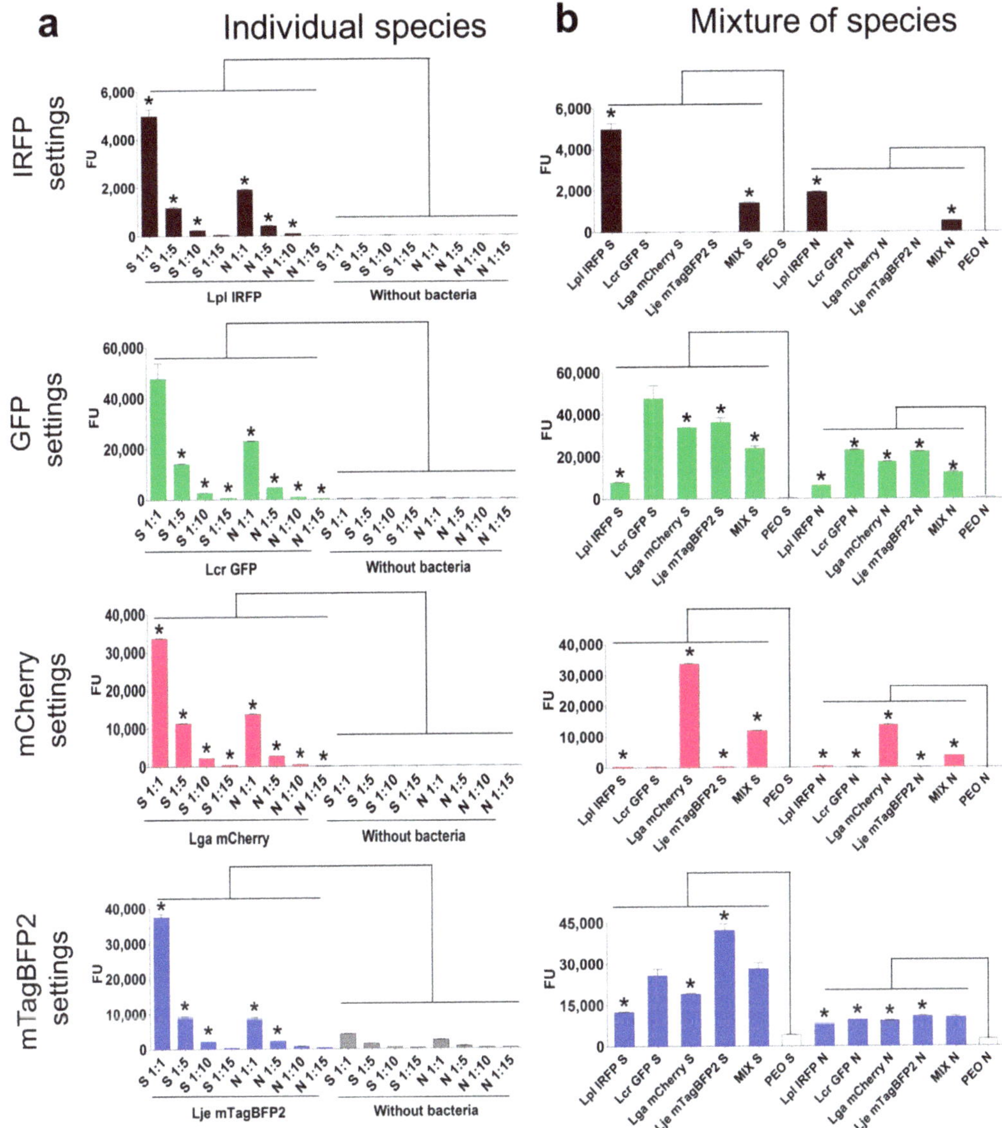

Figure 9. Fluorescence intensities of individual fluorescent *Lactobacillus* species (**a**) and for their mixtures containing all four of the *Lactobacillus* species (**b**), both as dispersed in 4% (*w*/*v*) PEO (S, suspension) prior to incorporation and after their release from the nanofibers (N). Ratios indicate different dilution factors (ratios of the aliquot volume to the final volume). * $p < 0.05$ (Student's *t* tests), comparison of suspensions or nanofibers containing fluorescent strain to its counterpart without bacteria (**a**), or comparison of suspensions and nanofibers containing fluorescent strains or their mixtures to corresponding PEO control (**b**). PEO—polyethylene oxide.

3. Discussion

To establish methods for imaging of the vaginal lactobacilli, three vaginal *Lactobacillus* species, *L. gasseri* ATCC 33323, *L. crispatus* ATCC 33820, and *L. jensenii* ATCC 25258, and the control *L. plantarum* ATCC 8014 were genetically modified to express fluorescent proteins with different spectral properties: IRFP, GFP, mCherry, and mTagBFP2. This genetic engineering of the vaginal lactobacilli was challenging, particularly for *L. crispatus* [44]. Here, it was performed by electrotransformation using modified and optimized previously published protocols [45–47].

Expression of the fluorescent proteins varied between the bacterial species and was highest in the control *L. plantarum*, regardless of the fluorescent protein used. This might be associated with the use of the *ldh* promoter for the control of the transcription of the fluorescent proteins, which originated from *L. plantarum*. The fluorescent intensities of the species were influenced by the growth conditions. According to expectation, higher fluorescence was observed when the lactobacilli were grown with aeration, as the presence of oxygen is crucial for post-translational maturation of the fluorescent proteins, resulting in exo-methylene double bond formation that prevents isomerization [43]. However, these conditions were not favorable for these lactobacilli, which are anaerobes or facultative anaerobes [48,49]. To further improve this approach, anaerobic fluorescent proteins could be considered. Expression of mTagBFP2 affected the growth of *L. plantarum* (not shown), which suggested possible phototoxicity. Nevertheless, all of these species expressed all of these fluorescent proteins, and the fluorescence measurements were proportional to the bacterial concentrations, thus also defining the suitability of this approach for quantifying these bacteria. Very little autofluorescence was seen for the nontransformed bacteria, except when using the mTagBFP2 settings. This contrasted with the expression of mTagBFP2 in *L. rhamnosus*, where autofluorescence was not an issue [42].

The expression of these different fluorescent proteins was also used to distinguish between these different bacterial species in mixtures containing species in equal ratios or in mixtures in which one of the species predominated. This resulted in a very useful tool to gain better insight into the behavior of the lactobacilli in future studies. The fluorescence signals correlated with the contents of the individual strains in the mixtures; however, the overlap between the different fluorescent proteins expressed by the same or different species was also observed. This was particularly evident for mTagBFP2; when using the mTagBFP2 settings, there was considerable fluorescence determined also for the other fluorescent proteins. Further, considerable overlap was observed between *L. crispatus* expressing GFP and *L. jensenii* expressing mTagBFP2, where the individual bacteria were detected in both fluorescence channels. We concluded that fluorescent proteins can be applied to distinguish between vaginal lactobacilli; however, proteins with longer excitation and emission wavelengths (IRFP, mCherry) are more appropriate due to the lower autofluorescence.

Most of the probiotic dosage forms are designed for oral application due to their beneficial effects on the gut. For therapeutic effects in the vagina, intravaginal administration of the probiotics is crucial [50]. To allow intravaginal applications of *Lactobacillus* probiotics, we incorporated these into small diameter fibers, i.e., nanofibers, which were produced using electrospinning [19]. Electrospun nanofibers represent a next-generation delivery system that can be used for biologicals, such as microorganisms, stem cells, proteins, and nucleic acids [23]. The incorporation of bacteria affects the characteristics of nanofibers [33]. Here, a change in diameter was observed, which may be due to the release of bacterial products that can influence the properties of the polymer suspension. Lactobacilli retained their fluorescence after incorporation into nanofibers, as well as after dissolution of the nanofibers, and the fluorescence intensities again correlated with the bacterial concentrations. In bacterial mixtures, the fluorescent proteins with longer excitation and emission wavelengths (i.e., mCherry, IRFP) were clearly distinguished, while for mTagBFP2 and GFP, autofluorescence and spectral overlap interfered with these measurements; this might be resolved with the appropriate compensation. The proposed approach represents a clear advantage over non-specific bacterial staining in pre-formulation and formulation studies of lactobacilli-containing nanofibers.

4. Materials and Methods

4.1. Bacterial Strains and Culturing

Four different strains from the genus *Lactobacillus* were used in this study: *L. crispatus* ATCC 33820; *L. gasseri* ATCC 33323; *L. jensenii* ATCC 25258; and *L. plantarum* ATCC 8014). *Lactococcus lactis* NZ9000 and *E. coli* DH5α were used as the cloning hosts. Lactobacilli were grown in De Man, Rogosa, and Sharpe (MRS) medium (Merck, Darmstadt, Germany) at 37 °C without and with aeration. *Lc. lactis* NZ9000 was grown in M-17 medium (Merck) supplemented with 0.5% (*v*/*v*) glucose (GM-17) at 30 °C, without aeration. *E. coli* was grown in a lysogeny broth medium at 37 °C, with aeration. All of the strains were kept frozen at −80 °C for long-term storage.

4.2. Plasmid Construction

KOD Hot Start DNA polymerase (Merck Millipore, Burlington, MA, USA) was used to fuse the fluorescent protein genes with the *ldh* promoter using overlap-extension PCR [51]. Plasmids pMEC276 [52], pNZ-IRFP713 [39], pCDLbu-1ΔEc-Ptuf34-mCherry [53], and pBAD-mTagBFP2 [54] were used as templates for *ldh*, *IRFP*, *mCherry*, and *mTagBFP2*, respectively. Three individual PCR reactions were prepared using the primers (Integrated DNA Technologies, Leuven, Belgium) specified in Supplementary Material: Table S2. In the first reaction, the promoter and fluorescent protein genes were amplified, and DNA overlaps were introduced (e.g., for *ldh-IRFP* fusion, primer pairs ldh-F/ldh-IRFP-R and IRFP-ldh-F/IRFP-R were used). These two DNA fragments were put in a new PCR mixture without primers to allow direct fusion of the two DNA fragments via complementary overlaps. Finally, in the third PCR reaction, the fused promoter and gene were amplified with a forward primer of the promoter and reverse primer of the fluorescent protein gene (e.g., ldh-F/IRFP-R for *ldh-IRFP*). Gene fusions *ldh-mCherry* and *ldh-mTagBFP2* were prepared in a similar fashion. The DNA products were then inserted into the pJET1.2/blunt vector and transformed into DH5α competent *E. coli* cells. Cloned products were digested using the XbaI/BglII restriction enzymes (Thermo Scientific, Waltham, MA, USA) and ligated into the pNZ8148 [55] plasmid using T4 DNA ligase (Thermo Scientific, Waltham, MA, USA). The three pNZ8148 derivatives thus obtained were: pNZ-ldh-IRFP, pNZ-ldh-mCherry, and pNZ-ldh-mTagBFP2. For the expression of GFP, the plasmid pMEC276 that contained *ldh-GFP* (for clarity, also indicated as pNZ-ldh-GFP) was kindly provided by C. Daniel [52]. The plasmids were transformed into *Lc. lactis* NZ9000 with electroporation using a Gene Pulser II apparatus (Biorad, Hercules, CA, USA), according to the manufacturer instructions (MoBiTec GmbH, Goettingen, Germany). Plasmids were isolated using peqGOLD Plasmid Miniprep Kit I (Peqlabs, Erlangen, Germany) and NucleoSpin Plasmid (Macherey and Nagel, Düren, Germany). Additional treatments with mixtures of lysozyme (50 mg/mL) and mutanolysin were performed when isolating from *Lc. lactis*. Genetic constructs were confirmed by nucleotide sequencing (Eurofins Genomics, Ebersberg, Germany).

4.3. Electrotransformation of Lactobacilli

The electrotransformation of lactobacilli was performed with a Gene Pulser II apparatus. Bacterial suspensions (100 μL) were mixed with 5 μL or 10 μL plasmid DNA (200–300 ng/μL) and added to an electroporation cuvette. The species-specific conditions for electroporation and electrocompetent cell preparation are as specified below. After electroporation, 1 mL fresh MRS medium was added to the bacteria, which were then left for 2–3 h to recover, at 30 °C or 37 °C. After the recovery, the bacteria were plated on MRS plates containing 10 μg/mL chloramphenicol (MRSC10). The plates were then placed into anaerobic bags (GasPak^{TM} EZ; Becton Dickinson, Franklin Lakes, NJ, USA) or jars (AnaeroGen^{TM} 2.5l; Thermo Scientific, Waltham, MA, USA) and incubated at 37 °C for 48 h to 72 h.

Electroporation-competent *L. plantarum* was prepared as previously described [45]. Fresh overnight cultures of *L. plantarum* were inoculated in 50 mL MRS medium at 1:50 and grown at 30 °C until an optical density at 600 nm (OD_{600}) of 0.4–0.6. The bacteria were then washed twice with 10 mL 10 mM $MgCl_2$ and once with 10 mL electroporation buffer (0.5 M sucrose, 10% (*v/v*) glycerol). The electroporation parameters for the transformation of *L. plantarum* were: 25 µF, 600 Ω, and 1.8 kV. After electroporation, the bacteria were left to recover at 30 °C for 2 h. A similar protocol was used for *L. jensenii*, with a higher growth and recovery temperature (37 °C) and a higher voltage applied during electroporation (2.4 kV).

To prepare electrocompetent *L. gasseri*, overnight cultures were inoculated in 50 mL MRS medium at 1:50 and grown at 37 °C until an OD_{600} of 0.4. Then, ampicillin (10 µg/mL) was added, and the bacteria were incubated to an OD_{600} of 0.8, followed by three washes with 0.5 M sucrose. *L. gasseri* was electroporated at 25 µF, 400 Ω, and 2.4 kV, and allowed to recover at 37 °C for 3 h, before plating on MRSC10 plates [47].

Overnight cultures of *L. crispatus* were inoculated into 10 mL sterile-filtered MRS medium containing 0.8% (*w/v*) glycine, and left for ~10 h at 37 °C until reaching an OD_{600} of 0.5. The bacteria were washed twice with 0.5 M sucrose, incubated on ice with 50 mM EDTA for 5 min, and again washed with 0.5 M sucrose. Electroporation was performed at 25 µF, 600 Ω, and 1.5 kV [46]. Before plating on MRSC10 plates, the bacteria were left to recover at 37 °C for 3 h.

The transformed bacteria were kept frozen at −80 °C in MRS with 20% (*v/v*) glycerol for long-term storage.

4.4. Culturing of Lactobacilli

Different growth conditions were used for the lactobacilli for the different experiments. Lactobacilli were transferred from frozen stocks to solid MRS media and grown anaerobically at 37 °C for 2–3 days. A single colony was picked and grown in liquid MRS media for 1 day. For measurement of fluorescence, overnight cultures of *Lactobacillus* species transformed with the fluorescent proteins encoded in the pNZ plasmids were inoculated in 5 mL MRSC10 medium and grown at 37 °C without and with aeration, and without and with shaking (180–200 rpm). Biliverdin (15.5 µg/mL; Sigma Aldrich, St. Louis, MO, USA) was added to the medium of species engineered to produce IRFP. The species were grown until late exponential or early stationary phase (OD_{600} 1.5–2.0) when the bacteria were centrifuged at 4400× *g* for 10 min at 4 °C (Centrifuge 5702 R; Eppendorf, St. Louis, MO, USA) and resuspended in phosphate-buffered saline (PBS, pH 7.4) to an OD_{600} of 3.0.

Prior to electrospinning, the engineered bacteria were grown in 400 mL at 37 °C with shaking until reaching OD_{600} of 2.0–3.0. The bacteria were washed twice with water and resuspended in 10 mL water. PEO powder (Mw 900 kDa; Sigma Aldrich, Darmstadt, Germany) was added to the lactobacilli suspensions and stirred at 400 rpm at room temperature until the polymer was completely dissolved to provide the polymer concentration of 4% (*w/v*).

4.5. Electrospinning of Lactobacilli

Bacterial cells (OD_{600} of 9.0–11.0) were mixed with PEO as described above. The homogenous bacterial-polymer suspensions were filled into a 5 mL syringe that was fixed to an electrospinning machine (Fluidnatek LE100; BioInicia SL, Valencia, Spain). A high voltage of 13 ± 2 kV was applied. The flow rate of the suspension in the syringe was 250–350 µL/h, and the distance between the needle and collector was 15 cm. The electrospinning process was conducted in a climate-controlled environment at 37 °C and 17% relative humidity.

4.6. Fluorescence Measurement

The fluorescence of the bacterial suspensions in PBS (200 µL) was measured using a microplate reader (Infinite M1000; Tecan, Männedorf, Switzerland) in 96-well black, flat-bottomed plates. All of the samples were measured in duplicate. Depending on the characteristics of the fluorescent protein, different excitation and emission wavelengths were applied: 402/457 nm for mTagBFP2; 488/509 nm for GFP; 587/610 nm for mCherry;

and 690/713 nm for IRFP. The excitation and emission spectra of these fluorescent proteins are provided in Supplementary Material: Figure S1.

To test for correlations between the bacterial concentrations and the fluorescence intensities, serial dilutions of the bacteria were used, corresponding to OD_{600} of 3.0, 2.0, 1.0, 0.5, and 0.25. Nontransformed bacteria were included as the control.

The overlap of the fluorescence signals of the four different fluorescent proteins was assessed by mixing fluorescent strains either in equal ratios (1:1:1:1) or in ratios where individual strain represented 50% of all the bacteria (e.g., 3:1:1:1). These were then compared to the same ratios of the nontransformed bacteria. Individual species that expressed different fluorescent proteins were compared to assign each species with a unique fluorescent protein; selected combinations were *L. plantarum* expressing IRFP, *L. gasseri* expressing mCherry, *L. crispatus* expressing GFP, and *L. jensenii* expressing mTagBFP2, with these compared in a similar fashion. These were also used for confocal microscopy imaging and incorporation into nanofibers.

To measure the fluorescence of the lactobacilli following electrospinning, 10 ± 3 mg PEO nanofibers with the incorporated fluorescent lactobacilli were dissolved in 900 µL PBS and diluted with PBS using the dilution factors (ratios of the aliquot volume to the final volume) of 1:1, 1:5, 1:10, and 1:15. The fluorescence intensities were compared to those of bacterial-polymer suspensions prior to electrospinning, where the concentration of bacteria was estimated to be ~3.6-fold higher. The same dilution factors were applied for the bacterial-polymer suspensions.

4.7. Confocal Microscopy

Lactobacilli were grown as described in section Culturing of lactobacilli, resuspended in PBS to an OD_{600} of 3.0, and fixed to a microscope slide with StatSpine Cytofuge 2 (Iris Sample Processing, Westwood, MA, USA) by centrifugation at maximum speed at 4400 rpm for 10 min. The samples were left at room temperature for 1 h to dry and then mounted with a mounting medium (Invitrogen, Waltham, MA, USA) with 4′,6-diamidino-2-phenylindole (DAPI), or IBIDI mounting medium without DAPI. Fluorescent bacteria were visualized with a confocal microscope (LSM-710; Carl Zeiss, Oberkochen, Germany), and images were acquired and processed with the ZEN 2010 B SP1 software (Carl Zeiss, Oberkochen, Germany). The strains were detected with different settings: brightfield, DAPI, Alexa 488, Alexa 543, and Alexa 647, using the 63× immersion oil objective.

For imaging of the nanofibers, a microscope slide was added to the collector of the electrospinning machine, and it was left there for nanofibers to be deposited onto it. A cover slip was added on top and glued with nail polish. Imaging was performed as above, using the 40× immersion oil objective. Nontransformed bacteria were included as the control.

4.8. Scanning Electron Microscopy

A scanning electron microscope (Supra 35 VP; Carl Zeiss, Oberkochen, Jena, Germany) was used to visualize the nanofibers with the incorporated bacteria, as well as free bacteria. Individual species and their mixtures were dispersed in water, and 3 µL of the suspension was pipetted onto a metal stub and air-dried. The double-sided conductive tape was used to attach the nanofiber mats to the metal stubs. The scanning electron microscopy was operated at an acceleration voltage of 1 kV, with a secondary electron detector. Bacterial and nanofiber size were analyzed using the ImageJ 1.51j8 software (National Institutes of Health, Bethesda, MD, USA), where the length and width of 30 randomly selected bacteria or nanofibers (regions without bacteria) were measured.

4.9. Statistical Analyses

Statistical analyses were performed using the GraphPad Prism 7.0 software, San Diego, CA, USA. Student's *t* tests were used to define the significances of the differences between the fluorescent bacteria and their respective controls. Calculation of slopes of the regression lines and their comparison was also performed with GraphPad Prism 7.0. All of the data are presented as means ± standard deviation (SD).

5. Conclusions

In this paper, we pursued two major goals for wider vaginal probiotics use, namely fluorescent labeling to allow for future distribution studies and designing of an appropriate delivery system. Three of the most important vaginal *Lactobacillus* species, *L. gasseri*, *L. crispatus*, and *L. jensenii*, and the control *L. plantarum* were engineered to produce compatible fluorescent proteins with different spectral properties. The fluorescence intensities were mostly dependent on lactobacilli species and growth conditions. The aeration during culturing promoted the expression of fluorescent proteins compared to samples grown without aeration. The four species were successfully incorporated into nanofibers by electrospinning, which indicated that this technique is appropriate for designing solid nanofiber-based vaginal delivery systems for probiotics. The lactobacilli retained their fluorescence after incorporation into these PEO nanofibers and after their release from them. This research presents a cutting-edge technology to accurately track, by fluorescence imaging, the release of lactobacilli from nanofibers and interactions with the indigenous vaginal microbiota in future in vitro and in vivo studies.

Supplementary Materials: The following are available online at https://www.mdpi.com/article/10.3390/ijms222413631/s1.

Author Contributions: S.S. carried out the experimental work and wrote the draft of the manuscript. T.V.P. assisted with confocal imaging. J.K. revised the manuscript and supervised the work. Š.Z. revised the manuscript, assisted with electrospinning, and performed scanning electron microscopy. A.B. revised the manuscript, planned, and supervised the work. All authors have read and agreed to the published version of the manuscript.

Funding: This work was funded by the Slovenian Research Agency (grants J4-9327, P1-0189 and P4-0127).

Institutional Review Board Statement: Not applicable.

Informed Consent Statement: Not applicable.

Data Availability Statement: The data presented in this study are available in the article or its Supplementary Material.

Acknowledgments: We thank Christopher Berrie for his critical reading of the manuscript. Our thanks also go to Catherine Daniel for providing pMEC276 and Vladislav Verkhusha for providing pBAD-mTagBFP2 (Addgene plasmid #34632).

Conflicts of Interest: The authors declare no conflict of interest. The funders had no role in the design of the study; in the collection, analyses, or interpretation of data; in the writing of the manuscript, or in the decision to publish the results.

References

1. Cribby, S.; Taylor, M.; Reid, G. Vaginal microbiota and the use of probiotics. *Interdiscip. Perspect. Infect. Dis.* **2008**, *2008*, 256490. [CrossRef]
2. Petrova, M.I.; Lievens, E.; Malik, S.; Imholz, N.; Lebeer, S. *Lactobacillus* species as biomarkers and agents that can promote various aspects of vaginal health. *Front. Physiol.* **2015**, *6*, 81. [CrossRef]
3. Kim, J.M.; Park, Y.J. Probiotics in the prevention and treatment of postmenopausal vaginal infections: Review article. *J. Menopausal Med.* **2017**, *23*, 139–145. [CrossRef]
4. Simoes, J.A.; Aroutcheva, A.A.; Shott, S.; Faro, S. Effect of metronidazole on the growth of vaginal lactobacilli in vitro. *Infect. Dis. Obstet. Gynecol.* **2001**, *9*, 41–45. [CrossRef]
5. Kumar, N.; Behera, B.; Sagiri, S.S.; Pal, K.; Ray, S.S.; Roy, S. Bacterial vaginosis: Etiology and modalities of treatment-a brief note. *J. Pharm. Bioallied Sci.* **2011**, *3*, 496–503. [CrossRef] [PubMed]
6. Stapleton, A.E.; Au-Yeung, M.; Hooton, T.M.; Fredricks, D.N.; Roberts, P.L.; Czaja, C.A.; Yarova-Yarovaya, Y.; Fiedler, T.; Cox, M.; Stamm, W.E. Randomized, placebo-controlled phase 2 trial of a *Lactobacillus crispatus* probiotic given intravaginally for prevention of recurrent urinary tract infection. *Clin. Infect. Dis.* **2011**, *52*, 1212–1217. [CrossRef]

7. Ghartey, J.P.; Smith, B.C.; Chen, Z.; Buckley, N.; Lo, Y.; Ratner, A.J.; Herold, B.C.; Burk, R.D. *Lactobacillus crispatus* dominant vaginal microbiome is associated with inhibitory activity of female genital tract secretions against *Escherichia coli*. *PLoS ONE* **2014**, *9*, e96659. [CrossRef]
8. Butler, D.S.C.; Silvestroni, A.; Stapleton, A.E. Cytoprotective effect of *Lactobacillus crispatus* CTV-05 against uropathogenic *E. coli*. *Pathogens* **2016**, *5*, 27. [CrossRef]
9. Nardini, P.; Nahui Palomino, R.A.; Parolin, C.; Laghi, L.; Foschi, C.; Cevenini, R.; Vitali, B.; Marangoni, A. *Lactobacillus crispatus* inhibits the infectivity of *Chlamydia trachomatis* elementary bodies, in vitro study. *Sci. Rep.* **2016**, *6*, 29024. [CrossRef] [PubMed]
10. Atassi, F.; Pho Viet Ahn, D.L.; Lievin-Le Moal, V. Diverse expression of antimicrobial activities against bacterial vaginosis and urinary tract Infection pathogens by cervicovaginal microbiota strains of *Lactobacillus gasseri* and *Lactobacillus crispatus*. *Front. Microbiol.* **2019**, *10*, 2900. [CrossRef]
11. Li, T.; Liu, Z.; Zhang, X.; Chen, X.; Wang, S. Local probiotic *Lactobacillus crispatus* and *Lactobacillus delbrueckii* exhibit strong antifungal effects against vulvovaginal candidiasis in a rat model. *Front. Microbiol.* **2019**, *10*, 1033. [CrossRef] [PubMed]
12. Charteris, W.P.; Kelly, P.M.; Morelli, L.; Collins, J.K. Antibacterial activity associated with *Lactobacillus gasseri* ATCC 9857 from the human female genitourinary tract. *World J. Microbiol. Biotechnol.* **2001**, *17*, 615–625. [CrossRef]
13. Spurbeck, R.R.; Arvidson, C.G. *Lactobacillus jensenii* surface-associated proteins inhibit *Neisseria gonorrhoeae* adherence to epithelial cells. *Infect. Immun.* **2010**, *78*, 3103–3111. [CrossRef]
14. Shim, Y.H.; Lee, S.J.; Lee, J.W. Antimicrobial activity of lactobacillus strains against uropathogens. *Pediatr. Int.* **2016**, *58*, 1009–1013. [CrossRef]
15. Morais, I.M.C.; Cordeiro, A.L.; Teixeira, G.S.; Domingues, V.S.; Nardi, R.M.D.; Monteiro, A.S.; Alves, R.J.; Siqueira, E.P.; Santos, V.L. Biological and physicochemical properties of biosurfactants produced by *Lactobacillus jensenii* P6A and *Lactobacillus gasseri* P65. *Microb. Cell Factories* **2017**, *16*, 155. [CrossRef] [PubMed]
16. Kim, J.; Muhammad, N.; Jhun, B.H.; Yoo, J.W. Probiotic delivery systems: A brief overview. *J. Pharm. Investig.* **2016**, *46*, 377–386. [CrossRef]
17. Borges, S.; Barbosa, J.; Teixeira, P. Drug delivery systems for vaginal infections. In *Frontiers in Clinical Drug Research: Anti-Infectives*; Rahman, A.U., Ed.; Bentham Science Publishers: Sharjah, United Arab Emirates, 2015; Volume 2, pp. 233–258. [CrossRef]
18. Torres-Martinez, E.J.; Cornejo Bravo, J.M.; Serrano Medina, A.; Perez Gonzalez, G.L.; Villarreal Gomez, L.J. A summary of electrospun nanofibers as drug delivery system: Drugs loaded and biopolymers used as matrices. *Curr. Drug Deliv.* **2018**, *15*, 1360–1374. [CrossRef] [PubMed]
19. Teo, W.E.; Ramakrishna, S. A review on electrospinning design and nanofibre assemblies. *Nanotechnology* **2006**, *17*, R89–R106. [CrossRef]
20. Sharma, A.; Pathak, D.; Patil, D.S.; Dhiman, N.; Bhullar, V.; Mahajan, A. Electrospun PVP/TiO2 nanofibers for filtration and possible protection from various viruses like COVID-19. *Technologies* **2021**, *9*, 89. [CrossRef]
21. Pathak, D.; Kumar, S.; Andotra, S.; Thomas, J.; Kaur, N.; Kumar, P.; Kumar, V. New tailored organic semiconductors thin films for optoelectronic applications. *Eur. Phys. J. Appl. Phys.* **2021**, *95*, 10201. [CrossRef]
22. Sofi, H.S.; Abdal-Hay, A.; Ivanovski, S.; Zhang, Y.S.; Sheikh, F.A. Electrospun nanofibers for the delivery of active drugs through nasal, oral and vaginal mucosa: Current status and future perspectives. *Mater. Sci. Eng. C Mater. Biol. Appl.* **2020**, *111*, 110756. [CrossRef] [PubMed]
23. Stojanov, S.; Berlec, A. Electrospun nanofibers as carriers of microorganisms, stem cells, proteins, and nucleic acids in therapeutic and other applications. *Front. Bioeng. Biotechnol.* **2020**, *8*, 130. [CrossRef] [PubMed]
24. Heunis, T.D.; Botes, M.; Dicks, L.M. Encapsulation of *Lactobacillus plantarum* 423 and its bacteriocin in nanofibers. *Probiotics Antimicrob. Proteins* **2010**, *2*, 46–51. [CrossRef]
25. Feng, K.; Zhai, M.Y.; Zhang, Y.; Linhardt, R.J.; Zong, M.H.; Li, L.; Wu, H. Improved viability and thermal stability of the probiotics encapsulated in a novel electrospun fiber mat. *J. Agric. Food Chem.* **2018**, *66*, 10890–10897. [CrossRef]
26. Luan, Q.; Zhou, W.; Zhang, H.; Bao, Y.; Zheng, M.; Shi, J.; Tang, H.; Huang, F. Cellulose-based composite macrogels from cellulose fiber and cellulose nanofiber as intestine delivery vehicles for probiotics. *J. Agric. Food Chem.* **2018**, *66*, 339–345. [CrossRef]
27. Škrlec, K.; Zupančič, S.; Prpar Mihevc, S.; Kocbek, P.; Kristl, J.; Berlec, A. Development of electrospun nanofibers that enable high loading and long-term viability of probiotics. *Eur. J. Pharm. Biopharm.* **2019**, *136*, 108–119. [CrossRef] [PubMed]
28. Yu, H.; Liu, W.; Li, D.; Liu, C.; Feng, Z.; Jiang, B. Targeting delivery system for *Lactobacillus plantarum* based on functionalized electrospun nanofibers. *Polymers* **2020**, *12*, 1565. [CrossRef]
29. Fung, W.Y.; Yuen, K.H.; Liong, M.T. Agrowaste-based nanofibers as a probiotic encapsulant: Fabrication and characterization. *J. Agric. Food Chem.* **2011**, *59*, 8140–8147. [CrossRef]
30. Ceylan, Z.; Meral, R.; Karakaş, C.Y.; Dertli, E.; Yilmaz, M.T. A novel strategy for probiotic bacteria: Ensuring microbial stability of fish fillets using characterized probiotic bacteria-loaded nanofibers. *Innov. Food Sci. Emerg. Technol.* **2018**, *48*, 212–218. [CrossRef]
31. Amna, T.; Hassan, M.S.; Pandeya, D.R.; Khil, M.S.; Hwang, I.H. Classy non-wovens based on animate *L. gasseri*-inanimate poly(vinyl alcohol): Upstream application in food engineering. *Appl. Microbiol. Biotechnol.* **2013**, *97*, 4523–4531. [CrossRef]
32. Soares, J.M.D.; Abreu, R.E.F.; da Costa, M.M.; de Oliveira, H.P. Investigation of *Lactobacillus paracasei* encapsulation in electrospun fibers of Eudragit® L100. *Polímeros* **2020**, *30*, e2020025. [CrossRef]
33. Zupančič, S.; Škrlec, K.; Kocbek, P.; Kristl, J.; Berlec, A. Effects of electrospinning on the viability of ten species of lactic acid bacteria in poly(ethylene oxide) nanofibers. *Pharmaceutics* **2019**, *11*, 483. [CrossRef]

34. Silva, J.A.; De Gregorio, P.R.; Rivero, G.; Abraham, G.A.; Nader-Macias, M.E.F. Immobilization of vaginal *Lactobacillus* in polymeric nanofibers for its incorporation in vaginal probiotic products. *Eur. J. Pharm. Sci.* **2021**, *156*, 105563. [CrossRef]
35. Nagy, Z.K.; Wagner, I.; Suhajda, A.; Tobak, T.; Harasztos, A.H.; Vigh, T.; Sóti, P.L.; Pataki, H.; Molnar, K.; Marosi, G. Nanofibrous solid dosage form of living bacteria preparedby electrospinning. *Express Polym. Lett.* **2014**, *8*, 352–361. [CrossRef]
36. Herbel, S.R.; Vahjen, W.; Wieler, L.H.; Guenther, S. Timely approaches to identify probiotic species of the genus *Lactobacillus*. *Gut Pathog.* **2013**, *5*, 27. [CrossRef] [PubMed]
37. Welch, J.L.M.; Hasegawa, Y.; McNulty, N.P.; Gordon, J.I.; Borisy, G.G. Spatial organization of a model 15-member human gut microbiota established in gnotobiotic mice. *Proc. Natl. Acad. Sci. USA* **2017**, *114*, E9105–E9114. [CrossRef] [PubMed]
38. Yu, Q.H.; Dong, S.M.; Zhu, W.Y.; Yang, Q. Use of green fluorescent protein to monitor *Lactobacillus* in the gastro-intestinal tract of chicken. *FEMS Microbiol. Lett.* **2007**, *275*, 207–213. [CrossRef] [PubMed]
39. Berlec, A.; Završnik, J.; Butinar, M.; Turk, B.; Štrukelj, B. In vivo imaging of *Lactococcus lactis*, *Lactobacillus plantarum* and *Escherichia coli* expressing infrared fluorescent protein in mice. *Microb. Cell Factories* **2015**, *14*, 181. [CrossRef]
40. Han, X.; Wang, L.; Li, W.; Li, B.; Yang, Y.; Yan, H.; Qu, L.; Chen, Y. Use of green fluorescent protein to monitor *Lactobacillus plantarum* in the gastrointestinal tract of goats. *Braz. J. Microbiol.* **2015**, *46*, 849–854. [CrossRef]
41. Karimi, S.; Ahl, D.; Vagesjo, E.; Holm, L.; Phillipson, M.; Jonsson, H.; Roos, S. In vivo and in vitro detection of luminescent and fluorescent *Lactobacillus reuteri* and application of red fluorescent mCherry for assessing plasmid persistence. *PLoS ONE* **2016**, *11*, e0151969. [CrossRef]
42. Spacova, I.; Lievens, E.; Verhoeven, T.; Steenackers, H.; Vanderleyden, J.; Lebeer, S.; Petrova, M.I. Expression of fluorescent proteins in *Lactobacillus rhamnosus* to study host-microbe and microbe-microbe interactions. *Microb. Biotechnol.* **2018**, *11*, 317–331. [CrossRef]
43. Tsuji, F.I. Oxidation of a cyclic tripeptide by molecular oxygen and the development of fluorescence in the Aequorea green fluorescent protein. *Int. Congr. Ser.* **2002**, *1233*, 37–44. [CrossRef]
44. Stephenson, D.P.; Moore, R.J.; Allison, G.E. Transformation of, and heterologous protein expression in, *Lactobacillus agilis* and *Lactobacillus vaginalis* isolates from the chicken gastrointestinal tract. *Appl. Environ. Microbiol.* **2011**, *77*, 220–228. [CrossRef]
45. Berthier, F.; Zagorec, M.; Champomier-Vergès, M.; Ehrlich, S.D.; Morel-Deville, F. Efficient transformation of *Lactobacillus sake* by electroporation. *Microbiol. Soc.* **1996**, *142*, 1273–1279. [CrossRef]
46. Beasley, S.S.; Takala, T.M.; Reunanen, J.; Apajalahti, J.; Saris, P.E. Characterization and electrotransformation of *Lactobacillus crispatus* isolated from chicken crop and intestine. *Poult. Sci.* **2004**, *83*, 45–48. [CrossRef] [PubMed]
47. Allain, T.; Mansour, N.M.; Bahr, M.M.; Martin, R.; Florent, I.; Langella, P.; Bermudez-Humaran, L.G. A new lactobacilli in vivo expression system for the production and delivery of heterologous proteins at mucosal surfaces. *FEMS Microbiol. Lett.* **2016**, *363*, fnw117. [CrossRef]
48. Talwalkar, A.; Kailasapathy, K. The role of oxygen in the viability of probiotic bacteria with reference to *L. acidophilus* and *Bifidobacterium* spp. *Curr. Issues Intest. Microbiol.* **2004**, *5*, 1–8.
49. Gupta, P.K.; McGrath, C. Microbiology, inflammation, and viral infections. In *Comprehensive Cytopathology*, 3rd ed.; Bibbo, M., Wilbur, D., Eds.; Elsevier Inc.: Amsterdam, The Netherlands, 2008; pp. 91–129. [CrossRef]
50. Nader-Macias, M.E.; Juarez Tomas, M.S. Profiles and technological requirements of urogenital probiotics. *Adv. Drug Deliv. Rev.* **2015**, *92*, 84–104. [CrossRef]
51. Kadkhodaei, S.; Memari, H.R.; Abbasiliasi, S.; Rezaei, M.A.; Movahedi, A.; Shun, T.J.; Ariff, A.B. Multiple overlap extension PCR (MOE-PCR): An effective technical shortcut to high throughput synthetic biology. *RSC Adv.* **2016**, *6*, 66682–66694. [CrossRef]
52. Salome-Desnoulez, S.; Poiretb, S.; Foligne, B.; Muharram, G.; Peucelle, V.; Lafont, F.; Daniel, C. Persistence and dynamics of fluorescent *Lactobacillus plantarum* in the healthy versus inflamed gut. *Gut Microbes* **2021**, *13*, 1–16. [CrossRef] [PubMed]
53. Tauer, C.; Heinl, S.; Egger, E.; Heiss, S.; Grabherr, R. Tuning constitutive recombinant gene expression in *Lactobacillus plantarum*. *Microb. Cell. Factories* **2014**, *13*, 150. [CrossRef] [PubMed]
54. Subach, O.M.; Cranfill, P.J.; Davidson, M.W.; Verkhusha, V.V. An enhanced monomeric blue fluorescent protein with the high chemical stability of the chromophore. *PLoS ONE* **2011**, *6*, e28674. [CrossRef] [PubMed]
55. Mierau, I.; Kleerebezem, M. 10 years of the nisin-controlled gene expression system (NICE) in *Lactococcus lactis*. *Appl. Microbiol. Biotechnol.* **2005**, *68*, 705–717. [CrossRef] [PubMed]

International Journal of
Molecular Sciences

Article

Modification of a Single Atom Affects the Physical Properties of Double Fluorinated Fmoc-Phe Derivatives

Moran Aviv [1,2,3,4,†], Dana Cohen-Gerassi [1,2,3,5,†], Asuka A. Orr [6], Rajkumar Misra [1,2,3], Zohar A. Arnon [1,2,3], Linda J. W. Shimon [7], Yosi Shacham-Diamand [5,8,9], Phanourios Tamamis [6,10] and Lihi Adler-Abramovich [1,2,3,*]

1 Department of Oral Biology, The Goldschleger School of Dental Medicine, Sackler Faculty of Medicine, Tel Aviv University, Tel Aviv 6997801, Israel; dr.moranaviv@gmail.com (M.A.); dana21cohen@gmail.com (D.C.-G.); rajkumarmisra96@gmail.com (R.M.); zoharnon@gmail.com (Z.A.A.)
2 The Center for Nanoscience and Nanotechnology, Tel Aviv University, Tel Aviv 6997801, Israel
3 The Center for the Physics and Chemistry of Living Systems, Tel Aviv University, Tel Aviv 6997801, Israel
4 School of Mechanical Engineering, Afeka Tel Aviv Academic College of Engineering, Tel Aviv 6910717, Israel
5 Department of Materials Science and Engineering, Iby and Aladar Fleischman Faculty of Engineering, Tel Aviv University, Tel Aviv 6997801, Israel; yosish@tauex.tau.ac.il
6 Artie McFerrin Department of Chemical Engineering, Texas A&M University, College Station, TX 77843-3122, USA; asukaorr@tamu.edu (A.A.O.); tamamis@tamu.edu (P.T.)
7 Department of Chemical Research Support, Weizmann Institute of Science, Rehovot 76132701, Israel; linda.Shimon@weizmann.ac.il
8 Department of Physical Electronics, School of Electrical Engineering, Faculty of Engineering, Tel Aviv University, Tel Aviv 69978, Israel
9 TAU/TiET Food Security Center of Excellence (T2FSCoE), Thapar Institute of Engineering and Technology, Patiala 147004, India
10 Department of Materials Science and Engineering, Texas A&M University, College Station, TX 77843-3003, USA
* Correspondence: LihiA@tauex.tau.ac.il
† These authors contributed equally to this work.

Citation: Aviv, M.; Cohen-Gerassi, D.; Orr, A.A.; Misra, R.; Arnon, Z.A.; Shimon, L.J.W.; Shacham-Diamand, Y.; Tamamis, P.; Adler-Abramovich, L. Modification of a Single Atom Affects the Physical Properties of Double Fluorinated Fmoc-Phe Derivatives. *Int. J. Mol. Sci.* **2021**, *22*, 9634. https://doi.org/10.3390/ijms22179634

Academic Editor: Raghvendra Singh Yadav

Received: 23 July 2021
Accepted: 30 August 2021
Published: 6 September 2021

Publisher's Note: MDPI stays neutral with regard to jurisdictional claims in published maps and institutional affiliations.

Copyright: © 2021 by the authors. Licensee MDPI, Basel, Switzerland. This article is an open access article distributed under the terms and conditions of the Creative Commons Attribution (CC BY) license (https://creativecommons.org/licenses/by/4.0/).

Abstract: Supramolecular hydrogels formed by the self-assembly of amino-acid based gelators are receiving increasing attention from the fields of biomedicine and material science. Self-assembled systems exhibit well-ordered functional architectures and unique physicochemical properties. However, the control over the kinetics and mechanical properties of the end-products remains puzzling. A minimal alteration of the chemical environment could cause a significant impact. In this context, we report the effects of modifying the position of a single atom on the properties and kinetics of the self-assembly process. A combination of experimental and computational methods, used to investigate double-fluorinated Fmoc-Phe derivatives, Fmoc-3,4F-Phe and Fmoc-3,5F-Phe, reveals the unique effects of modifying the position of a single fluorine on the self-assembly process, and the physical properties of the product. The presence of significant physical and morphological differences between the two derivatives was verified by molecular-dynamics simulations. Analysis of the spontaneous phase-transition of both building blocks, as well as crystal X-ray diffraction to determine the molecular structure of Fmoc-3,4F-Phe, are in good agreement with known changes in the Phe fluorination pattern and highlight the effect of a single atom position on the self-assembly process. These findings prove that fluorination is an effective strategy to influence supramolecular organization on the nanoscale. Moreover, we believe that a deep understanding of the self-assembly process may provide fundamental insights that will facilitate the development of optimal amino-acid-based low-molecular-weight hydrogelators for a wide range of applications.

Keywords: self-assembly; low-molecular-weight hydrogelator; phase-transition; molecular-dynamics

1. Introduction

Supramolecular self-assembly based on noncovalent interactions between monomeric building blocks is a powerful strategy for the design of well-ordered functional architectures [1–4]. Due to their unique mechanical and physicochemical properties, such assembled systems show great promise in a wide range of applications, including electro-optics [5], biomedicine [6,7], chemical separation [8], and material science [9]. These supramolecular assemblies form a rich variety of nano- and micro-architectures including tubes, fibers, films, plates, and vesicles [10–12].

Low-molecular-weight peptides and amino acid derivatives that self-assemble to form ordered nanostructures have attracted significant attention in recent years. Their natural ability to form supramolecular networks can be utilized for biomedical applications including controlled drug delivery [13], vaccine development [14], tissue engineering [15], and regenerative medicine [16]. The orderly nanostructures resulting from the self-assembly process contain large internal cavities, capable of trapping large quantities of water molecules, and allow these building blocks to serve as gelators for self-supporting hydrogels [17–19].

Since the early work of Janmey and co-workers who developed hydrogels based on Fmoc-Leu-Asp [20], and the later work of Xu and co-workers who developed nanofibrous hydrogels based on Fmoc-D-Ala-D-Ala [11], there has been a keen interest in fluorenylmethoxycarbonyl (Fmoc) protected amino acids and dipeptides and their possible applications [5,21,22]. The main advantages of these materials are their ease of synthesis, low cost, similarity to the natural extracellular matrix, and good biocompatibility.

Among the molecular hydrogelators reported to date, those based on Fmoc-phenylalanine (Fmoc-Phe) have attracted particular interest due to the wide variety of unique properties that can readily be obtained by chemical and biological decoration. In addition, the assembly properties of Fmoc-Phe can be greatly enhanced by the incorporation of various substituents, including halogens, on the benzyl side chain [23–25]. Incorporation of single halogen substituents on the aromatic side chain of Fmoc-Phe was shown to enhance the efficient self-assembly of these amino acid derivatives dramatically (relative to Fmoc-Phe), and to produce hydrogel fibril networks that promote hydrogelation in aqueous solvents [23–25]. Other studies have reported that the position of halogen substitution (ortho, meta, para) and the halogen itself (F, Cl, Br) have a strong influence on the self-assembly and hydrogelation rate, and the emergent viscoelasticity of the resulting hydrogels [24,26–28]. These findings indicate that the transducer type and position have a profound influence on the molecular self-assembly process and the resulting properties.

Another well-studied Fmoc-Phe derivative is Fmoc-pentafluoro-phenylalanine (Fmoc-F_5-Phe) that was previously shown to form a hydrogel [25]. Recently, we demonstrated that Fmoc-F_5-Phe undergoes a phase transition from spheres to a fibrillary gel, and ultimately to crystals [29]. When co-assembled with either Fmoc-F_5-Phe-PEG or Fmoc-Phe-Phe, this building block exhibits ideal stress-responsive behavior [30] or synergistic improvement of the mechanical properties [31], respectively. Moreover, we recently demonstrated the antibacterial activity of Fmoc-F_5-Phe against *Streptococcus mutans* in dental composite restoratives [32] and described its use as an antibacterial coating for different surfaces [33].

The present study reports the use of a combination of experimental and computational methods to investigate the molecular self-assembly process and phase transition during gelation of two previously unreported Fmoc-Phe derivatives: Fmoc-3,4-difluoro-phenylalanine (Fmoc-3,4F-Phe), and Fmoc-3,5-difluoro-phenylalanine (Fmoc-3,5F-Phe). In addition, we used X-ray diffraction (XRD) and molecular dynamics (MD) simulations to examine how changing the position of a single fluorine in the aromatic ring of Fmoc-Phe hydrogels, affects their physical properties, including self-assembly kinetics, morphology, physical characterization, phase transition as well as structural characterization at the atomic and molecular level.

2. Results and Discussion

2.1. Kinetic Analysis and Morphology of Double-Fluorinated Fmoc-Phe Hydrogels

Inspired by the self-assembly of Fmoc-Phe and the formation of hydrogel, we studied two new double fluorinated Fmoc-Phe building blocks: Fmoc-3,4F-Phe and Fmoc-3,5F-Phe

(Figure 1a–c). Whereas Fmoc-3,4F-Phe consist of two adjacent fluorine atoms, the fluorine atoms in the Fmoc-3,5F-Phe are farther apart, resulting in a lower electron repulsion. We examined the propensity of the building block to form self-supporting 3D hydrogel in response to a thermal switch, pH switch, or solvent switch using ethanol (EtOH) and dimethyl sulfoxide (DMSO) (Figure S1, Supporting Information).

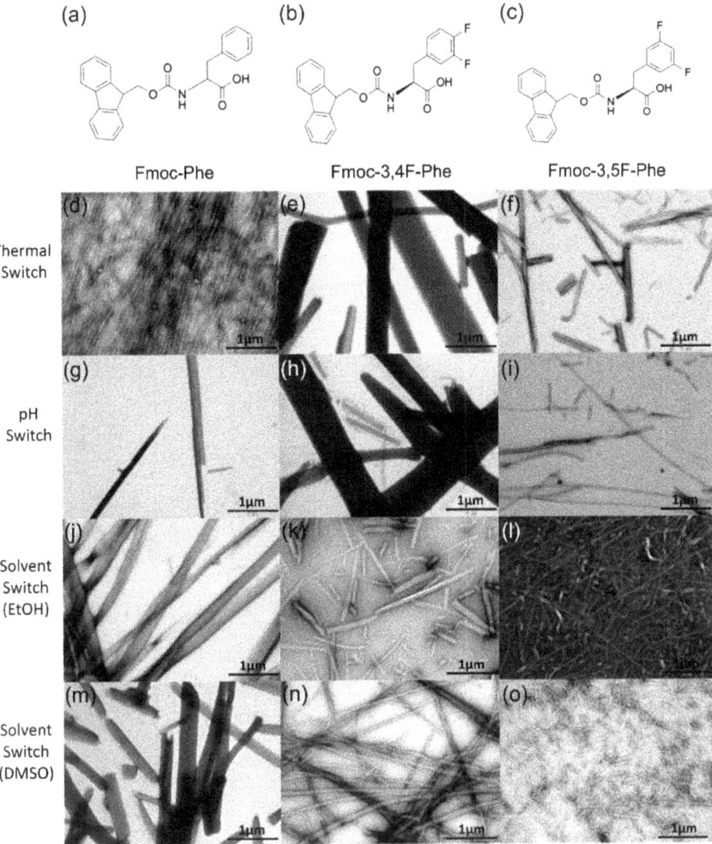

Figure 1. Formation of double fluorinated Fmoc-Phe hydrogels. Molecular structure of (**a**) Fmoc-Phe, (**b**) Fmoc-3,4F-Phe, and (**c**) Fmoc-3,5F-Phe. (**d–o**) TEM images of the structures formed by the various methods (**d–f**) Thermal Switch (**g–i**) pH Switch (**j–l**) Solvent Switch (EtOH) (**m–o**) Solvent Switch (DMSO), scale bar 1 μm.

In order to investigate the morphological structure and the alterations between the derivatives, we performed transmission electron microscopy (TEM) analysis (Figure 1). All the conditions tested exhibited fibrillary network morphology typical of supramolecular gels (Figure 1d–o). However, rapid formation of stable and rigid hydrogels (within a few minutes) was seen only when the solvent switch method, using DMSO as the organic solvent, was employed. For this reason, further comprehensive study of the Fmoc-Phe derivatives employed only this method. Comparing the nanostructures of the different hydrogels revealed that while Fmoc-Phe forms thick fibrils with an average diameter of 175 nm (Figure 1m), Fmoc-3,4F-Phe exhibits thinner and partly tangled fibrils, approximately 45 nm in diameter (Figure 1n), and the fibrils formed by Fmoc-3,5F-Phe were the thinnest with an average diameter of 30 nm (Figure 1o).

The gelation process of low-molecular-weight building blocks in the solvent switch method is usually characterized by an optical change from an opaque solution to a more transparent hydrogel [24,34] as organized nanostructures are formed [34]. While Fmoc-Phe formed an opaque hydrogel, both Fmoc-3,4F-Phe and Fmoc-3,5F-Phe formed transparent hydrogels (Figure 2a). Neither the Fmoc-Phe nor the Fmoc-3,4F-Phe hydrogel was stable, and phase separation was observed after one week, whereas the Fmoc-3,5F-Phe hydrogel was stable over time and maintained a clear 3D self-supporting hydrogel structure for at least 1 month.

The kinetics of the self-assembly process to nanostructures was monitored by measuring the turbidity of the solution at a wavelength of 350 nm over time (Figure 2b). This revealed that the absorbance of Fmoc-Phe remained high and stable over 16 h with an optical density (OD) of more than 2. In contrast, the absorbances of the double-fluorinated Fmoc-Phe hydrogels were much lower with a minimum OD value of 0.3 reached by Fmoc-3,4F-Phe after 3 min, and by Fmoc-3,5F-Phe after 44 min (Figure 2c). Interestingly, the absorbance of Fmoc-3,4F-Phe subsequently increased gradually to reach an OD of 1, while the absorbance of the Fmoc-3,5F-Phe hydrogel remained stable for 16 h. Although the building blocks of both Fmoc-3,4F-Phe and Fmoc-3,5F-Phe can be self-assembled into nanofibers, the chemical bonds, arrangement and kinetics are significantly different. Apparently, in the Fmoc-3,4F-Phe the electron clouds create repulsion resulting in an unstable hydrogel which decays over time. In contrast, the lower electron rejection in the Fmoc-3,5F-Phe results in a more stable structure for a longer period of time.

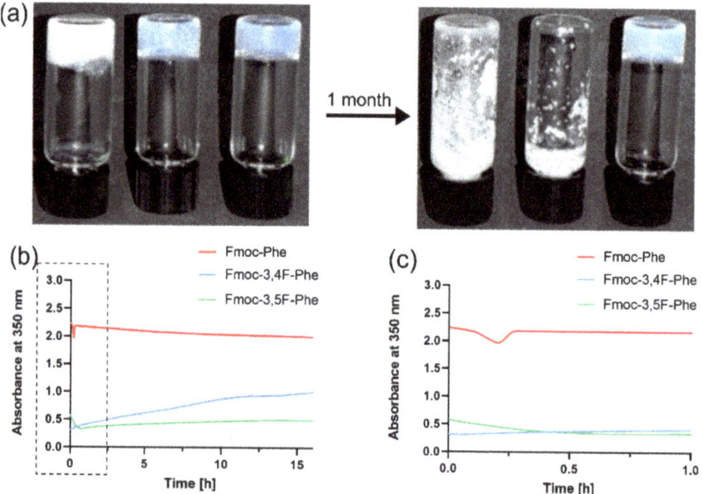

Figure 2. Kinetic characterization of the assemblies. (**a**) Hydrogels formed from Fmoc-Phe derivatives one-hour post gelation (left side) and one-month post gelation (right side). Vials from left to right: Fmoc-Phe, Fmoc-3,4F-Phe, and Fmoc-3,5F-Phe. (**b**) OD kinetics at 350 nm for the first 16 h of hydrogel formation. (**c**) Higher scale enlargement of the OD kinetics for the first hour (dashed area).

2.2. Physical Characterization of Double-Fluorinated Fmoc-Phe Hydrogels

Rheological analysis was performed at 25 °C to evaluate the mechanical properties of the hydrogels. First, dynamic strain sweep (5 Hz) and frequency sweep (0.5% strain) oscillatory measurements were performed to identify the appropriate conditions (Figure S2, Supporting Information). Based on the frequency sweep and oscillatory strain sweep analysis, the in-situ kinetics of the hydrogels formation and their mechanical properties were characterized by time sweep measurements at a fixed strain of 0.5% and frequency of 5 Hz, over 6 h (Figure 3). The storage modulus, G', is typically significantly higher than the loss modulus, G'', of these hydrogels, which is indicative of a viscoelastic gel. In all cases,

Tanδ (G″/G′) values after gelation were in the range of 0.05-0.015, i.e., <0.1, indicating stable hydrogel formation [35] (Figure S3, Supporting Information).

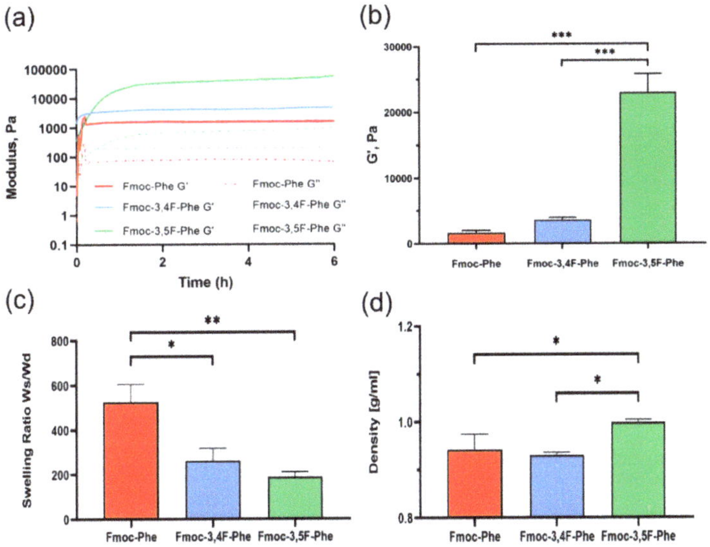

Figure 3. Physical Properties of Fmoc-Phe and the double fluorinated Fmoc-Phe hydrogels. (**a**) In situ time sweep oscillation measurements of storage and loss modulus. (**b**) The averaged storage modulus, G′, one-hour post-gelation. (**c**) Swelling ratio. (**d**) Density measurement. Representative results from three independent experiments are presented as mean ± SD; * $p < 0.05$, ** $p < 0.01$ and *** $p < 0.001$ as measured using one-way ANOVA.

The results of the rheological analysis revealed that the G′ of the Fmoc-Phe hydrogel, is relatively low, at 1600 Pa, whereas the Fmoc-3,4F-Phe has a slightly higher G′ value of 4800 Pa. In contrast, the Fmoc-3,5F-Phe hydrogel exhibits high rigidity, with a G′ value of more than 50,000 Pa. Significant differences in G′ values were observed by 1-hour post gelation (Figure 3b). At this time, the Fmoc-Phe hydrogel was entirely gelated with a G′ value of 1600 Pa, while the double fluorinated hydrogels had only achieved 40–70% of their end point storage modulus values of 3500 Pa and 23,000 Pa for Fmoc-3,4F-Phe and Fmoc-3,5F-Phe, respectively. The high storage modulus of the Fmoc-3,5F-Phe is probably due to stronger π–π interactions which contribute to the hydrogel stability, which caused by the higher distance between the fluorine atoms in comparison to Fmoc-3,4F-Phe. The relatively high mechanical rigidity of Fmoc-3,5F-Phe, combined with its high stability make the material interesting for tissue engineering and cell culture applications, where the mechanical properties are essential for controlling processes, such as stem cell differentiation. In this context, mesenchymal stem cells have been shown to undergo stiffness-directed fate differentiation into osteogenic lineages on rigid hydrogels [36].

The uptake of water into a hydrogel is very important because it can determine the overall permeation of nutrients and the excretion of cellular waste out of the hydrogel [37]. Whereas after 24 h in ddH_2O, the Fmoc-3,5F-Phe hydrogel retained its initial 3D-shape with a swelling ratio of 187 Ws/Wd (Figure 3c), the Fmoc-3,4F-Phe disintegrated slightly and presented a higher swelling ratio of 262 Ws/Wd, and the Fmoc-Phe hydrogel broke apart completely with a very high swelling ratio of 525 Ws/Wd. Despite the importance of water absorption, high swelling can be a disadvantage and can make the material unsuitable for use as a scaffold in aqueous environments because it will not retain its 3D structure. The density of the double fluorinated Fmoc-Phe hydrogels was examined in order to provide information about the significant differences seen in water absorption. As expected, the Fmoc-3,5F-Phe hydrogel had a higher density than the others with a value of approximately 1 g mL^{-1} (Figure 3d). While Fmoc-3,4F-Phe and Fmoc-Phe have almost the same density of 0.93-0.94 g mL^{-1}, the significant differences in swelling properties are probably due to the elasticity of the chains composing the hydrogel. It is possible that the chains in Fmoc-Phe are more elastic and flexible, which improves the water interaction and diffusion properties, while Fmoc-3,4F-Phe might form a more rigid fibrous matrix, which allows an exchange of water molecule at a steady state but limits the swelling [38]. Additionally, the lower electron rejection in the Fmoc-3,5F-Phe might allow for closer and denser interactions, contributing a denser packing which results in less swelling and higher mechanical properties.

The physical characterization highlights the advantages of these two double-fluorinated Fmoc-Phe hydrogels. Fmoc-Phe form a weak hydrogel which does not retain shape in an aqueous environment. Moreover, the Fmoc-F_5-Phe form a weak hydrogel that is unstable over time, probably due to the large electron clouds derived from its five fluorine atoms. We have previously reported the improvement of the Fmoc-F_5-Phe stability by forming a hybrid hydrogel with additional peptide [31]. Here we present how a single atom modification can form hydrogels which exhibit higher rigidity and stability over time, even in aqueous solution. It is shown that Fmoc-3,5F-Phe, in particular, remains very stable and shows improved physical properties compared to the other hydrogels.

2.3. Phase Transition and Morphological Characterization of the Assemblies

Time-lapse optical microscopy measurements in real-time were used to monitor the kinetics and dynamics of the self-assembly process. The samples were sealed in a glass capillary to prevent evaporation, and based on our previous report of the Fmoc-F_5-Phe phase transition [29], the experiment was performed in 50:50 DMSO:ddH_2O solution. Interestingly, we also observed a phase transition for Fmoc-Phe and the double fluorinated Fmoc-Phe under these conditions (Figure 4, Movies S1–S3). Figure 4 presents real-time images and a schematic illustration of the phase transition of Fmoc-Phe and the two derivatives. At t_0, all three building blocks exhibited a spherical structure, which over the next few minutes was gradually replaced by a fibrillary network (t_{mid}). Interestingly, from that point on, the kinetics and morphologies of the various building blocks diverged, with Fmoc-Phe exhibiting multiple nucleation centers (Figure 4a,d and Movie S1). Initially, a similar process was observed for Fmoc-3,4F-Phe; however, a few minutes later, this continued to a third phase involving the growth of needle-like crystals (Figure 4b,e and Movie S2). In contrast to the other two building blocks, the assembled fibrils of Fmoc-3,5F-Phe are aligned along the capillary (Figure 4c,f and Movie S3). The disassembly and replacement of the spherical structures seen suggests that they are thermodynamically metastable compared to the gel and crystal phases. This is in accordance with previous reports concerning other supramolecular systems [29,39–43]. Our results demonstrate that the three different building blocks all exhibit a phase transition under the same conditions. These phase transitions are not reversible process, as the final phase is the preferred and stable thermodynamic state. This behavior appears to follow Ostwald's rule of stages whereby monomers of metastable structural species can be replaced by a more stable phase such as fibrils and crystals. However, further analysis will be needed to obtain a deeper understanding of the kinetics as well as the thermodynamics of this process.

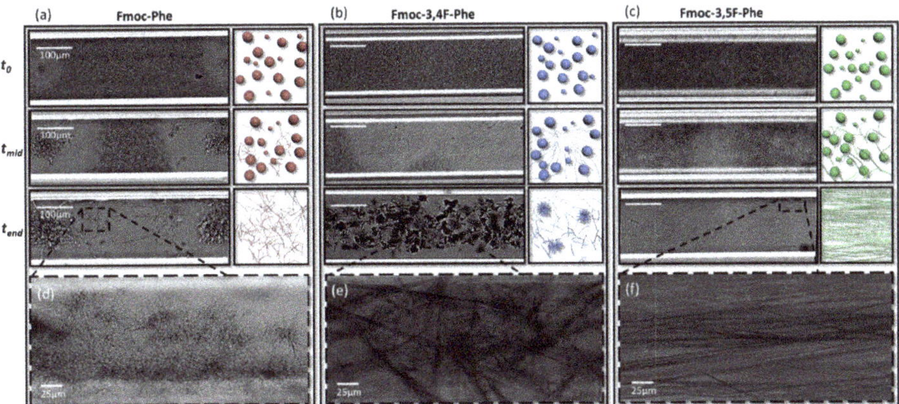

Figure 4. Real-time monitoring of the phase transition over time, and a schematic illustration of the different structures. Phase transition at three different time points; (t_0) spherical structures, (t_{mid}) the spheres disassembly slowly and are replaced by a fibrillary gel, (t_{end}) final morphology of each building block (**a**) Fmoc-Phe (**b**) Fmoc-3,4F-Phe (**c**) Fmoc-3,5F-Phe. The final stage is shown at a magnified scale to enable visualization of the differences in morphology for (**d**) Fmoc-Phe (**e**) Fmoc-3,4F-Phe (**f**) Fmoc-3,5F-Phe.

The results of the microscopy analysis are in accordance with the physical characterization of the structures. The low storage modulus obtained for the Fmoc-Phe hydrogel, as well as the low density and high swelling ratio, might be explained by the thick fibrillar morphology. In contrast, the thinner and more entangled fibrils of Fmoc-3,4F-Phe probably contribute to the higher storage modulus of this hydrogel. In addition, the curled fibrils could make it difficult for water to penetrate and thereby explain the lower swelling. Similarly, the thin, long, aligned nanofibrils seen in Fmoc-3,5F-Phe, might be directly responsible for the high storage modulus [44]. The straight orientation of the fibrils could be difficult to fold and reduce the flexibility, explaining how Fmoc-3,5F-Phe can be organized in a denser structure that exhibits less swelling.

2.4. Structural Analysis by Powder and Single Crystal XRD

Powder XRD (PXRD) was used to study the molecular organization of the different building blocks further. The PXRD pattern of Fmoc-Phe, Fmoc-3,4F-Phe and Fmoc-3,5F-Phe presented in Figure 5a includes peaks for all the samples, indicating their crystalline nature. However, Fmoc-3,4F-Phe and Fmoc-3,5F-Phe exhibit sharper and higher intensity peaks, which indicate a higher degree of crystallization. The diffraction patterns of Fmoc-3,4F-Phe and Fmoc-3,5F-Phe are different, which implies two distinct crystalline arrangements.

Using single crystal XRD, we were able to solve the crystal structure of Fmoc-3,4F-Phe for the first time, and detailed crystallographic data are presented in Table S1 in the Supporting Information. This analysis revealed that the Fmoc-3,4F-Phe crystallizes in the orthorhombic space group $P2_12_12_1$ with one Fmoc-3,4F-Phe and one co-crystallized DMSO solvent molecule in the asymmetric unit (Figure 5b). Careful observation of the crystal structure indicated that the backbone torsional angle φ and ψ have a value near $-130°$ and $165°$, which fall in the extended sheet region of the Ramachandran plot. More importantly, the carbamate group of Fmoc-3,4F-Phe is connected to the neighboring molecule through an intermolecular hydrogen bond (N–H···O). In addition, the carboxylate group of each Fmoc-3,4F-Phe forms an intermolecular hydrogen bond (O–H···O) with the DMSO solvent molecule. The face-to-face π–π interactions between the Fmoc-Fmoc groups of the subunit, together with π–π interactions between the side chain 3,4F-Phe groups and a neighboring molecule, stabilize the overall packing (Figure 5c).

Interestingly, weak intermolecular C–H···F hydrogen bonds were observed between the subunits, and the molecule further self-organized to produce a helical like architecture along the crystallographic c-axis in higher-order packing (Figure 5d). This helical like architecture correlates with the presence of chiral carbons in the Fmoc-3,4F-Phe structure, and represents a generic structural motif in self-assembled amino acid and peptide nanostructures [45]. In contrast to the orthorhombic space group $P2_12_12_1$ crystals seen for Fmoc-3,4F-Phe, Fmoc-Phe crystallized in the monoclinic space group $P2_1$ [26,27,46,47]. The Fmoc-Phe crystal structure exhibits strong intermolecular hydrogen bonding N1–H1···O2 between Fmoc-Phe molecules, and strong π–π stacking interactions between the large bulky Fmoc group, as well as between phenylalanines, in a similar scenario to that observed in the Fmoc-3,4F-Phe crystal structure. However, while the carboxylic acid group of Fmoc-Phe forms intermolecular hydrogen bonds with the adjacent carboxylic group of another Fmoc-Phe molecule, the carboxylic acid group of Fmoc-3,4F-Phe undergoes intermolecular hydrogen bonding with a DMSO solvent molecule. Importantly, this difference in intermolecular hydrogen bonding, means that Fmoc-3,4F-Phe has significantly higher order packing than Fmoc-Phe.

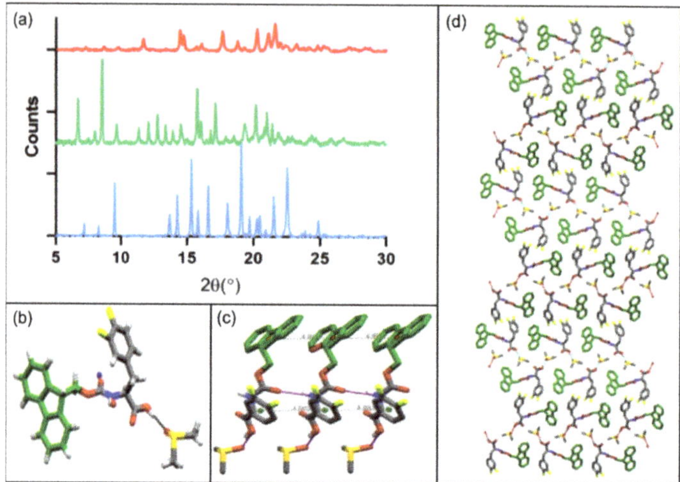

Figure 5. PXRD and Single crystal XRD structures. (**a**) PXRD of Fmoc-Phe (red) and the double fluorinated Fmoc-Phe building blocks; Fmoc-3,4F-Phe (blue), and Fmoc-3,5F-Phe (green). (**b**) Single crystal XRD structure of Fmoc-3,4F-Phe, demonstrating the asymmetric unit. (**c**) Intermolecular backbone hydrogen bonding and π–π stacking of the Fmoc-3,4F-Phe molecule, with a distance between two centroids of 4.983 Å. (**d**) Fmoc-3,4F-Phe crystal packing along the c-axis.

2.5. Structural Analysis by MD Simulations

After solving the structure of Fmoc-3,4F-Phe by single crystal XRD analysis, and without success in crystalizing Fmoc-3,5F-Phe, we further explored the structural differences between the two molecules through MD simulations. The MD simulations were used to independently study and compare the properties of Fmoc-3,4F-Phe and Fmoc-3,5F-Phe during the first moments of self-assembly without any bias from XRD results. Five-replicate explicit solvent (water) MD simulations, followed by structural analyses of the resulting MD simulation trajectories, for each derivative were performed. Both derivatives exhibited gradual formation of aggregates, which may represent the initial organization of Fmoc-3,4F-Phe and Fmoc-3,5F-Phe. We note that the interactions and statistical trends described below are reproducible across all MD simulations despite the formation of different aggregate structures within MD simulation replicates.

The results of the MD simulations indicate that the aggregates formed by self-assembly of Fmoc-3,4F-Phe or Fmoc-3,5F-Phe are significantly stabilized by π–π interactions between aromatic groups as well as F-F and F-Phe contacts (Figure 6a–c). In addition, both derivatives interacted occasionally through hydrogen bonds between terminal carboxyl groups or backbone amide and carbonyl groups of opposing monomers.

Our analysis revealed similarities and differences in the structural properties of the two derivatives as a result of the interactions formed within the simulated clusters (Figure 6a). While both Fmoc-3,4F-Phe and Fmoc-3,5F-Phe formed structures reminiscent of an antiparallel β-sheet at a similar rate (Figure 6a), Fmoc-3,4F-Phe formed a parallel β-sheet-like structure, in which two monomers are bonded through a hydrogen bond between their backbone amide and carbonyl groups as well as face-to-face π–π interactions between the Fmoc-Fmoc or Phe-Phe groups (Figure 6b, boxed in red dotted lines) more often than Fmoc-3,5F-Phe. Importantly, this ordered structure is reminiscent of the crystal structure of Fmoc-3,4F-Phe (Figures 5c and 6b boxed in red dotted lines). The preferential ability of Fmoc-3,4F-Phe to form face-to-face π–π interactions between Fmoc-Phe and Phe-Phe groups (Figure 6b, boxed in black and blue dotted lines, respectively) could facilitate the formation of the parallel β-sheet-like structures observed in the MD simulations and crystal structure (Figures 5c and 6b boxed in red dotted lines). In contrast, Fmoc-3,5F-Phe aggregates were more frequently stabilized by Fmoc-Fmoc π-stacking interactions, and contacts between fluorine of the 3,5F-Phe group and Fmoc, in the absence of Fmoc-Phe π-stacking, as well as interactions between the fluorine of the 3,5F-Phe group and the terminal O of opposing monomers (Figure 6c boxed in red, black, and blue dotted lines respectively).

In addition to the differences in the frequency of interactions formed within their aggregates, we also observed that the calculated radius of gyration of the monomers within the clusters formed by Fmoc-3,5F-Phe was consistently lower than for those formed by Fmoc-3,4F-Phe. This was true across aggregates of different sizes (Figure 6d). Thus, for clusters containing the same number of building block-monomers, clusters formed by Fmoc-3,5F-Phe were more densely packed. This supports the experimental density measurements and provides additional evidence for differences in the self-assembly properties of the Fmoc-3,4F-Phe and Fmoc-3,5F-Phe systems.

Solvent exposure calculations of the Fmoc moiety and Phe sidechain of Fmoc-3,4F-Phe and Fmoc-3,5F-Phe within the independent detected clusters indicated that the Fmoc moiety is generally less solvent exposed in the clusters formed by Fmoc-3,5F-Phe compared to those formed by Fmoc-3,4F-Phe (Figure S4, orange and blue data points, respectively). In addition, the Phe sidechain is generally more solvent exposed in the clusters formed by Fmoc-3,5F-Phe compared to those formed by Fmoc-3,4F-Phe (Figure S4, yellow and grey data points, respectively). Thus, the Fmoc moiety is buried more deeply within the clusters of Fmoc-3,5F-Phe than in the clusters of Fmoc-3,4F-Phe, and the Phe sidechain is less buried within the clusters of Fmoc-3,5F-Phe than in the clusters of Fmoc-3,4F-Phe. This suggests that the different position of the fluorine atom in Fmoc-3,4F-Phe compared to Fmoc-3,5F-Phe may influence the hydrophobicity of the Phe sidechain. Specifically, as 3,4F-Phe is less solvent exposed than 3,5F-Phe within the detected clusters, 3,4F-Phe appears to be more hydrophobic than 3,5F-Phe. This is supported by polar desolvation energy calculations performed in AMSOL [48] for the isolated 3,4F-Phe and 3,5F-Phe sidechains, predicting that 3,5F-Phe should be less hydrophobic than 3,4F-Phe [49]. This difference in hydrophobicity between 3,4F-Phe and 3,5F-Phe provides an additional potential explanation for the differences in the interactions formed within the clusters of Fmoc-3,4F-Phe and the clusters of Fmoc-3,5F-Phe (Figure 6a,b).

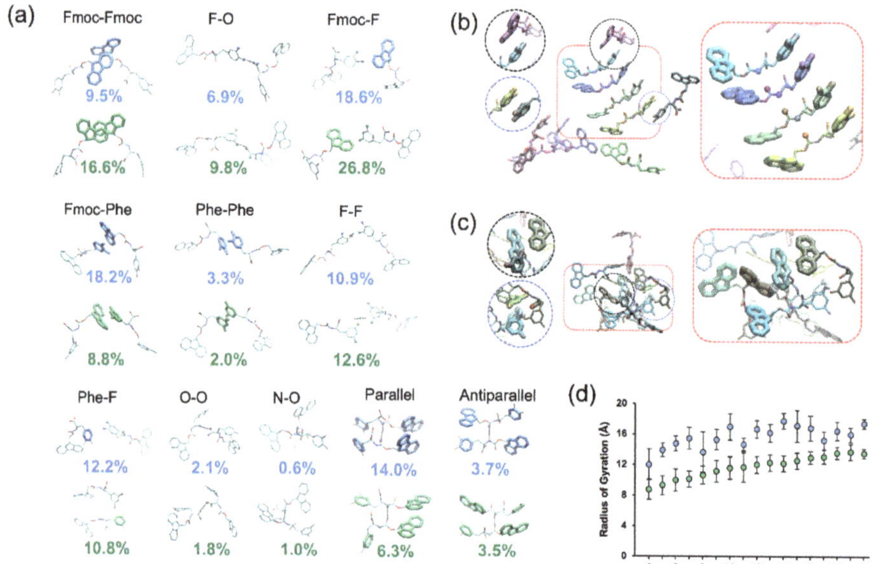

Figure 6. MD simulations depicting the structural properties of the Fmoc-3,4F-Phe and Fmoc-3,5F-Phe assemblies. (**a**) Molecular graphics images of interactions between (top row, blue) Fmoc-3,4F-Phe and (bottom row, green) Fmoc-3,5F-Phe pairs observed in simulations and their average percent frequency across all MD simulations. Percent frequency was calculated as the total number of instances for which a given interaction was observed, divided by the total number of interactions observed in each simulation. The average standard deviation of the average percent frequencies is $0.9 \pm 0.6\%$ with a minimum and maximum standard deviation of 0.1 and 2.6%, respectively; the relatively small values indicate good reproducibility across different simulation runs per system. (**b**,**c**) Representative aggregate of 10 (**b**) Fmoc-3,4F-Phe and (**c**) Fmoc-3,5F-Phe building-block monomers observed in the simulations with key differences in interactions encircled in dotted lines and zoomed-in in the nearby panels. Molecular graphics images in (**a**–**c**) were produced using Visual Molecular Dynamics (VMD) version 1.9.4 [50]. (**d**) Radius of gyration (Å) of building block-monomers within the clusters observed in the simulations of Fmoc-3,4F-Phe (blue) and Fmoc-3,5F-Phe (green)..

3. Materials and Methods

3.1. Materials

Fmoc-L-Phe was purchased from Sigma-Aldrich (Israel), Fmoc-3,4F-Phe and Fmoc-3,5F-Phe were purchased from Chem-Implex Int'l inc (IL, USA).

3.2. Preparation of Fmoc-Phe Derivatives Self-Assemblies

Three different techniques were used for the preparation of Fmoc-Phe derivatives self-assemblies: Thermal-Switch, pH-Switch and Solvent-Switch. ***Thermal Switch***: Preformed structures were assembled by dissolving Fmoc-Phe derivatives in ddH_2O at concentration of 5 g L^{-1} and heating to 90 °C. Structures were visible when samples were slowly cooled down to room temperature. ***pH Switch***: Fmoc-Phe derivatives were mixed in ddH_2O at concentration of 5 g L^{-1} and sonicated until dissolved. A solution of 0.5 M NaOH was slowly added to the peptide until pH 7.5 was measured. The solution was left undisturbed until gelation was observed. ***Solvent Switch*** [51]: Stock solution were prepared in dimethyl sulfoxide (DMSO) or absolute ethanol (EtOH) at a concentration of 100 g L^{-1} and 10 g L^{-1}, respectively. The hydrogels were formed by diluting the stock solution with ddH_2O, resulting in a final concentration of 5 g L^{-1}.

3.3. OD Kinetics Analysis

Immediately after hydrogels preparation, 100 μL samples were placed into a 96-well plate. Absorbance and kinetics were measured at a wavelength of 350 nm using a TECAN Infinite M200PRO plate reader for 16 h.

3.4. Optical Microscopy Analysis

Samples of the three building blocks were prepared at a ratio of 50:50 DMSO:ddH_2O and transferred into a thin glass capillary (0.2 mm inner diameter, 0.1 mm wall) and sealed to avoid evaporation. The capillary was attached to a glass slide and observed under an optical microscope. Bright-field imaging was performed using an Eclipse Ti-E inverted microscope (Nikon, Japan), equipped with a Zyla scMOS camera (Andor, UK). For the phase transition kinetics experiment, time-lapse image series were acquired using a 20× objective over time, with 10 s interval.

3.5. TEM

Hydrogels samples were placed on a 400-mesh copper grid. After 1 min, the piece of gel as well as the excess fluid was removed. Negative staining was obtained by covering the grid with 10 μL of 2% uranyl acetate in water. After 2 min, excess uranyl acetate solution was removed. Samples were viewed using a JEOL 1200EX electron microscope operating at 80 kV.

3.6. Rheology Analysis

In situ hydrogel formation, mechanical properties, and kinetics were characterized using an AR-G2 rheometer (TA Instruments). Time-sweep oscillatory tests in 20 mm parallel plate geometry were performed on 220 μL fresh solution (resulting in a gap size of 0.6 mm) at room temperature. Oscillatory strain (0.01–100%) and frequency sweeps (0.1–100 Hz) were conducted to find the linear viscoelastic region, in which the time-sweep oscillatory tests were performed (Figure S2, Supporting information). G′ and G″, the storage and loss moduli, respectively, were obtained at 5 Hz oscillation and 0.5% strain deformation for each sample.

3.7. Swelling

Identical volumes of hydrogel samples were placed on plates. Five samples were used for each hydrogel. The initial weight (Wi) was recorded, and the hydrogels were placed in ddH_2O. To allow equilibration and swelling, all samples were left to swell for 24 h on an orbital shaker (50 rpm) at room temperature. The equilibrated swollen mass (Ws) was recorded after gently absorbing excess water from each sample. The hydrogel samples were subsequently lyophilized, and their dry weight (Wd) was measured. The equilibrated swelling ratio (Q) was defined as the ratio of Ws to Wd.

3.8. Density

The density measurement was conducted using a pycnometer at 23 °C. First, we filled the pycnometer with ddH_2O and calculated the exact volume of each 5 ml pycnometer (Vp). Then we placed 500 μL hydrogel sample inside the pycnometer, weighed it (Ms) and waited an hour until complete gelation. We gently added ddH_2O and calculate the ddH_2O volume inside the pycnometer (Vd). Sample density was calculating by dividing Ms to (Vp-Vd).

3.9. Crystallization, Data Collection and Structure Determination

Fmoc-3,4F-Phe 100 g L^{-1} stock solutions were prepared in DMSO and diluted into ddH_2O at a 70:30 ratio. The samples were left, half sealed, on the bench at room temperature, and crystals grew within 4 months. For data collection, a crystal was coated in Paratone oil (Hampton Research), mounted on a MiTeGencryo-loop, and flash-frozen in liquid nitrogen. Crystal data were measured at 100 K on a RigakuXtaLab^{Pro} diffractometer equipped with CuKα radiation (λ = 1.54184 Å) and a Dectris Pilatus3R 200K-A detector. The data were processed with CrysAlisPro programs (RigakuOD). The structure was solved by direct methods with SHELXT-2016/4 and refined with full-matrix least squares refinement based on F2 with SHELXL-2016/4. The crystallographic data have been deposited in the Cambridge Crystallographic Data Centre (CCDC) under no. 2043733 and are presented in supplementary information Table S1, Supporting Information.

3.10. Molecular Modeling of Investigated Systems

Fmoc-3,4F-Phe and Fmoc-3,5F-Phe were computationally modeled independently for use as initial structures in MD simulations, described in the next section, analogously to refs [52,53]. For each system, 36 monomers (either Fmoc-3,4F-Phe or Fmoc-3,5F-Phe) were embedded in a $40 \times 40 \times 60$ Å^3 grid within a $90 \times 90 \times 90$ Å^3 water box, with an equal spacing of approximately 20 Å between each monomer and random configurations and orientations. The random configurations were extracted from short simulations of each monomer at infinite dilution. The derivative concentration within each system was higher than the concentration used in the experimental studies, aiming to artificially accelerate the self-assembly within the frame of the simulations [52,53].

3.11. MD Simulations

The self-assembly of Fmoc-3,4F-Phe and Fmoc-3,5F-Phe were independently investigated through MD simulations analogously to refs [52,53]. Prior to the execution of MD simulations, the simulation systems were first energetically minimized and equilibrated. In the energetic minimization stage, 50 steps of steepest descent followed by 50 steps of Adopted Basis Newton-Raphson energy minimization were first performed with the monomers initially fixed to their initial conformations to alleviate clashes primarily between the monomers and their surrounding environment. Subsequently, an additional 100 steps of steepest descent followed by 100 steps of Adopted Basis Newton-Raphson energy minimization were performed with the monomer heavy atoms were constrained with 0.1 kcal mol^{-1} Å^{-2}. Under the same constraints, after the energetic minimization, the system was equilibrated for 1 ns. Finally, five replicate MD simulation production runs with different initial velocities were performed for each system. Multiple serial simulation runs were preferred over other single simulation runs using enhanced sampling methods, as routinely used by Tamamis and co-authors [52,54–58], at the same computational cost as the former were used advantageously to check reproducibility of the results in the current study. Here, reproducibility was considered important to delineate the differences in self-assembly between the two systems with subtle structural differences at the monomeric state. In the MD simulation production runs, no constraints were imposed on the system. All energy minimization and MD simulations were performed using periodic boundary conditions and the CHARMM36 force field [59] in CHARMM [60]. Parameters and topologies for the derivatives were generated using CGenFF [61]. The temperature and pressure of the simulation systems was maintained at 300 K and 1.0 atm using the dual Nosé-Hoover thermostat and the Andersen-Hoover barostat, respectively. The bond lengths of covalently bonded hydrogens were constrained using the SHAKE algorithm [62].

3.12. Structural Analysis of MD Simulations

Both Fmoc-3,4F-Phe and Fmoc-3,5F-Phe were observed to form aggregates within their respective sets of MD simulations. We aimed to investigate the early-stage gradual formation of structural morphologies in the two systems, and thus, we simulated and analyzed 80 ns of each replicate in both systems under investigation, which corresponds to the time at which the systems started forming rather stable aggregates based on radius of gyration calculations. Simulation snapshots were extracted in 0.2 ns intervals per simulation production run, resulting in 2000 analyzed snapshots per system. Structural analysis programs were to understand and compare the structural organization properties of the two derivatives, independently, in the first moments of self-assembly. Crystallographic data of Fmoc-3,4F-Phe and visual inspection of the MD simulations were used to guide the definitions of key interactions that are formed in ordered assemblies of Fmoc-3,4F-Phe and Fmoc-3,5F-Phe, independently, in their respective simulations. The interactions detected are shown in Figure 6a. A uniform distance cutoff value of 4.0 Å was used as a criterion to define an interaction between the atoms described in Figure 6a. The same set of interactions was tracked for both the simulations of Fmoc-3,4F-Phe and Fmoc-3,5F-Phe for a fair comparison of the two derivatives' self-assembly properties. A number of n monomers were defined to form a cluster when each building block-monomer was in the vicinity of at least another one based on the 4.0 Å distance criterion defined above.

We additionally calculated the radius of gyration of the building block-monomers within each observed cluster to determine and compare the compactness of the clusters formed by Fmoc-3,4F-Phe to those formed by Fmoc-3,5F-Phe (Figure 6d). The radius of gyration of the building block-monomers within each cluster was calculated using the following equation in Wordom: [63,64]

$$R_g = \sqrt{\frac{1}{N}\sum_{k=1}^{N}(r_k - \bar{r})^2}$$

In the aforementioned equation, the radius of gyration, R_g, is calculated as the square root of the average deviation of N atoms, r_k, from the geometric center, r. The radius of gyration calculations was performed considering only building block-monomers within each cluster, with all other atoms omitted. Larger relative radius of gyration values indicates lower compactness, or lower packing density of building blocks in the clusters, while lower relative radius of gyration values indicates higher compactness, or higher density of building blocks in the clusters [55].

Finally, we calculated the solvent exposure of the Fmoc-moiety and modified Phe side chains within each observed cluster to determine and compare on how the aromatic rings of Fmoc-3,4F-Phe and Fmoc-3,5F-Phe were positioned (in terms of "burial" or not) within each of their respective clusters (Figure S4). The solvent exposition of the Fmoc-moiety and modified Phe side chains, independently, of the building block-monomers within each cluster was calculated as the solvent accessible surface area (SASA) of the Fmoc-moiety or modified Phe side chain divided by the total molecular surface area (TSA) of the same Fmoc-moiety or modified Phe side chain. The SASA and TSA values for each Fmoc-moiety or modified Phe side chain was calculated in Wordom [63,64]. A larger percent solvent exposure of a Fmoc-moiety or modified Phe side chain indicates that the moiety or side chain is more exposed to the solvent and more likely to be at the surface of the cluster; a smaller the percent solvent exposure of a Fmoc-moiety or modified Phe side chain indicates that the moiety or side chain is more "buried" and more likely to be encapsulated in the interior of the cluster.

3.13. Statistical Analysis

Statistical significance was examined using one-way ANOVA to determine the p-value. $p < 0.05$ was considered a statistically significant difference

4. Conclusions

The current study provides the first description of the behavior of the double-fluorinated Fmoc-Phe derivatives, Fmoc-3,4F-Phe and Fmoc-3,5F-Phe, and the effect that the position of a single fluorine has on the self-assembly process, and physical properties that the material produces. Our results reveal substantial differences between the derivatives. While Fmoc-3,5F-Phe is transparent and remains clear and stable over time, both Fmoc-Phe and Fmoc-3,4F-Phe disintegrate and undergo phase separation. Moreover, Fmoc-3,5F-Phe presents a more orderly and aligned microstructure of nanofibrils and poses a higher storage modulus, and lower swelling due to a higher density. Although all three building blocks exhibit a spontaneous phase transition process from metastable spheres to fibrils,

Fmoc-3,4F-Phe undergoes an additional self-assembly event, resulting in the formation of crystals. Single crystal XRD of the Fmoc-3,4F-Phe crystal structure revealed that π–π interactions and hydrogen bonding contribute to the crystal stability. MD simulations provided additional evidence of differences between the structural properties of Fmoc-3,4F-Phe and Fmoc-3,5F-Phe. While Fmoc-3,4F-Phe aggregates are more frequently stabilized through interactions reminiscent of the crystal structure, the stability of Fmoc-3,5F-Phe aggregates rely more on Fmoc-Fmoc π stacking, contacts between the F of the 3,5F-Phe group and Fmoc, as well as interactions between the F of the 3,5F-Phe group and the terminal O of the opposing monomer. The experimental data and the simulations both indicate that Fmoc-3,5F-Phe forms more compact aggregates than Fmoc-3,4F-Phe. These results highlight the effect of the position of a single amino acid on the self-assembly process and demonstrate that fluorination is an effective strategy to influence nanoscale supramolecular organization. Our results are consistent with the changes in the Phe fluorination pattern observed in previously published work and emphasize the often-unpredicted consequences of minor change in the building block structure that complicate rational design of such materials. They also demonstrate how the macroscale properties of a material can be modified by atomic scale changes in the constituent molecules. Furthermore, they provide fundamental insights that will facilitate the development of optimal amino-acid-based low-molecular-weight hydrogelators for a wide range of applications in various fields.

Supplementary Materials: The following are available online at https://www.mdpi.com/article/10.3390/ijms22179634/s1.

Author Contributions: Conceptualization, M.A., D.C.-G. and L.A.-A.; Data Curation, M.A., D.C.-G. and A.A.O.; MD simulation, A.A.O. and P.T.; formal analysis, M.A., D.C.-G., A.A.O., R.M., Z.A.A., L.J.W.S. and L.A.-A.; writing—original draft preparation, M.A., D.C.-G., A.A.O., R.M. and L.A.-A.; writing—review and editing, Y.S.-D., P.T. and L.A.-A.; supervision, Y.S.-D., P.T. and L.A.-A.; funding acquisition, P.T. and L.A.-A. All authors have read and agreed to the published version of the manuscript.

Funding: This work was supported by the European Research Council (ERC), under the European Union's Horizon 2020 research and innovation program (grant agreement No. 948102) (L.A.-A.) and the ISRAEL SCIENCE FOUNDATION (grant No. 1732/17) (L.A.-A.).

Institutional Review Board Statement: Not applicable.

Informed Consent Statement: Not applicable.

Data Availability Statement: CCDC 2043733 contains the supplementary crystallographic data for this paper. These data can be obtained free of charge from the Cambridge Crystallographic Data Centre via www.ccdc.cam.ac.uk/data_request/cif (accessed on 30 August 2021).

Acknowledgments: The authors acknowledge the Chaoul Center for Nanoscale Systems of Tel Aviv University for the use of instruments and staff assistance. The authors acknowledge the Marian Gertner Institute for Medical Nano systems at Tel Aviv University (D.C.-G). All MD simulations and computational analysis were conducted using the Ada supercomputing cluster at the Texas A&M High Performance Research Computing Facility, and additional facilities at Texas A&M University. A.A.O. acknowledges the Texas A&M University Graduate Diversity Fellowship from the TAMU Office of Graduate and Professional Studies. We thank Davide Levy for PXRD analysis and instrumental assistance and are grateful to the members of the Adler-Abramovich group for the helpful discussions.

Conflicts of Interest: The authors declare no conflict of interest.

References

1. Ariga, K.; Leong, D.T.; Mori, T. Nanoarchitectonics for Hybrid and Related Materials for Bio-Oriented Applications. *Adv. Funct. Mater.* **2018**, *28*, 1702905. [CrossRef]
2. Dhiman, S.; George, S.J. Temporally Controlled Supramolecular Polymerization. *Bull. Chem. Soc. Jpn.* **2018**, *91*, 687–699. [CrossRef]

3. Freeman, R.; Han, M.; Álvarez, Z.; Lewis, J.A.; Wester, J.R.; Stephanopoulos, N.; McClendon, M.T.; Lynsky, C.; Godbe, J.M.; Sangji, H.; et al. Reversible self-assembly of superstructured networks. *Science* **2018**, *362*, 808–813. [CrossRef]
4. Gazit, E. Self-assembled peptide nanostructures: The design of molecular building blocks and their technological utilization. *Chem. Soc. Rev.* **2007**, *36*, 1263. [CrossRef]
5. Xu, Y.; Wang, B.; Kaur, R.; Minameyer, M.B.; Bothe, M.; Drewello, T.; Guldi, D.M.; von Delius, M. A Supramolecular [10]CPP Junction Enables Efficient Electron Transfer in Modular Porphyrin–[10]CPP⊃ ∩Fullerene Complexes. *Angew. Chem. Int. Ed.* **2018**, *57*, 11549–11553. [CrossRef] [PubMed]
6. Wei, G.; Su, Z.; Reynolds, N.P.; Arosio, P.; Hamley, I.W.; Gazit, E.; Mezzenga, R. Self-assembling peptide and protein amyloids: From structure to tailored function in nanotechnology. *Chem. Soc. Rev.* **2017**, *46*, 4661–4708. [CrossRef] [PubMed]
7. Zhang, W.; Yu, X.; Li, Y.; Su, Z.; Jandt, K.D.; Wei, G. Protein-mimetic peptide nanofibers: Motif design, self-assembly synthesis, and sequence-specific biomedical applications. *Prog. Polym. Sci.* **2018**, *80*, 94–124. [CrossRef]
8. Weingarten, A.S.; Kazantsev, R.V.; Palmer, L.C.; McClendon, M.; Koltonow, A.R.; Samuel, A.P.S.; Kiebala, D.J.; Wasielewski, M.R.; Stupp, S.I. Self-assembling hydrogel scaffolds for photocatalytic hydrogen production. *Nat. Chem.* **2014**, *6*, 964–970. [CrossRef] [PubMed]
9. Kasotakis, E.; Mitraki, A. Designed Self-Assembling Peptides as Templates for the Synthesis of Metal Nanoparticles. *Methods Mol. Biol.* **2013**, *996*, 195–202. [CrossRef]
10. Aida, T.; Meijer, E.W.; Stupp, S.I. Functional Supramolecular Polymers. *Science* **2012**, *335*, 813–817. [CrossRef] [PubMed]
11. Zhang, Y.; Gu, H.; Yang, Z.; Xu, B. Supramolecular Hydrogels Respond to Ligand-Receptor Interaction. *J. Am. Chem. Soc.* **2003**, *125*, 13680–13681. [CrossRef] [PubMed]
12. Knowles, T.P.J.; Oppenheim, T.W.; Buell, A.K.; Chirgadze, D.Y.; Welland, M.E. Nanostructured films from hierarchical self-assembly of amyloidogenic proteins. *Nat. Nanotechnol.* **2010**, *5*, 204–207. [CrossRef] [PubMed]
13. Roth-Konforti, M.E.; Comune, M.; Halperin-Sternfeld, M.; Grigoriants, I.; Shabat, D.; Adler-Abramovich, L. UV Light-Responsive Peptide-Based Supramolecular Hydrogel for Controlled Drug Delivery. *Macromol. Rapid Commun.* **2018**, *39*, 1800588. [CrossRef] [PubMed]
14. Wen, Y.; Collier, J.H. Supramolecular peptide vaccines: Tuning adaptive immunity. *Curr. Opin. Immunol.* **2015**, *35*, 73–79. [CrossRef]
15. Koutsopoulos, S. Self-assembling peptide nanofiber hydrogels in tissue engineering and regenerative medicine: Progress, design guidelines, and applications. *J. Biomed. Mater. Res. Part A* **2016**, *104*, 1002–1016. [CrossRef] [PubMed]
16. Choe, S.; Bond, C.W.; Harrington, D.A.; Stupp, S.I.; McVary, K.T.; Podlasek, C.A. Peptide amphiphile nanofiber hydrogel delivery of sonic hedgehog protein to the cavernous nerve to promote regeneration and prevent erectile dysfunction. *Nanomed. Nanotechnol. Biol. Med.* **2017**, *13*, 95–101. [CrossRef] [PubMed]
17. Du, X.; Zhou, J.; Shi, J.; Xu, B. Supramolecular Hydrogelators and Hydrogels: From Soft Matter to Molecular Biomaterials. *Chem. Rev.* **2015**, *115*, 13165–13307. [CrossRef]
18. Dou, X.-Q.; Feng, C.-L. Amino Acids and Peptide-Based Supramolecular Hydrogels for Three-Dimensional Cell Culture. *Adv. Mater.* **2017**, *29*, 1604062. [CrossRef]
19. Karoyo, A.; Wilson, L. Physicochemical Properties and the Gelation Process of Supramolecular Hydrogels: A Review. *Gels* **2017**, *3*, 1. [CrossRef]
20. Vegners, R.; Shestakova, I.; Kalvinsh, I.; Ezzell, R.M.; Janmey, P.A. Use of a gel-forming dipeptide derivative as a carrier for antigen presentation. *J. Pept. Sci.* **1995**, *1*, 371–378. [CrossRef]
21. Aviv, M.; Halperin-Sternfeld, M.; Grigoriants, I.; Buzhansky, L.; Mironi-Harpaz, I.; Seliktar, D.; Einav, S.; Nevo, Z.; Adler-Abramovich, L. Improving the Mechanical Rigidity of Hyaluronic Acid by Integration of a Supramolecular Peptide Matrix. *ACS Appl. Mater. Interfaces* **2018**, *10*, 41883–41891. [CrossRef]
22. Ghosh, M.; Halperin-Sternfeld, M.; Grinberg, I.; Adler-Abramovich, L. Injectable Alginate-Peptide Composite Hydrogel as a Scaffold for Bone Tissue Regeneration. *Nanomaterials* **2019**, *9*, 497. [CrossRef]
23. Ryan, D.M.; Doran, T.M.; Nilsson, B.L. Complementary π–π Interactions induce multicomponent coassembly into functional fibrils. *Langmuir* **2011**, *27*, 11145–11156. [CrossRef] [PubMed]
24. Ryan, D.M.; Anderson, S.B.; Nilsson, B.L. The influence of side-chain halogenation on the self-assembly and hydrogelation of Fmoc-phenylalanine derivatives. *Soft Matter* **2010**, *6*, 3220–3231. [CrossRef]
25. Ryan, D.M.; Anderson, S.B.; Senguen, F.T.; Youngman, R.E.; Nilsson, B.L. Self-assembly and hydrogelation promoted by F5-phenylalanine. *Soft Matter* **2010**, *6*, 475–479. [CrossRef]
26. Rajbhandary, A.; Brennessel, W.W.; Nilsson, B.L. Comparison of the Self-Assembly Behavior of Fmoc-Phenylalanine and Corresponding Peptoid Derivatives. *Cryst. Growth Des.* **2018**, *18*, 623–632. [CrossRef]
27. Liyanage, W.; Nilsson, B.L. Substituent Effects on the Self-Assembly/Coassembly and Hydrogelation of Phenylalanine Derivatives. *Langmuir* **2016**, *32*, 787–799. [CrossRef] [PubMed]
28. Ryan, D.M.; Doran, T.M.; Anderson, S.B.; Nilsson, B.L. Effect of C-terminal modification on the self-assembly and hydrogelation of fluorinated Fmoc-Phe derivatives. *Langmuir* **2011**, *27*, 4029–4039. [CrossRef]
29. Cohen-Gerassi, D.; Arnon, Z.A.; Guterman, T.; Levin, A.; Ghosh, M.; Aviv, M.; Levy, D.; Knowles, T.P.J.; Shacham-Diamand, Y.; Adler-Abramovich, L. Phase Transition and Crystallization Kinetics of a Supramolecular System in a Microfluidic Platform. *Chem. Mater.* **2020**, *32*, 8342–8349. [CrossRef]

30. Ryan, D.M.; Doran, T.M.; Nilsson, B.L. Stabilizing self-assembled Fmoc–F 5 –Phe hydrogels by co-assembly with PEG-functionalized monomers. *Chem. Commun.* **2011**, *47*, 475–477. [CrossRef]
31. Halperin-Sternfeld, M.; Ghosh, M.; Sevostianov, R.; Grigoriants, I.; Adler-Abramovich, L. Molecular co-assembly as a strategy for synergistic improvement of the mechanical properties of hydrogels. *Chem. Commun.* **2017**, *53*, 9586–9589. [CrossRef]
32. Schnaider, L.; Ghosh, M.; Bychenko, D.; Grigoriants, I.; Ya'ari, S.; Shalev Antsel, T.; Matalon, S.; Sarig, R.; Brosh, T.; Pilo, R.; et al. Enhanced Nanoassembly-Incorporated Antibacterial Composite Materials. *ACS Appl. Mater. Interfaces* **2019**, *11*, 21334–21342. [CrossRef]
33. Ya'ari, S.; Halperin-Sternfeld, M.; Rosin, B.; Adler-Abramovich, L. Surface Modification by Nano-Structures Reduces Viable Bacterial Biofilm in Aerobic and Anaerobic Environments. *Int. J. Mol. Sci.* **2020**, *21*, 7370. [CrossRef] [PubMed]
34. Jayawarna, V.; Ali, M.; Jowitt, T.A.; Miller, A.F.; Saiani, A.; Gough, J.E.; Ulijn, R.V. Nanostructured hydrogels for three-dimensional cell culture through self-assembly of fluorenylmethoxycarbonyl-dipeptides. *Adv. Mater.* **2006**, *18*, 611–614. [CrossRef]
35. Colquhoun, C.; Draper, E.R.; Schweins, R.; Marcello, M.; Vadukul, D.; Serpell, L.C.; Adams, D.J. Controlling the network type in self-assembled dipeptide hydrogels. *Soft Matter* **2017**, *13*, 1914–1919. [CrossRef]
36. Engler, A.J.; Sen, S.; Sweeney, H.L.; Discher, D.E. Matrix Elasticity Directs Stem Cell Lineage Specification. *Cell* **2006**, *126*, 677–689. [CrossRef]
37. Hoffman, A.S. Hydrogels for biomedical applications. *Adv. Drug Deliv. Rev.* **2002**, *54*, 3–12. [CrossRef]
38. Pelham, R.J.; Wang, Y.-l. Cell locomotion and focal adhesions are regulated by substrate flexibility. *Proc. Natl. Acad. Sci. USA* **1997**, *94*, 13661–13665. [CrossRef]
39. Fichman, G.; Guterman, T.; Damron, J.; Adler-Abramovich, L.; Schmidt, J.; Kesselman, E.; Shimon, L.J.W.; Ramamoorthy, A.; Talmon, Y.; Gazit, E. Spontaneous structural transition and crystal formation in minimal supramolecular polymer model. *Sci. Adv.* **2016**, *2*, e1500827. [CrossRef]
40. Levin, A.; Mason, T.O.; Adler-Abramovich, L.; Buell, A.K.; Meisl, G.; Galvagnion, C.; Bram, Y.; Stratford, S.A.; Dobson, C.M.; Knowles, T.P.J.; et al. Ostwald's rule of stages governs structural transitions and morphology of dipeptide supramolecular polymers. *Nat. Commun.* **2014**, *5*, 5219. [CrossRef]
41. Orbach, R.; Adler-Abramovich, L.; Zigerson, S.; Mironi-Harpaz, I.; Seliktar, D.; Gazit, E. Self-assembled Fmoc-peptides as a platform for the formation of nanostructures and hydrogels. *Biomacromolecules* **2009**, *10*, 2646–2651. [CrossRef] [PubMed]
42. Chen, L.; Raeburn, J.; Sutton, S.; Spiller, D.G.; Williams, J.; Sharp, J.S.; Griffiths, P.C.; Heenan, R.K.; King, S.M.; Paul, A.; et al. Tuneable Mechanical Properties in Low Molecular Weight Gels. *Soft Matter* **2011**, *7*, 9721–9727. [CrossRef]
43. Dudukovic, N.A.; Zukoski, C.F. Mechanical Properties of Self-Assembled Fmoc-Diphenylalanine Molecular Gels. *Langmuir* **2014**, *30*, 4493–4500. [CrossRef]
44. Xie, Y.Y.; Zhang, Y.W.; Qin, X.T.; Liu, L.P.; Wahid, F.; Zhong, C.; Jia, S.R. Structure-Dependent Antibacterial Activity of Amino Acid-Based Supramolecular Hydrogels. *Colloids Surf. B Biointerfaces* **2020**, *193*, 111099. [CrossRef]
45. Bera, S.; Mondal, S.; Xue, B.; Shimon, L.J.W.; Cao, Y.; Gazit, E. Rigid helical-like assemblies from a self-aggregating tripeptide. *Nat. Mater.* **2019**, *18*, 503–509. [CrossRef] [PubMed]
46. Draper, E.R.; Morris, K.L.; Little, M.A.; Raeburn, J.; Colquhoun, C.; Cross, E.R.; McDonald, T.O.; Serpell, L.C.; Adams, D.J. Hydrogels formed from Fmoc amino acids. *CrystEngComm* **2015**, *17*, 8047–8057. [CrossRef]
47. Liyanage, W.; Brennessel, W.W.; Nilsson, B.L. Spontaneous Transition of Self-assembled Hydrogel Fibrils into Crystalline Microtubes Enables a Rational Strategy to Stabilize the Hydrogel State. *Langmuir* **2015**, *31*, 9933–9942. [CrossRef] [PubMed]
48. Hawkins, G.D.; Giesen, D.J.; Lynch, G.C.; Chambers, C.C.; Rossi, I.; Storer, J.W.; Li, J.; Zhu, T.; Thompson, J.D.; Winget, P.; et al. Version 7.1 Manual Based in Part on AMPAC, Version 2.1. Available online: https://comp.chem.umn.edu/amsol/amsol_Manual_v7.1.2010.05.11.pdf (accessed on 30 August 2021).
49. Sterling, T.; Irwin, J.J. ZINC 15–Ligand Discovery for Everyone. *J. Chem. Inf. Model.* **2015**, *55*, 2324–2337. [CrossRef] [PubMed]
50. Humphrey, W.; Dalke, A.; Schulten, K. VMD: Visual molecular dynamics. *J. Mol. Graph.* **1996**, *14*, 33–38. [CrossRef]
51. Mahler, A.; Reches, M.; Rechter, M.; Cohen, S.; Gazit, E. Rigid, Self-Assembled Hydrogel Composed of a Modified Aromatic Dipeptide. *Adv. Mater.* **2006**, *18*, 1365–1370. [CrossRef]
52. Chen, Y.; Orr, A.A.; Tao, K.; Wang, Z.; Ruggiero, A.; Shimon, L.J.W.; Schnaider, L.; Goodall, A.; Rencus-Lazar, S.; Gilead, S.; et al. High-Efficiency Fluorescence through Bioinspired Supramolecular Self-Assembly. *ACS Nano* **2020**, *14*, 2798–2807. [CrossRef]
53. Tao, K.; Chen, Y.; Orr, A.A.; Tian, Z.; Makam, P.; Gilead, S.; Si, M.; Rencus-Lazar, S.; Qu, S.; Zhang, M.; et al. Enhanced Fluorescence for Bioassembly by Environment-Switching Doping of Metal Ions. *Adv. Funct. Mater.* **2020**, *30*, 1909614. [CrossRef]
54. Kokotidou, C.; Jonnalagadda, S.V.R.; Orr, A.A.; Seoane-Blanco, M.; Apostolidou, C.P.; Raaij, M.J.; Kotzabasaki, M.; Chatzoudis, A.; Jakubowski, J.M.; Mossou, E.; et al. A novel amyloid designable scaffold and potential inhibitor inspired by GAIIG of amyloid beta and the HIV-1 V3 loop. *FEBS Lett.* **2018**, *592*, 1777–1788. [CrossRef]
55. Tamamis, P.; Adler-Abramovich, L.; Reches, M.; Marshall, K.; Sikorski, P.; Serpell, L.; Gazit, E.; Archontis, G. Self-assembly of phenylalanine oligopeptides: Insights from experiments and simulations. *Biophys. J.* **2009**, *96*, 5020–5029. [CrossRef]
56. Tamamis, P.; Kasotakis, E.; Mitraki, A.; Archontis, G. Amyloid-like self-assembly of peptide sequences from the adenovirus fiber shaft: Insights from molecular dynamics simulations. *J. Phys. Chem. B* **2009**, *113*, 15639–15647. [CrossRef] [PubMed]
57. Deidda, G.; Jonnalagadda, S.V.R.; Spies, J.W.; Ranella, A.; Mossou, E.; Forsyth, V.T.; Mitchell, E.P.; Bowler, M.W.; Tamamis, P.; Mitraki, A. Self-Assembled Amyloid Peptides with Arg-Gly-Asp (RGD) Motifs As Scaffolds for Tissue Engineering. *ACS Biomater. Sci. Eng.* **2017**, *3*, 1404–1416. [CrossRef] [PubMed]

58. Jonnalagadda, S.V.R.; Kokotidou, C.; Orr, A.A.; Fotopoulou, E.; Henderson, K.J.; Choi, C.H.; Lim, W.T.; Choi, S.J.; Jeong, H.K.; Mitraki, A.; et al. Computational Design of Functional Amyloid Materials with Cesium Binding, Deposition, and Capture Properties. *J. Phys. Chem. B* **2018**, *122*, 7555–7568. [CrossRef] [PubMed]
59. Best, R.B.; Zhu, X.; Shim, J.; Lopes, P.E.M.; Mittal, J.; Feig, M.; MacKerell, A.D. Optimization of the additive CHARMM all-atom protein force field targeting improved sampling of the backbone φ, ψ and side-chain χ1 and χ2 Dihedral Angles. *J. Chem. Theory Comput.* **2012**, *8*, 3257–3273. [CrossRef]
60. Brooks, B.R.; Brooks, C.L.; Mackerell, A.D.; Nilsson, L.; Petrella, R.J.; Roux, B.; Won, Y.; Archontis, G.; Bartels, C.; Boresch, S.; et al. CHARMM: The biomolecular simulation program. *J. Comput. Chem.* **2009**, *30*, 1545–1614. [CrossRef] [PubMed]
61. Vanommeslaeghe, K.; Hatcher, E.; Acharya, C.; Kundu, S.; Zhong, S.; Shim, J.; Darian, E.; Guvench, O.; Lopes, P.; Vorobyov, I.; et al. CHARMM general force field: A force field for drug-like molecules compatible with the CHARMM all-atom additive biological force fields. *J. Comput. Chem.* **2010**, *31*, 671–690. [CrossRef]
62. Ryckaert, J.P.; Ciccotti, G.; Berendsen, H.J.C. Numerical integration of the cartesian equations of motion of a system with constraints: Molecular dynamics of n-alkanes. *J. Comput. Phys.* **1977**, *23*, 327–341. [CrossRef]
63. Seeber, M.; Cecchini, M.; Rao, F.; Settanni, G.; Caflisch, A. Wordom: A program for efficient analysis of molecular dynamics simulations. *Bioinformatics* **2007**, *23*, 2625–2627. [CrossRef] [PubMed]
64. Seeber, M.; Felline, A.; Raimondi, F.; Muff, S.; Friedman, R.; Rao, F.; Caflisch, A.; Fanelli, F. Software news and updates Wordom: A user-friendly program for the analysis of molecular structures, trajectories, and free energy surfaces. *J. Comput. Chem.* **2011**, *32*, 1183–1194. [CrossRef] [PubMed]

MDPI
St. Alban-Anlage 66
4052 Basel
Switzerland
Tel. +41 61 683 77 34
Fax +41 61 302 89 18
www.mdpi.com

International Journal of Molecular Sciences Editorial Office
E-mail: ijms@mdpi.com
www.mdpi.com/journal/ijms

www.ingramcontent.com/pod-product-compliance
Lightning Source LLC
LaVergne TN
LVHW071023170726
843515LV00006B/1188

* 9 7 8 3 0 3 6 5 6 5 8 7 3 *